Industrial Coating and Lining Application Specialist

Level Two

Trainee Guide

PEARSON

Boston Columbus Indianapolis New York San Francisco Amsterdam
Cape Town Dubai London Madrid Milan Munich Paris Montreal Toronto Delhi
Mexico City Sao Paulo Sydney Hong Kong Seoul Singapore Taipei Tokyo

NCCER

President: Don Whyte
Director of Product Development: Daniele Dixon
Industrial Coating and Lining Application Specialist Project Manager: Jennifer Wilkerson
Production Manager: Tim Davis
Quality Assurance Coordinator: Debie Ness
Editors: Rob Richardson, Chris Wilson
Desktop Publishing Coordinator: James McKay
Production Assistant: Laura Wright

NCCER would like to acknowledge the contract service provider for this curriculum:
Topaz Publications, Syracuse, New York.

10 9 8 7 6 5 4 3 2 1

ISBN 10: 0-13-604510-3
ISBN 13: 978-0-13-604510-6

Preface

To the Trainee

Congratulations! If you are training under an NCCER Accredited Training Sponsor, you have successfully completed *Industrial Coating and Lining Application Specialist Level One*. Now you are well on your way to more specific skills training in the field of coatings.

In this level, you will learn about the identification of different types of corrosion; the steps to follow to execute a work plan; and how to construct a safe and appropriate containment system for specific applications. You will also expand your knowledge of surface preparations, industrial coatings, and coating applications.

Most importantly, you will learn how to work independently and therefore spend more time demonstrating your abilities and understanding of the concepts presented in this text.

We wish you success as you progress through this training program. Should you have any comments on how NCCER might improve upon this textbook, please complete the User Update form located at the back of each module and send it to us. We will always consider and respond to input from our customers.

We also invite you to visit the NCCER website at **www.nccer.org** for the latest releases, training information, newsletter, and much more. You can also reference the Pearson product catalog online at **www.crafttraining.com**. Your feedback is welcome. You may email your comments to **curriculum@nccer.org** or send general comments and inquiries to **info@nccer.org**.

NCCER Standardized Curricula

NCCER is a not-for-profit 501(c)(3) education foundation established in 1996 by the world's largest and most progressive construction companies and national construction associations. It was founded to address the severe workforce shortage facing the industry and to develop a standardized training process and curricula. Today, NCCER is supported by hundreds of leading construction and maintenance companies, manufacturers, and national associations. The NCCER Standardized Curricula was developed by NCCER in partnership with Pearson Education, Inc., the world's largest educational publisher.

Some features of NCCER's Standardized Curricula are as follows:

- An industry-proven record of success
- Curricula developed by the industry for the industry
- National standardization providing portability of learned job skills and educational credits
- Compliance with the Office of Apprenticeship requirements for related classroom training (CFR 29:29)
- Well-illustrated, up-to-date, and practical information

NCCER also maintains a Registry that provides transcripts, certificates, and wallet cards to individuals who have successfully completed modules of NCCER's Standardized Curricula. *Training programs must be delivered by an NCCER Accredited Training Sponsor in order to receive these credentials.*

NACE International

NACE International has more than 25,000 members in 100 countries networking on many corrosion control technologies. NACE's largest program is the NACE Coating Inspector Program, providing comprehensive training and certification for coating inspectors who often work alongside applicators to ensure lasting success of protective coating systems. Paint applicators considering a move into the coating inspection field are encouraged to consider coating inspection training and certification, described at **www.nace.org**. NACE International also is a leader in the development of written standards for the coatings industry, which can be accessed by NACE members at no charge at **www.nace.org**.

NCCER Standardized Curricula

NCCER's training programs comprise more than 80 construction, maintenance, pipeline, and utility areas and include skills assessments, safety training, and management education.

Boilermaking
Cabinetmaking
Carpentry
Concrete Finishing
Construction Craft Laborer
Construction Technology
Core Curriculum: Introductory Craft Skills
Drywall
Electrical
Electronic Systems Technician
Heating, Ventilating, and Air Conditioning
Heavy Equipment Operations
Highway/Heavy Construction
Hydroblasting
Industrial Coating and Lining Application Specialist
Industrial Maintenance Electrical and Instrumentation Technician
Industrial Maintenance Mechanic
Instrumentation
Insulating
Ironworking
Masonry
Millwright
Mobile Crane Operations
Painting
Painting, Industrial
Pipefitting
Pipelayer
Plumbing
Reinforcing Ironwork
Rigging
Scaffolding
Sheet Metal
Signal Person
Site Layout
Sprinkler Fitting
Tower Crane Operator
Welding

Maritime

Maritime Industry Fundamentals
Maritime Pipefitting
Maritime Structural Fitter

Green/Sustainable Construction

Building Auditor
Fundamentals of Weatherization
Introduction to Weatherization
Sustainable Construction Supervisor
Weatherization Crew Chief
Weatherization Technician
Your Role in the Green Environment

Energy

Alternative Energy
Introduction to the Power Industry
Introduction to Solar Photovoltaics
Introduction to Wind Energy
Power Industry Fundamentals
Power Generation Maintenance Electrician
Power Generation I&C
Maintenance Technician
Power Generation Maintenance Mechanic
Power Line Worker
Power Line Worker: Distribution
Power Line Worker: Substation
Power Line Worker: Transmission
Solar Photovoltaic Systems Installer
Wind Turbine Maintenance Technician

Pipeline

Control Center Operations, Liquid
Corrosion Control
Electrical and Instrumentation
Field Operations, Liquid
Field Operations, Gas
Maintenance
Mechanical

Safety

Field Safety
Safety Orientation
Safety Technology

Supplemental Titles

Applied Construction Math
Tools for Success

Management

Fundamentals of Crew Leadership
Project Management
Project Supervision

Spanish Titles

Acabado de concreto: nivel uno
Aislamiento: nivel uno
Albañilería: nivel uno
Andamios
Carpintería:
Formas para carpintería, nivel tres
Currículo básico: habilidades introductorias del oficio
Electricidad: nivel uno
Herrería: nivel uno
Herrería de refuerzo: nivel uno
Instalación de rociadores: nivel uno
Instalación de tuberías: nivel uno
Instrumentación: nivel uno, nivel dos, nivel tres, nivel cuatro
Orientación de seguridad
Paneles de yeso: nivel uno
Seguridad de campo

Acknowledgments

This curriculum was revised as a result of the farsightedness and leadership of the following sponsors:

Basic Industries
Brand Energy
International Paint
Greenman-Pedersen, Inc.
Altus Energy Services

This curriculum would not exist were it not for the dedication and unselfish energy of those volunteers who served on the Authoring Team. A sincere thanks is extended to the following:

Kat Coronado
Thomas Gibbons
David Hurtado
Richard McGinty
Cathey Perkins
Chester Ramirez
Gil Rogers
John Troeger

NCCER Partners

American Fire Sprinkler Association
Associated Builders and Contractors, Inc.
Associated General Contractors of America
Association for Career and Technical Education
Association for Skilled and Technical Sciences
Construction Industry Institute
Construction Users Roundtable
Construction Workforce Development Center
Design Build Institute of America
GSSC – Gulf States Shipbuilders Consortium
ISN
Manufacturing Institute
Mason Contractors Association of America
Merit Contractors Association of Canada
NACE International
National Association of Minority Contractors
National Association of Women in Construction
National Insulation Association
National Technical Honor Society
NAWIC Education Foundation
North American Crane Bureau
North American Technician Excellence
Pearson
Pearson Qualifications International
Prov
SkillsUSA®
Steel Erectors Association of America
U.S. Army Corps of Engineers
University of Florida, M. E. Rinker School of Building Construction
Women Construction Owners & Executives, USA

NCCER Business Partners

ACT
Prov
Pearson
Judgment Index
NACB
ConstructNet International

Contents

Industrial Coating Safety

69201-10

INDUSTRIAL COATING AND LINING APPLICATION SPECIALIST LEVEL TWO

- 69207-10 Coating Applications Two
- 69206-10 Industrial Coatings Two
- 69205-10 Surface Preparation Two
- 69204-10 Containment
- 69203-10 Work Planning and Quality Control
- 69202-10 Corrosion Protection
- 69201-10 Industrial Coating Safety
- Industrial Coating and Lining Application Specialist Level One
- Core Curriculum: Introductory Craft Skills

This course map shows all of the modules in *Industrial Coating and Lining Application Specialist Level Two*. The suggested training order begins at the bottom and proceeds up. Skill levels increase as you advance on the course map. The local Training Program Sponsor may adjust the training order.

Objectives

When you have completed this module, you will be able to do the following:

1. Identify appropriate health and safety regulatory bodies pertaining to surface preparation and coating work.
2. Determine access requirements to structures needing coating work.
3. Identify site-specific lockout/tagout procedures and demonstrate verification of lockout.
4. Select the correct fall-prevention system based on the job-site conditions.
5. Recognize intrinsically safe/explosion-proof equipment and devices.
6. Identify proper personal protective equipment (PPE) for surface preparation, coating mixing and thinning, and coating application operations.
7. Describe safety procedures associated with the following:
   - Pre-cleaning
   - Hand and power tool cleaning
   - Dry and wet abrasive blast cleaning
   - Ultra-high-pressure waterjetting
   - Chemical stripping
   - Disturbance of coatings containing toxic metals
   - Material handling, storage, disposal, and waste minimization
   - Mixing, thinning, and coating operations

Trade Terms

Accidents
Acute effect
Acute exposure
Anchorage
Apparatus
Arc
Atmospheric hazards
Best practices plan
Body harness
Bonding
Chronic effect
Chronic exposure
Combustible
Competent person
Compliance
Confined space
Connectors
Decibels (dBA)
Duty rating
Electrically powered tools
Energy source
Energy-isolating device
Free fall
Ground fault circuit interrupter (GFCI)
Grounded
Guards
Hazard Communication (HazCom) Standard
Hypothermia
Impact tools
Lanyard
Leading edges
Lifelines
Lockout
Lockout devices
Material safety data sheet (MSDS)
Non-permit required confined space
Oxygen-deficient
Oxygen-enriched
Permit required confined space
Personal fall-arrest system
Platforms
Pneumatic tools
Power tools
Scaffolding
Self-retracting lifeline
Silica
Tagout
Tagout devices
Two-point swing scaffolding

Prerequisites

Before you begin this module, it is recommended that you successfully complete *Industrial Coating and Lining Application Specialist Level One.*

Contents

Topics to be presented in this module include:

Contents *(continued)*

Contents *(continued)*

Figures and Tables

Figures and Tables *(continued)*

1.0.0 INTRODUCTION

Industrial coating work can be dangerous if technicians do not follow safe work practices. Most **accidents** on the job are caused by worker carelessness, poor safety planning, lack of training, failure on the part of the employer or employee to follow safety regulations, or a lack of accountability.

To help prevent accidents, a company must have a safety program. Safety programs provide the process needed to work safely according to company rules and safeguards. Safety must be incorporated into all phases of the job and involve all employees at every level, including management.

Regulations require that a company-appointed, **competent person** be on site before any job is started. A competent person is someone who has experience and training for the job and knows the hazards. He or she knows the rules and regulations and has the authority to correct a job if work is not being done safely.

In addition to their own rules, companies must comply with local, state, and national regulations. Safety regulations are created to make work sites safe and accident-free. It is important to remember that there is a good reason why each regulation exists. Following good safety practices helps to save lives.

> **NOTE**
> This module has been developed to comply with U.S. safety regulations. Users from other countries must refer to applicable local regulations. In addition, work in specialized environments such as nuclear power plants, oil extraction, or refinery operations requires additional safety training and certification.

The following ten basic rules of safety should become second nature to everyone on the job site:

1. Use tools, equipment, and PPE the way they were designed. Always maintain positive control of the tools you are using.
2. Wear a hard hat and safety equipment at all times where required.
3. Inspect equipment daily.
4. Only perform tasks for which you have been trained.
5. Understand company safety rules and policies.
6. Take responsibility for yourself and co-workers.
7. Correct or report all unsafe conditions to a supervisor immediately.
8. Adopt the zero-accident philosophy.
9. Get involved with the company safety program.
10. When in doubt, stop and ask!

Always report unsafe conditions or acts, equipment defects, and other hazardous situations. Never assume that a poor condition has already been reported or that an unsafe procedure is okay because that's the way it has always been done. When a co-worker performs an unsafe act, speak up; it could save his or her life.

1.1.0 Industrial and Governmental Organizations Involved with Safety

Career activities may require working with a number of industry organizations, such as NACE International, the Society for Protective Coating (SSPC), the National Painting Contractors' Association (NPCA), the Painting and Decorating Contractors' Association (PDCA) and ASTM International (originally known as the American Society for Testing and Materials), as well as United States government agencies, such as the Environmental Protection Agency (EPA) and the Occupational Safety and Health Administration (OSHA). These organizations can provide a wealth of information about how to do jobs safely. Information from these organizations can be accessed from their websites, and printed materials can be ordered from them. Most likely, an employer will have some information on hand.

Many of the professional organizations have been around for a number of years—in some cases, more than 100 years. This reflects the depth of commitment that coating professionals have for corrosion prevention, coating safety, and business ethics.

This course is co-sponsored by NACE International (originally known as the National Association of Corrosion Engineers). As of this publication, NACE International has more than 23,000 members in 109 countries. The NACE International mission is "to protect people, assets, and the environment from the effect of corrosion." They do this by setting standards for the corrosion industry and distributing the latest information worldwide in journals and technical papers. They also hold a number of conferences and meetings every year.

SSPC began in 1950 as the Steel Structures Painting Council and, as the name implies, their main concern was the use of coatings to protect industrial steel structures. In 1997, the name of the association was changed to The Society for Protective Coatings because they broadened their mission to include the protection of concrete, steel, and other structures. SSPC is a good source of information for surface preparation, coating selection, coating application, environmental regulations, and health and safety issues that concern the industrial coatings industry.

NPCA is a trade association that was created in 1888 as the National Paint, Oil, and Varnish Association. Its primary role today is to serve as an industrial coating industry voice at the federal, state and local levels—this includes when a law is being discussed that will change the regulations or make new regulations for products used in industrial coating operations. In addition, NPCA has an interest in legal processes involving products used in the industry.

The Painting and Decorating Contractors of America (PDCA) began in 1884 as the Master House Painters Association of the United States and Canada. PDCA's mission statement reads: "PDCA exists to lead the industry by providing quality products, programs, services, and opportunities essential to the success of our members." PDCA's goals are to support advancements in the industrial coating industry; promote communication among industry professionals; and promote business ethics.

ASTM International was founded in 1898. It is a voluntary standards development organization. These standards cover a variety of materials, products, industries, and services, such as metals, paints, plastics, petroleum, construction, aviation, energy, the environment, consumer products, and electronics. In addition, ASTM offers continuing technical education and training programs for industry and government.

The U.S. EPA was created in 1970 as an independent regulatory agency responsible for the implementation of federal laws designed to protect the environment. EPA requirements must be followed for virtually every aspect of a job. The EPA's mission includes the following:

> The establishment and enforcement of environmental protection standards consistent with national environmental goals... The conduct of research on the adverse effects of pollution and on methods and equipment for controlling it; the gathering of information on pollution; and the use of this information in strengthening environmental protection programs and recommending policy changes... assisting others, through grants, technical assistance and other means, in arresting pollution of the environment... assisting the Council on Environmental Quality in developing and recommending to the President new policies for the protection of the environment.

OSHA, part of the U.S. Department of Labor, is a government agency that helps protect millions of workers each year. Their mission is to ensure a safe and healthy environment in the workplace. Since the agency was created in 1971, workplace deaths have been cut in half and injury and illness rates have declined 40 percent. At the same time, there has been an increase in the number of people working. In fact, since 1971, the number of people working in the United States has doubled from 56 million workers at 3.5 million work sites to over 115 million workers at 7.5 million sites.

In Canada, workers are covered by provincial, territorial, or federal labor codes. Workers covered by federal legislation (including those in mining, transportation, and federal employment) are covered by the Canada Labor Code (CLC). All other workers are covered by the various health and safety codes of the province or territories where they work. The Canadian Centre for Occupational Health and Safety (CCOHS), an agency of the government of Canada, was created in 1978 and is mandated to promote safe and healthy workplaces to help prevent work-related injuries and illnesses. Environmental protection is administered by a department called Environment Canada. This department administers nearly two dozen governmental acts either in whole or in part. The department uses regulations to place strict controls on areas governed by these acts. The Canadian Department of Justice provides the full text of federal acts and regulations. The Canadian Standards Association (CSA) is a non-profit organization that creates consensus standards. They publish environmental standards or guidelines and occupational health and safety (OHS) standards. When any of the standards are referenced in federal, provincial, or territorial regulations/codes as being necessary for compliance, they have the force of law.

In Mexico, labor and environmental laws are defined in great detail in the Mexican Constitution. Each law designates which Secretary (Cabinet Branch) is responsible for oversight. Regulations define the laws and are created by the Secretaries to detail the procedures and processes for compliance. Standards for occupational health and safety are created by the Secretaría del Trabajo y Previsión Social (STPS). Standards for environmental protection are created by the Secretaría de Medio Ambiente y Recursos Naturales (SEMARNAT). The standards are published as Norma Oficial Mexicanas (NOMs) and have the force of law. The NOMs include the abbreviation of the applicable Secretariat as part of the coding (NOM-STPS-xxxxxx or NOM-SEMARNATxxxxxxxx). Standards include the detail for the forms and format of the documentation that must be maintained for laws and regulations. They also detail fines and sanctions. Normas Mexicanas (NMXs) are voluntary standards that usually serve as reference guides; however, they may be referenced in a NOM as obligatory for compliance of that NOM.

In Europe, based on Article 137 of the European Community (EC) Treaty, a variety of community measures in the field of environmental protection, safety, and health have been adopted. The resulting European directives are legally binding and have to be transposed into the national laws of the member countries. Article 137 and other directives on specific environmental, safety, and health issues form the basis of European environmental, safety, and health legislation. The standardization and resultant development of European standards (ENs) for health, safety, and environmental protection is handled by the European Committee for Standardisation (CEN). These standards are incorporated into the regulations and codes of the member countries. Each of the member countries may have additional and/or more stringent codes as well.

1.1.1 OSHA

OSHA adopts and enforces safety regulations known as standards; some, like the ***Hazard Communications (HazCom) Standard***, apply to all industries. Other standards only apply to a specific industry.

Twenty-six states have their own OSHA programs. The state regulations must be at least as strict as the national standards. Some state OSHA plans make employee noncompliance a criminal act. Find out which regulations apply to your specific job site. The federal and state OSHA programs keep workplaces safe through accident investigation, inspections, and outreach.

If there is a serious accident at a job site in which three or more workers are hospitalized or someone is killed, OSHA must be notified. OSHA will then investigate the accident. Accident investigation is OSHA's tool for uncovering hazards that were missed during routine or previous inspections. Accident investigation is really about accident prevention. A lot can be learned from an accident, most importantly how to avoid another one. Investigations are very useful when they discover every contributing factor to the accident. This helps correct the unsafe condition or activity and helps prevent it from happening again.

OSHA inspects workplaces to ensure **compliance** with minimum safety standards. If OSHA compliance officers find any violations on a site, they may issue a citation and a penalty. Citations inform the employer and employees of the regulations and standards that have been broken. Penalties are fines the company must pay for each violation. Even though the company is paying the fine, employees will still feel the effects. Fines affect employees because the company will have less money for raises, bonuses, and hiring more workers to replace those who may have been injured.

OSHA plays an important role in preventing on-the-job injuries and illnesses. This is accomplished through outreach, education, and compliance assistance. OSHA provides various types of training materials to assist employers and workers in understanding the standards.

The OSHA website (www.osha.gov) has nearly 450,000 pages of safety information. The website also offers assistance and interactive e-tools to help both employers and employees understand and comply with the standards.

In addition to the website, OSHA offices provide employers and employees with information and training to help them understand and comply with safety regulations. OSHA also offers free workplace consultations. These consultations are available to any business that needs on-site help in establishing safety and health programs, as well as identifying and correcting workplace hazards. These offices can also answer questions for employees who want additional information but do not have a complaint.

There is a difference between compliance and best practices. OSHA determines compliance; individual company policies determine best practices. OSHA sets minimum safety requirements, investigates serious accidents, and inspects work sites. A company is in compliance when it meets all minimum safety regulations.

In addition to specific safety guidelines, OSHA regulations include a general duty clause. Each employer has a general duty to provide employees with a workplace free from recognized hazards that may cause death or serious harm. Companies are required to set practices and procedures to make sure all workers are safe. A safety program that includes the best available practices and equipment to ensure worker safety is known as a **best practices plan**. Best practices often exceed the minimum safety requirements. Best practices can change as new safety equipment and procedures are developed.

NOTE

Keep in mind that if an employee gets injured and did not follow the company's safety program or OSHA standards, the employee can be held responsible for the accident.

1.1.2 EPA

The U.S. Environmental Protection Agency was established in 1970 to consolidate in one agency a variety of federal research, monitoring, standard-setting, and enforcement activities to ensure environmental protection. The EPA's mission is to protect human health and to safeguard the natural environment including the air, water, and land upon which life depends. Therefore, it is important to understand how work being done on a construction site affects the environment. This includes knowing the major environmental regulations covering the release, treatment, storage, and disposal of potentially hazardous materials into the air, water, and soil.

Industrial coatings work generates a lot of unwanted material—paint chips, solvents, excess coatings, dirty rags, and contaminated protective clothing. All of this is called a waste stream. The waste stream may be hazardous or non-hazardous. Either way, the waste needs to be disposed of, and depending on the type of waste, it can be costly. A lot of money and effort goes into disposing of waste, so the first step in waste management is to prevent waste. Naturally, some waste is unavoidable but some is caused by carelessness.

Control waste by mixing only the amount of coatings required to properly do the job. This means carefully calculating the amounts based on the DFT specified for the job. Another way to control waste is to use only the amount of solvent and rags necessary to do a good surface preparation job. When using dry abrasives, blast only until obtaining the correct surface profile—doing more just blasts metal and contaminates the abrasive.

Minimizing waste decreases costs in two ways. It saves money on materials, and it also saves money on disposal fees. Minimizing waste is also good for the environment. Waste deliberately or accidentally released into the environment can pollute water and air, kill animals, and destroy vegetation. In addition, hefty fines may be associated with improper waste disposal.

Non-hazardous waste must be disposed of in accordance with the employer's policy. These policies are usually created based on local laws about waste disposal, so be sure to follow them.

Hazardous waste is a waste that is dangerous or potentially dangerous to human health or the environment. Hazardous wastes can be liquids, solids, contained gases, or sludge. They can be the by-products of manufacturing processes or discarded products, like solvents. These products must be disposed of in accordance with the EPA *Resource Conservation and Recovery Act (RCRA) 40 CFR 261* or with local law. Local law can be stricter than the RCRA but it cannot be more lenient. Look on the MSDS for the product to find out how to safely dispose it.

The EPA has determined that a number of specific wastes are hazardous. These are defined in EPA document *Title 40, Protection of Environment Part 261—Identification and Listing of Hazardous Waste.* Other wastes can also be classified as hazardous, including those that are spontaneously combustible or have a low flash point (like some solvents), and those that are toxic (like some paints).

As you learned in the *Industrial Coatings* module in *Level One*, any person or site creating hazardous waste is a generator, and based on how much waste they produce per month, they may need to have a special waste disposal program. Even companies exempt from following the rules that large and small generators must follow need to properly dispose of hazardous waste. All large- and small-quantity generators are required to do the following:

- Obtain an EPA identification number from the state
- Comply with the manifest system
- Handle wastes properly before shipment; for example, packaging, labeling, marking, and accumulation time
- Comply with recordkeeping and reporting requirements

Information about determining whether a work site needs an EPA identification number and how to get one is found in *Notification of Regulated Waste Activity Instructions and Form*, EPA Form 8700-12 (OMB# 2050-0028). Helpful tips about what to do with hazardous waste are listed in the Frequently Asked Questions (FAQ) section of the EPA website (www.epa.gov).

It is very important to know what waste is hazardous and what is not. Hazardous waste costs more to dispose of than non-hazardous waste. Hazardous and non-hazardous waste need to be stored and disposed of separately. Both must be placed in appropriate containers in accordance with company policy. Hazardous waste must be disposed of by a contractor authorized to dispose of it. Sometimes waste needs to be tested to determine its exact contents. Experienced certified safety professionals (CSPs) and certified industrial hygienists (CIHs) are often involved in testing.

Should a hazardous material ever be spilled at the job site, check the MSDS for cleanup instructions. If required, inform the EPA of a spill (depending on the type of material and how large of an area is involved). Some states may have more stringent rules about handling hazardous materials and cleaning up spills.

> **CAUTION**
>
> Always report spills to a supervisor. Some states require that a report be made within one hour of a spill.

Always be certain to store, mix, and use materials properly. Place any waste in the proper receptacle to await disposal. When the local law indicates waste needs to be tested to determine whether it is hazardous, be aware of the correct operation of any testing equipment and the procedure for testing. Know what the results mean, don't guess. Always follow the employer's documentation procedure for waste.

1.2.0 Counterfeit and Knock-Off Parts, Materials, and Equipment

> **WARNING**
>
> Counterfeit and knock-off parts, materials, and equipment can cause death or serious injury when intentionally or unintentionally used in or as personal safety equipment, rigging equipment, high-pressure tanks or tools, high-speed tools, and other types of equipment where sudden failure will be hazardous to personnel.

Due to globalization, many counterfeit and knock-off parts, materials, and equipment are entering the markets. These products are currently plaguing the auto, aircraft, and medical industries, but no market is immune. The U.S. Department of Energy (DOE) has discovered counterfeit and knock-off fasteners, valves, and other components or products being supplied for the nuclear power industry.

Counterfeit goods violate patents and copyrights in an attempt to duplicate the look of original equipment manufacturer (OEM) goods, including the packaging. Knock-off goods are similar to counterfeit goods, but evade patent and trademark laws by avoiding exact OEM brand names, sometimes varying only by one letter or by the addition of a dash, slash, or period(s). Knock-off packaging is almost identical to the OEM packaging. Even the labeling/marking identification, including part numbers, is often the same as OEM goods.

Both counterfeit and knock-off parts, materials, and equipment have no traceability of quality or integrity and no warranty. When obtaining replacement parts, materials, and equipment that will be hazardous if it fails, always obtain the items directly from the OEM or through the OEM's authorized distributor. In addition, always examine any goods and their packaging for signs of counterfeit or knock-off items.

For additional information, a copy of a suspect (knock-off) and counterfeit training booklet can be downloaded from the DOE HS-32 Suspect/Counterfeit-Defective Items website (eh.doe.gov/sci) along with other pertinent information.

2.0.0 Basic Components of a Safety Program

There are certain basic elements common to most effective safety programs. The degree and extent to which these elements are included in a program will vary depending on the size and nature of the operations. The most basic components of a safety program include the following:

- A management support and policy statement
- A policy on alcohol and drug abuse
- Safety rules
- Orientation and training
- Access to first aid and follow-up medical care
- Pre-project and pre-task safety planning
- Safety meetings and employee involvement
- Subcontractor management
- Emergency action plans and methods for dealing with the media
- Inspections, employee observations, and audits
- Accident investigation and analysis
- Recordkeeping
- Accountability system
- Program evaluation and follow-up

In order to comply with specific OSHA regulations, these additional safety program elements may be required:

- Hazard communication
- Bloodborne pathogens
- Welding, burning, and cutting

- Respiratory protection
- PPE
- Elevated work and fall protection
- Hearing conservation
- Confined-space entry
- Lockout/tagout
- Cranes, hoists, and other lifting devices
- Mobile equipment operations

2.1.0 Policy Statement

A policy statement outlines management's philosophies and goals toward safety and loss prevention. It is signed by the president or chief executive officer (CEO) and communicated to all affected employees. A good policy statement recognizes management's responsibility for safety and loss prevention without diminishing each employee's role. It expresses a strong, positive, and realistic commitment to safety. Management support and commitment are the most important factors in the success of any safety program. Employees should be aware of the company's management policy statement.

2.2.0 Alcohol and Drug Abuse

Alcohol and drug abuse are national problems that result in the following:

- Increased worker's compensation and healthcare costs
- Decreased productivity
- Increased tools and materials costs
- Increased absenteeism

Many owners and contractors have recognized this problem and have implemented programs to combat the situation. Employees should be aware of the company's drug and alcohol abuse programs. Some larger companies sponsor employee assistance programs (EAPs).

2.3.0 Safety Rules

Safety rules are a necessary part of any program. They must be logical, and they should be prepared and presented in terms that are easily understood. Rules are of no value unless they are communicated and understood, and workers buy into the safety process. Everyone on site must also understand the consequences of breaking safety rules.

Safety rules are generally divided into the following categories:

- General/company safety rules that apply to all employees
- Client-specific, job-specific, or site-specific safety rules
- Craft or special safety rules that apply to a specific type of work operation

Safety rules can be presented on posters at a site, during orientation, safety meetings, or safety training. Progressive contractors plan safety into the job at the bid stage.

2.4.0 Safety Training

Safety training is an important tool for preventing accidents. Training can be broken down into two major categories: new employee orientation and job-specific safety training. Statistics have shown that untrained employees who have been on the job 30 days or less account for 25 percent of all injuries. This clearly shows the need for an effective orientation program. Safety training should be conducted when a new employee is hired, an existing employee is assigned to a new job, and when new jobs are started.

In order for safety training to be effective, it should cover the following:

- Hazard assessment
- Correct work procedures
- Care, use, maintenance, and limitations of any required PPE
- The hazards and safeguards associated with any harmful materials used
- Where to go for help
- Site emergency reporting and response procedures

Safety meetings are often used to maintain employee interest and involvement. They are a key element in any safety program. Safety meetings can vary from a formal presentation to short five-minute toolbox talks.

Safety meetings, when properly conducted, can be used to do the following:

- Exchange information regarding specific safety matters
- Provide an outlet for workers to discuss critical issues
- Provide a written record of the actions taken
- Establish an effective communication link between management and employees
- Focus on the hazards and safeguards associated with a planned task

Both federal and state governmental regulations require employers to train their employees on the hazards and safeguards associated with the work they are doing. Safety meetings can be used as training sessions.

2.5.0 Emergency Response

Every job site should have a plan for responding to emergencies. Planning is especially important if the accident happens in a remote area that does not have a telephone or cellular phone service. Be aware of what the employer's emergency plan involves. Find out who needs to be called in case of an emergency and what protection is required.

Emergency action plans can cut down response time to job-site emergencies. In many cases, advance planning can even reduce the severity of the incident. For example, if the names, addresses, and phone numbers of the nearest medical, fire, police, and emergency response agencies are posted at each job-site phone, it is easier to get the help needed. Also, as part of the emergency action plan, everyone should be familiar with the site emergency reporting and response procedures. Another part of the emergency action plan is the availability of prompt access to first aid and follow-up medical care. Emergency action plans may also require that one or more people on each job site be trained in basic first aid and cardiopulmonary resuscitation (CPR). Lives are saved when all of these elements are incorporated into emergency action plans.

Emergency-response teams are the first line of defense in emergencies. Only employees who are physically capable of performing the duties and who have been properly trained can be assigned to an emergency-response team. Depending on the size of the operation, there may be one or several teams trained in the following areas:

- Use of various types of fire extinguishers (required for all employees)
- First aid, including CPR
- Shutdown procedures
- Evacuation procedures
- Chemical spill-control procedures
- Use of self-contained breathing apparatus (SCBA)
- Search and emergency-rescue procedures

The type and extent of the emergency will depend on site operations. The response will vary according to the type of process, the material handled, the number of employees, and the availability of outside resources.

Emergency-response teams must be trained in the types of emergencies that could happen, and must know the emergency actions that need to be performed. They should be informed about special hazards, such as storage and use of flammable materials, toxic chemicals, and radioactive sources to which they may be exposed during fire and other emergencies.

It is important to determine when not to intervene. For example, team members must be able to determine if the fire is too large for them to handle or whether it is safe to perform an emergency rescue. If emergency-response team members could be injured or incapacitated, they should wait for professional firefighters and/or emergency-response groups.

2.6.0 Accident Investigation

All accidents, injuries, illnesses, and near misses must be reported and promptly investigated. The primary purpose of an accident investigation is to prevent re-occurrences. Accident investigations are important because they do the following:

- Determine the causes of accidents
- Help prevent re-occurrences
- Satisfy insurance company requirements and government regulations
- Document facts and preserve evidence for legal purposes
- Detect trends and identify potential problem areas

The front-line supervisor generally performs the primary accident investigation function. This is because it is generally agreed that front-line supervisors are the most familiar with the work area and employees.

2.7.0 Recordkeeping

Recordkeeping is a critical part of a company's safety and health program. The data that is collected helps keep track of work-related injuries and illnesses. Once this data is gathered and analyzed, it can be used to help identify and correct problem areas. This will help prevent future illnesses and injuries. Recordkeeping not only provides information to management about illness and injury, it also informs workers about incidents that happen in the work area. When workers are aware of injuries, illnesses, and hazards in the workplace, they are more likely to follow safe work practices and report workplace hazards.

Recordkeeping is also required by OSHA. Specific illness and injury information is reported to OSHA as part of the agency's site-specific inspection targeting program. The Bureau of Labor Statistics (BLS) also uses injury and illness records as the source data for the Annual Survey of Occupational Injuries and Illnesses. This report shows safety and health trends nationwide and industry wide.

2.8.0 Best Practices

Best practices are the rules and safety procedures that the industry considers to be the safest method to get a job done. Best practices take a pro-active approach to safety, identifying possible safety issues in advance and establishing procedures to avoid potential accidents. Although a company sets best practices, they are often based on OSHA regulations. Best practices are designed to help employees comply with required safety standards.

Confined-space safety is a good example of best practices and compliance. Industrial sites are required to follow the confined-space standard.

2.9.0 Guidelines for a Safe Workplace

Employees have a responsibility to follow the company's rules and safety procedures. These are some general safety rules that promote a safe workplace:

- Follow all rules, policies, and procedures, both written and spoken, for the job being done.
- Never start a job until the appropriate procedure for the job has been reviewed and there is an understanding of how to do the task properly is known.
- Never perform unauthorized jobs.
- Never start a job that appears unsafe or that will be unnecessarily risky.
- Never start a job that requires additional assistance without that assistance.
- Never engage in any unsafe behaviors.
- Stay focused on the job at all times. Anticipate potential hazards and take steps to avoid them.
- Immediately report all job-related incidents, near misses, and injuries to the supervisor.
- Immediately report all hazards to the supervisor. Correct those conditions that are controllable.
- Tell co-workers to stop what they are doing when they are being unsafe or creating a hazardous situation.

2.10.0 HazCom Program

Using hazardous materials improperly can cause health problems. These can include minor problems such as skin or eye irritation, or serious conditions such as cancer. These materials also present physical hazards such as fire, corrosion, and reactivity. You have the right to know the hazards of all the chemicals you will be exposed to on the job. OSHA directs employers to tell workers about hazardous materials on the job site through the *HazCom Standard*. You may have heard of it as the worker right-to-know program.

On-Site

Ask Questions

Find out about all potential job hazards before beginning work. Ask the supervisor or competent person on site to answer the following questions before starting a job:

- What are the hazards associated with this job?
- Where is the energy source and how can it be shut down or controlled?
- What could possibly go wrong?
- What must be done to protect everyone from personal injury?

Employers must educate their workers about the hazardous materials they might use on the job. This is done through a written HazCom program and training. In addition, all materials must have proper labels and **material safety data sheets (MSDSs)**. Labels and MSDSs provide information about health hazards, safety precautions, and emergency responses. Workers have the following responsibilities when it comes to HazCom:

- Learn to recognize hazardous materials labels.
- Know where MSDSs are kept on the job site.
- Report any hazards spotted on the job site to the supervisor.
- Know the physical and health hazards of the materials being used and the protection required.
- Know what to do in an emergency.
- Understand the employer's HazCom program.

There are many types of hazardous materials on a work site. These materials and related hazards are shown in *Table 1*.

Table 1 Typical Hazardous Materials

Substance*	Hazard
Glue and adhesives	Toxic vapors, explosion, fire
Gasoline	Toxic vapors, explosion, fire
Paints	Toxic vapors, explosion, fire
Concrete and mortar	Toxic vapors, lung disease, cement dermatitis, chemical burns
Fuels and oils	Toxic vapors, explosion, fire
Wood	Flying particles
Compressed gases	Toxic gases, explosion, fire
Corrosive liquids	Burns, respiratory hazards
Radioactive materials	Burns, radiation sickness

* Each of these materials must have an MSDS. Do not use them unless you have read and understand the MSDS. The MSDS will help you determine the hazards that you may be exposed to and how to protect yourself.

201T01.EPS

Hazardous materials come in many different forms. Chemicals are not only liquids. They can also be solids, gases, vapors, fumes, dusts, and mists. Many common products contain several chemicals. For example, some existing paints contain cadmium, chromium, and lead.

Many chemicals pose health hazards, such as diseases or burns. Others pose physical hazards, including fire or explosion. Some have both health and physical hazards. The MSDS identifies the hazards and lists the required safety precautions and PPE.

Radioactive materials can be a serious hazard on a job site. They are used to test welds on piping and vessels, which may result in some contact. All radioactive materials must be properly labeled and warning signs must be posted in the work area. Only workers with special training and PPE are allowed in these areas. If signs like those shown in *Figure 1* are posted, stay away from that area.

Radioactive materials are a unique type of physical hazard. Radiation cannot be seen, but it affects the body. Excessive radiation exposure can cause skin burns, nausea, vomiting, infertil-

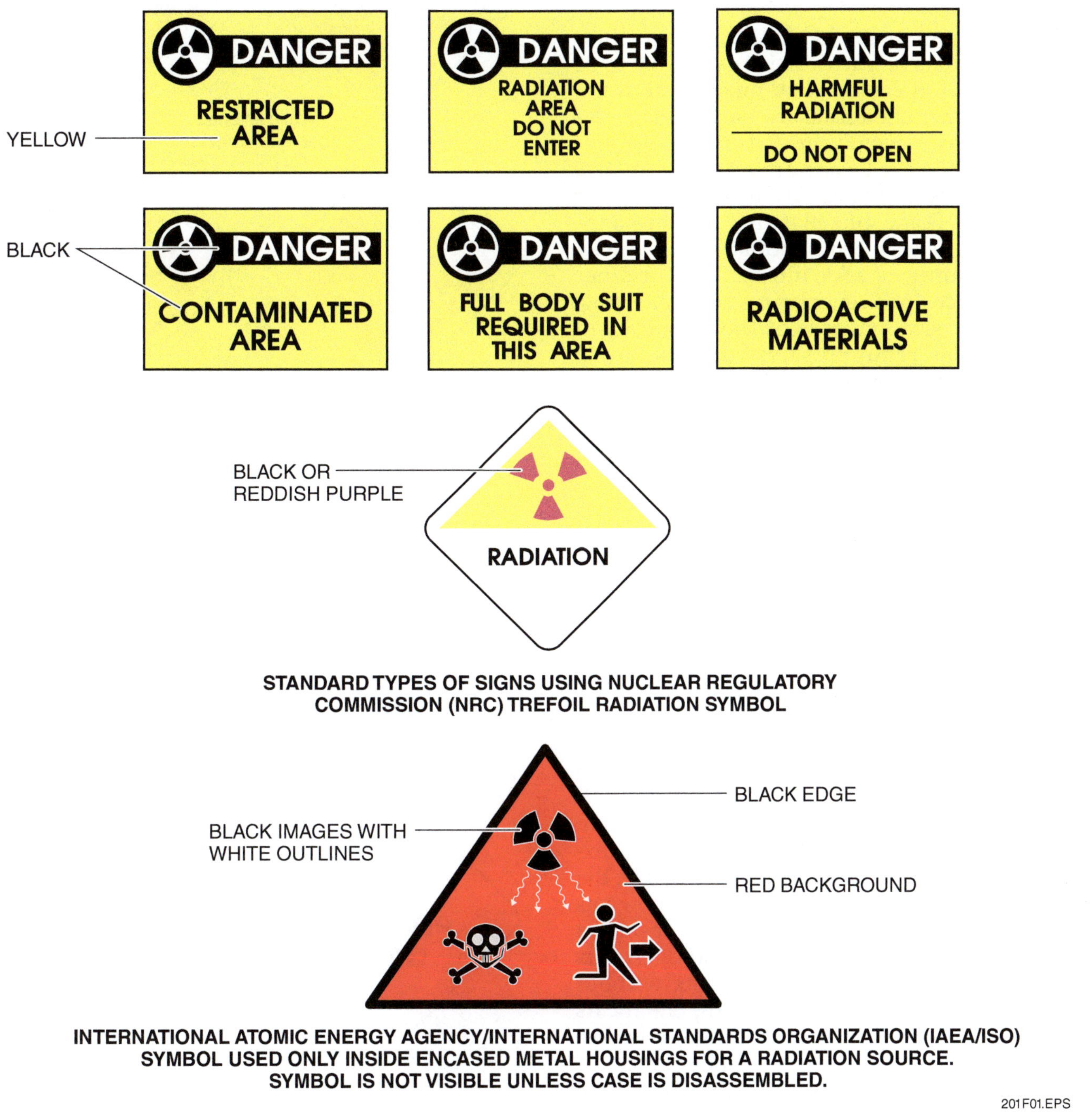

Figure 1 Warning signs for radiation hazards.

ity, and cancer. Radiation exposure can be minimized by limiting the amount of exposure time and/or increasing the distance from the source. Use the proper shielding or PPE. Check with the supervisor to find out if there are any radioactive hazards and how to avoid them.

> **WARNING**
>
> Radioactive materials require special handling, training, and PPE. Never enter an area where radioactive materials are being used unless the proper training is received and the appropriate PPE is used.

Chemicals can pose physical hazards, including acid reactions or burns. Acids can create toxic vapors or react violently when mixed with other chemicals. Other chemicals are flammable, **combustible**, or explosive (*Figure* 2). For example, solvents and compressed gases are often flammable. These materials can catch fire in their liquid or gaseous state. Some, like blasting caps, are dangerous solids. These items should be kept away from open flames, sparks, intense heat, or other ignition sources to prevent fires or explosions. Know the physical hazards of the chemicals in use, and follow required safety precautions to prevent fires and chemical reactions

Compressed gas is under very high pressure in a tank. Compressed gases must be handled carefully so that the tank is not damaged. A damaged tank may leak hazardous gases or vapors. If the tank is severely damaged, it may explode.

2.10.1 Chemical Hazards

Many chemicals are a health hazard. For example, solvents and paints can be poisonous. Even mild exposure can cause skin irritation, breathing problems, or allergic reactions. Hazardous products are marked TOXIC or POISON (*Figure* 3). A skull and crossbones symbol is frequently used to label poisons. Exposure to poisons can result in serious injury or death.

Chemicals can cause harm when they get into the body. This is known as chemical exposure. Exposure to toxic chemicals can occur through the skin, mouth, eyes, and lungs. These are known as routes of exposure. Chemicals can be absorbed through direct contact with skin, and open cuts increase exposure. Smoking or eating without washing your hands after handling chemicals or spent abrasives may cause ingestion of toxins. Chemicals can enter the eyes by rubbing them or by splashes, mists, or vapors. Breathing chemical vapors, mists, or dusts brings chemicals into the lungs. The PPE being used should reduce or eliminate the routes of chemical exposure.

The severity of chemical exposure depends on the individual, the toxicity, and the dose. Each person reacts to chemicals differently. Some are

Figure 2 Common labels for flammable, combustible, and explosive materials.

Figure 3 Common labels for toxic materials.

more sensitive than others. Chemical sensitivity can change over time. Toxicity is a measurement of how poisonous a chemical is. Some chemicals are very poisonous while others are only moderately irritating. The dose is how much of the chemical gets into the body over a certain time period. A big dose of a mild toxin can be deadly. Be aware of the dangers of the chemicals being used, and know what protection is required.

If a whole bottle of aspirin is swallowed, it will cause serious sickness. However, taking one or two aspirins for a headache can relieve minor pain. Different exposures have different results. Health officials calculate safety limits for various routes of exposure (eyes, skin, and mouth) over different time periods (15 minutes, 1 hour, 8 hours, 10 hours, and 12 hours). Generally, two extremes can occur: acute and chronic. A short-term, intense contact with a chemical is known as **acute exposure**. Some chemical exposure can affect a person right away. This is known as an **acute effect**. Dizziness, headaches, and skin rashes are common acute effects. For example, inhaling acetone vapors (exposure) will cause immediate dizziness (effect).

Some effects caused by chemicals do not appear for many months or even years. Low-level exposure over a long period of time can also result in health problems. This is known as **chronic exposure**. Often people do not realize they are being affected. Breathing asbestos dust over many years will cause asbestosis. This is a **chronic effect**.

It is important to know what types of chemicals you are working with and how to protect yourself. You also need to know what to do when you are exposed or chemicals are spilled. The product MSDS can help you. Always follow all established safety guidelines, wear the proper PPE, and ask the safety person on site if unsure of how to protect against chemical hazards.

2.11.0 Material Safety Data Sheets (MSDSs)

MSDSs are fact sheets prepared by the chemical manufacturer or importer. Each product used on a construction site must have an MSDS. An MSDS describes the substance, hazards, safe handling, first aid, and emergency spill procedures. OSHA does not have a mandatory form, but they do require inclusion of specific information. The American National Standards Institute (ANSI) has developed a standard form that meets national and international standards. Most chemical manufacturers use this form, which has the following sections:

- Chemical product and company information
- Composition/information on ingredients
- Hazard identification
- First aid measures
- Fire-fighting information
- Accidental release measures
- Handling and storage
- Exposure controls/personal protection
- Physical and chemical properties
- Stability and reactivity
- Toxicological properties
- Ecological properties
- Disposal considerations
- Transport information
- Regulatory information
- Other information

An MSDS can be difficult to read. The scientific information is fairly technical. *Figure 4* shows a sample MSDS for a common paint thinner. The most important things to look for on an MSDS are the specific hazards, personal protection, handling procedures, and first aid information. Most MSDSs have a 24-hour emergency-response number.

Read the sample MSDS in *Figure 4* and look for the safety information for this paint thinner. First, locate the hazards. Section 7 of the MSDS shows that the thinner is flammable, and relevant precautions are provided.

Section 8 shows that mechanical ventilation is needed to reduce hazardous vapors. This can be a fan in an open window. If ventilation cannot control exposure within acceptable limits, respiratory protection is needed. Section 8 also explains how to protect the eyes and skin.

Finally, find out how to dispose of the product. Section 13 lists requirements for safe disposal.

On a job site, all materials in containers must have a label. Labels describe what is in a container. They also warn of the chemical hazards. The *HazCom Standard* states that hazardous material containers must be labeled, tagged, or marked. The label must include the name of the material, the appropriate hazard warnings, and the name and address of the manufacturer. OSHA does not require a specific label format. Label information can be any type of message using words, pictures, or symbols. However, labels must describe the hazards present in a way that is readable and easily seen.

Material Safety Data Sheet
MATERIAL SAFETY DATA SHEET
R7K145
04 00

==

Section 1 -- PRODUCT AND COMPANY IDENTIFICATION

PRODUCT NUMBER HMIS CODES
Health 2*
R7K145 Flammability 3
Reactivity 0
PRODUCT NAME
Reducer No. 145
MANUFACTURER'S NAME EMERGENCY TELEPHONE NO.
THE SHERWIN–WILLIAMS COMPANY (216) 566–2917
101 Prospect Avenue N.W.
Cleveland, OH 44115
DATE OF PREPARATION INFORMATION TELEPHONE NO.
02–APR–07 (216) 566–2902

==

Section 2 -- COMPOSITION/INFORMATION ON INGREDIENTS
% by WT CAS No. INGREDIENT UNITS VAPOR PRESSURE

----

7 100–41–4 Ethylbenzene
ACGIH TLV 100 ppm 7.1 mm
ACGIH TLV 125 ppm STEL
OSHA PEL 100 ppm
OSHA PEL 125 ppm STEL
41 1330–20–7 Xylene
ACGIH TLV 100 ppm 5.9 mm
ACGIH TLV 150 ppm STEL
OSHA PEL 100 ppm
OSHA PEL 150 ppm STEL
51 107–98–2 1–Methoxy–2–propanol
ACGIH TLV 100 ppm 10.9 mm
ACGIH TLV 150 ppm STEL
OSHA PEL 100 ppm
OSHA PEL 150 ppm STEL

==

Section 3 -- HAZARDS IDENTIFICATION

ROUTES OF EXPOSURE
INHALATION of vapor or spray mist.
EYE or SKIN contact with the product, vapor or spray mist.
Contains alcohols and acetates which can be absorbed through the skin.
EFFECTS OF OVEREXPOSURE
EYES: Irritation.
SKIN: Prolonged or repeated exposure may cause irritation.
INHALATION: Irritation of the upper respiratory system.
May cause nervous system depression. Extreme overexposure may result in unconsciousness and possibly death.
SIGNS AND SYMPTOMS OF OVEREXPOSURE
Headache, dizziness, nausea, and loss of coordination are indications of excessive exposure to vapors or spray mists.
Redness and itching or burning sensation may indicate eye or excessive skin exposure.

201F04A.EPS

Figure 4 Sample MSDS (1 of 5).

R7K145 page 2

==

MEDICAL CONDITIONS AGGRAVATED BY EXPOSURE
None generally recognized.
CANCER INFORMATION
For complete discussion of toxicology data refer to Section 11.

==

Section 4 -- FIRST AID MEASURES

EYES: Flush eyes with large amounts of water for 15 minutes.
Get medical attention.
SKIN: Wash affected area thoroughly with soap and water.
Remove contaminated clothing and launder before reuse.
INHALATION: If affected, remove from exposure. Restore breathing. Keep warm and quiet.
INGESTION: Do not induce vomiting. Get medical attention immediately.

==

Section 5 -- FIRE FIGHTING MEASURES

FLASH POINT LEL UEL
80 F PMCC 1.0 13.7
FLAMMABILITY CLASSIFICATION
RED LABEL -- Flammable, Flash below 100°F (38°C)
EXTINGUISHING MEDIA
Carbon Dioxide, Dry Chemical, Foam
UNUSUAL FIRE AND EXPLOSION HAZARDS
Closed containers may explode when exposed to extreme heat.
Application to hot surfaces requires special precautions.
During emergency conditions overexposure to decomposition products may cause a health hazard. Symptoms may not be immediately apparent. Obtain medical attention.
SPECIAL FIRE FIGHTING PROCEDURES
Full protective equipment including self-contained breathing apparatus should be used.
Water spray may be ineffective. If water is used, fog nozzles are preferable. Water may be used to cool closed containers to prevent pressure buildup and possible auto ignition or explosion when exposed to extreme heat.

==

Section 6 -- ACCIDENTAL RELEASE MEASURES

STEPS TO BE TAKEN IN CASE MATERIAL IS RELEASED OR SPILLED
Remove all sources of ignition. Ventilate the area.
Remove with inert absorbent.

==

Section 7 -- HANDLING AND STORAGE

STORAGE CATEGORY
DOL Storage Class IC

201F04B.EPS

Figure 4 Sample MSDS (2 of 5).

R7K145 page 3

==

PRECAUTIONS TO BE TAKEN IN HANDLING AND STORAGE

Contents are FLAMMABLE. Keep away from heat, sparks, and open flame. During use and until all vapors are gone: Keep area ventilated – Do not smoke – Extinguish all flames, pilot lights, and heaters – Turn off stoves, electric tools and appliances, and any other sources of ignition.

Consult NFPA Code. Use approved Bonding and Grounding procedures.

Keep container closed when not in use. Transfer only to approve containers with complete and appropriate labeling. Do not take internally. Keep out of the reach of children.

==

Section 8 -- EXPOSURE CONTROLS/PERSONAL PROTECTION

PRECAUTIONS TO BE TAKEN IN USE

Use only with adequate ventilation.

Avoid contact with skin and eyes. Avoid breathing vapor and spray mist.

Wash hands after using.

VENTILATION

Local exhaust preferable. General exhaust acceptable if the exposure to materials in Section 2 are maintained below applicable exposure limits.

Refer to OSHA Standards 1910.94, 1910.107, 1910.108.

RESPIRATORY PROTECTION

If personal exposure cannot be controlled below applicable limits by ventilation, wear a properly fitted organic vapor/particulate respirator approved by NIOSH/MSHA for protection against materials in Section 2.

PROTECTIVE GLOVES

Wear gloves which are recommended by glove supplier for protection against materials in Section 2.

EYE PROTECTION

Wear safety spectacles with imperforated side shields.

OTHER PROTECTIVE EQUIPMENT

Use of barrier cream on exposed skin is recommended.

OTHER PRECAUTIONS

Intentional misuse by deliberately concentrating and inhaling the contents can be harmful or fatal.

==

Section 9 -- PHYSICAL AND CHEMICAL PROPERTIES

PRODUCT WEIGHT 7.41 lb/gal 887 g/l

SPECIFIC GRAVITY 0.89

BOILING POINT 248 – 292 F 120 – 144 C

MELTING POINT Not Available

VOLATILE VOLUME 100 %

EVAPORATION RATE Slower than ether

VAPOR DENSITY Heavier than air

SOLUBILITY IN WATER N.A.

VOLATILE ORGANIC COMPOUNDS (VOC Theoretical – As Packaged)

7.40 lb/gal 887 g/l Less Water and Federally Exempt Solvents

7.40 lb/gal 887 g/l Emitted VOC

201F04C.EPS

Figure 4 Sample MSDS (3 of 5).

R7K145 page 4

===

Section 10 -- STABILITY AND REACTIVITY

STABILITY -- Stable
CONDITIONS TO AVOID
None known.
INCOMPATIBILITY
None known.
HAZARDOUS DECOMPOSITION PRODUCTS
By fire: Carbon Dioxide, Carbon Monoxide
HAZARDOUS POLYMERIZATION
Will not occur

===

Section 11 -- TOXICOLOGICAL INFORMATION

CHRONIC HEALTH HAZARDS
Ethylbenzene is classified by IARC as possibly carcinogenic to humans (2B) based on inadequate evidence in humans and sufficient evidence in laboratory animals. Lifetime inhalation exposure of rats and mice to high ethylbenzene concentrations resulted in increases in certain types of cancer, including kidney tumors in rats and lung and liver tumors in mice.
These effects were not observed in animals exposed to lower concentrations.
There is no evidence that ethylbenzene causes cancer in humans.
Prolonged overexposure to solvent ingredients in Section 2 may cause adverse effects to the liver, urinary and reproductive systems.
Reports have associated repeated and prolonged overexposure to solvents with permanent brain and nervous system damage.

TOXICOLOGY DATA
CAS No. Ingredient Name

100–41–4 Ethylbenzene
LC50 RAT 4HR Not Available
LD50 RAT 3500 mg/kg
1330–20–7 Xylene
LC50 RAT 4HR 5000 ppm
LD50 RAT 4300 mg/kg
107–98–2 1–Methoxy–2–propanol
LC50 RAT 4HR Not Available
LD50 RAT 6600. mg/kg

===

Section 12 -- ECOLOGICAL INFORMATION

ECOTOXICOLOGICAL INFORMATION
No data available.

201F04D.EPS

Figure 4 Sample MSDS (4 of 5).

R7K145 page 5

==

Section 13 -- DISPOSAL CONSIDERATIONS

WASTE DISPOSAL METHOD
Waste from this product may be hazardous as defined under the Resource Conservation and Recovery Act (RCRA) 40 CFR 261.
Waste must be tested for ignitability to determine the applicable EPA hazardous waste numbers.
Incinerate in approved facility. Do not incinerate closed container. Dispose of in accordance with Federal, State/Provincial, and Local regulations regarding pollution.

==

Section 14 -- TRANSPORT INFORMATION

US Ground (DOT)
1 Gallon and Less may be Classed as CONSUMER COMMODITY, ORM-D
Larger Containers are Regulated as:
UN1263, PAINT RELATED MATERIAL, 3, PG III, (ERG#128)
DOT (Dept of Transportation) Hazardous Substances & Reportable Quantities
Ethyl benzene 1000 lb RQ
Xylenes (isomers and mixture) 100 lb RQ
Bulk Containers may be Shipped as (check reportable quantities):
RQ, UN1263, PAINT RELATED MATERIAL, 3, PG III, (XYLENES (ISOMERS AND MIXTURE)), (ERG#128) Canada (TDG)
UN1263, PAINT RELATED MATERIAL, CLASS 3, PG III, LIMITED QUANTITY, (ERG#128) IMO
UN1263, PAINT RELATED MATERIAL, CLASS 3, PG III, (27 C c.c.), EmS F-E, S-E

==

Section 15 -- REGULATORY INFORMATION

SARA 313 (40 CFR 372.65C) SUPPLIER NOTIFICATION
CAS No. CHEMICAL/COMPOUND % by WT % Element

100-41-4 Ethylbenzene 7
1330-20-7 Xylene 41
CALIFORNIA PROPOSITION 65
WARNING: This product contains chemicals known to the State of California to cause cancer and birth defects or other reproductive harm.
TSCA CERTIFICATION
All chemicals in this product are listed, or are exempt from listing, on the TSCA Inventory.

R7K145 page 6

==

Section 16 -- OTHER INFORMATION

This product has been classified in accordance with the hazard criteria
of the Canadian Controlled Products Regulations (CPR) and the MSDS contains all of the information required by the CPR.
The above information pertains to this product as currently formulated, and is based on the information available at this time. Addition of reducers or other additives to this product may substantially alter the composition and hazards of the product. Since conditions of use are outside our control, we make no warranties, express or implied, and assume no liability in connection with any use of this information.

201F04E.EPS

Figure 4 Sample MSDS (5 of 5).

Two common labeling systems are from the National Fire Protection Association (NFPA) and the U.S. Department of Transportation (DOT). The NFPA developed *NFPA 704* for stationary storage vessels, also called the NFPA hazard warning diamond (*Figure 5*). The four-color diamond can be a container label. It is also used on doors to note the hazard in a room or building. Each section and color represents a hazard: health, flammability, instability, and specific hazards. Numbers from zero to four indicate increasing hazards. The DOT also requires labels for hazardous material shipments (*Figure 6*). Either of these labels can be part of a company's HazCom labeling program. Be sure you understand the labeling method used by your company.

If the hazardous material is transferred from a labeled container, the new container must also be labeled. The new label must have all of the information from the original label. Make sure that any materials on the job site are clearly labeled.

WARNING

Never use a product or material from an unlabeled container.

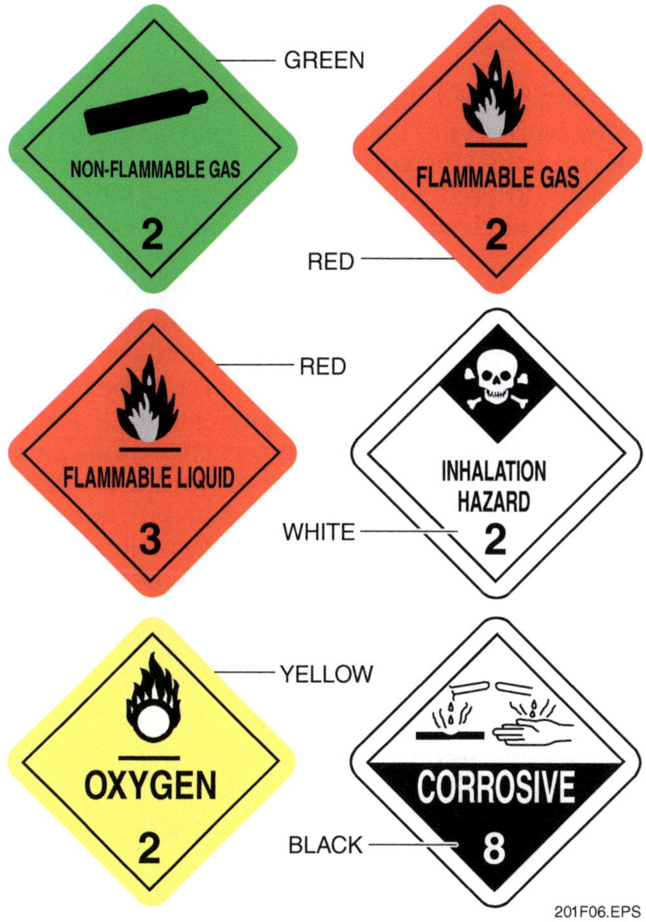

Figure 6 DOT hazardous materials labels.

FLAMMABILITY HAZARD (RED)

4 Extremely flammable – Rapidly vaporizes at normal pressure and temperature or is readily dispersed in air and will burn readily.
3 Flammable – Ignites at normal temperatures.
2 Ignites when moderately heated.
1 Ignites when preheated.
0 Will not burn.

HEALTH HAZARD (BLUE)

4 Extreme – Fatal with very short exposure. Wear special full protective suit and breathing apparatus.
3 Serious – Serious injury with short exposure. Wear full protective suit and breathing apparatus.
2 Moderate – Continued exposure can cause injury. Use breathing apparatus.
1 Slight – Exposure can cause irritation. Breathing apparatus may be worn.
0 Normal – No hazard.

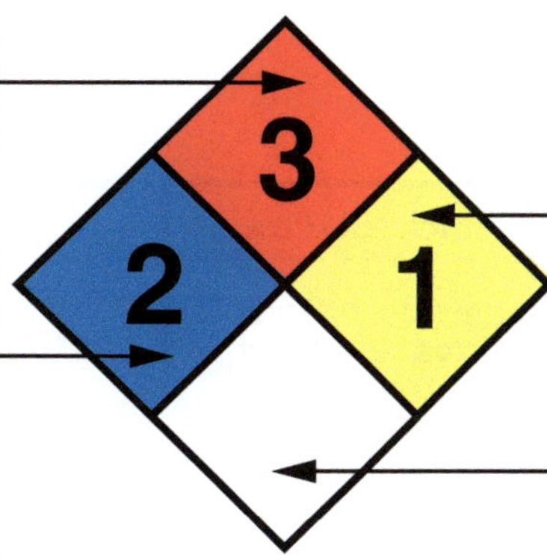

INSTABILITY (YELLOW)

4 Readily capable of detonation or of explosive decomposition or reaction at normal temperatures and pressures.
3 Capable of detonation or explosive reaction if shocked, heated under confinement, or mixed with water.
2 Violent chemical change possible but does not detonate. May react violently or form explosive mixtures with water.
1 Normally stable. Can become unstable at high temperatures and pressure. May react with water but not violently.
0 Normally stable. Not reactive to water.

SPECIFIC HAZARD (WHITE)

OXY	Oxidizer
ACID	Acid
ALK	Alkali
COR	CORROSIVE
~~W~~	Use NO WATER
☢	Radioactive

201F05.EPS

Figure 5 NFPA hazard warning diamond.

2.12.0 Personal Protective Equipment (PPE)

PPE is designed to protect workers from injury. Many workers are injured on the job because they are not using the correct PPE. In one incident, a worker suffered a serious eye injury after debris from the work site blew into his eyes. This worker had been told several times by his supervisor to wear safety glasses. Had he worn the proper safety glasses, this worker would not have been injured, nor would he have had to take time off from work to recover. See *Table 2* for some activities that require eye and face protection.

All the potentially dangerous conditions cannot be seen just by looking around a job site. Check the MSDSs for all of the products being used, and consider what type of accidents could happen before performing any job. Using common sense and knowing how to use PPE will greatly reduce any chances of getting hurt.

The best PPE is of no use unless it is:

- Regularly inspected.
- Properly cared for.
- Used properly when it is needed.
- Never altered or modified in any way.

PPE commonly used on work sites includes the following (see *Figures 7* and *8*):

- Hard hats
- Eye protection
- Safety harnesses
- Gloves
- Safety-toed shoes and boots
- Hearing protection
- Respiratory protection
- Protective clothing

CAUTION

Don't take chances with safety; use PPE every time the job requires it.

2.12.1 Respiratory Protection

Wherever there is danger of suffocation or other breathing hazards, a respirator must be used. Federal law specifies which type of respirator to use for various breathing hazards. Find out which respirator is required for the product being used by reading the site-specific respiratory plan the employer is required to prepare.

A respirator must be properly selected for the contaminant present and its concentration level. It must be properly fitted and used in accordance with the manufacturer's instructions. It must be worn during all times of exposure.

Employers must have a site-specific respiratory protection program with the following components:

- Standard operating procedures for selection and use
- Employee training
- Regular cleaning and disinfecting of equipment
- Sanitary storage
- Regular inspection
- Annual fit testing
- Medical qualifications, usually including pulmonary function testing

An employee is responsible for wearing respiratory protection when needed. In certain concentrations, vapors or fumes can be eliminated by the

Table 2 Eye and Face Protection Selection Chart

Source	Assessment of Hazard	Protection
IMPACT – Chipping, grinding, drilling, chiseling, sanding, etc.	Flying fragments, objects, large chips, particles, sand, dirt, etc.	Glasses with side protection, goggles, face shields. For severe exposure, use face shield over primary eye protection.
CHEMICALS – Acid and chemical handling	Splash	Goggles, eyecup, and cover types. For severe exposure, use face shield over primary eye protection.
	Irritating mists	Special purpose goggles.
DUST – Sanding, buffing, general dusty conditions	Nuisance dust	Goggles, eyecup, and cover types.

201T02.EPS

Figure 7 Examples of personal protective equipment.

REUSABLE HALF-MASK RESPIRATOR

POWERED AIR-PURIFYING RESPIRATOR

REUSABLE FULL FACEPIECE RESPIRATOR

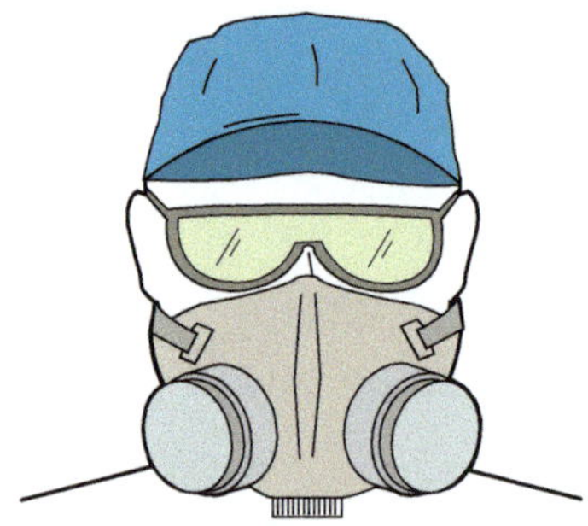

BEFORE USING A RESPIRATOR YOU MUST DETERMINE THE FOLLOWING:

1. THE TYPE OF CONTAMINANT(S) FOR WHICH THE RESPIRATOR IS BEING SELECTED
2. THE CONCENTRATION LEVEL OF THE CONTAMINANT(S)
3. WHETHER THE RESPIRATOR CAN BE PROPERLY FITTED ON THE WEARER'S FACE

ALL RESPIRATOR INSTRUCTIONS, WARNINGS, AND USE LIMITATIONS CONTAINED ON EACH PACKAGE MUST ALSO BE READ AND UNDERSTOOD BY THE WEARER BEFORE USE.

SELF-CONTAINED BREATHING APPARATUS

SUPPLIED-AIR RESPIRATOR

201F08.EPS

Figure 8 Respirators.

use of air-purifying devices, as long as the oxygen levels are acceptable. Smoke billowing from a fire and the fumes generated when welding are examples of fumes. Always check the cartridge on the respirator to ensure that it is the correct type for the air conditions and contaminants found on the job site (*Table 3*).

When selecting a respirator to wear while working with specific materials, first determine the hazardous ingredients contained in the material and their exposure levels, and then choose the proper respirator for protection at those levels. Always read the product's MSDS. It identifies the hazardous ingredients and should list the type of respirator and cartridge recommended for use with the product.

Limitations that apply to all half-mask (air-purifying) respirators are as follows:

- These respirators do not completely eliminate exposure to contaminants, but they will reduce the exposure to an acceptable level.
- These respirators do not supply oxygen and must not be used in areas where the oxygen level is below 19.5 percent.
- These respirators must not be used in areas where chemicals have poor warning signs, such as no taste or odor.
- Cartridges must have an indicator that shows when their useful life has ended or a schedule of how long they will protect a worker.

Be aware of breathing hazards. Effects of exposure include difficulty breathing, dizziness or nausea, and a chemical smell or taste. If these symptoms occur, leave the area immediately, return to a fresh air area, and seek any necessary assistance.

Respirators are useless unless properly fit-tested to each individual. To obtain the best protection from a respirator, perform positive and negative fit checks each time before wearing it. These fit checks must be repeated until a good face seal is obtained.

Table 3 Respirator Cartridge Color Code

Cartridge	Color Code
Organic Vapor	Black
Acid Gas	White
Organic Vapor/Acid Gas	Yellow
Ammonia	Green
Particle (HEPA)	Magenta
Multi-Gas/Vapor	Olive

201T03.EPS

To perform a positive fit check, do the following:

Step 1 Adjust the face piece for the best fit, then adjust the head and neck straps to ensure good fit and comfort.

WARNING

Do not overtighten the head and neck straps. Tighten them only enough to stop leakage. Overtightening can cause face-piece distortion and dangerous leaks.

Step 2 Block the exhalation valve with a hand or other material.

Step 3 Breathe out into the mask.

Step 4 Check for air leakage around the edges of the face piece.

Step 5 If the face piece puffs out slightly for a few seconds, a good face seal has been obtained.

To perform a negative fit check, do the following:

Step 1 Block the inhalation valve with a hand or other material.

Step 2 Attempt to inhale.

Step 3 Check for air leakage around the edges of the face piece.

Case History

Breathing Hazards

An employee sitting in a looped chain was lowered approximately 17 feet (5.2 m) into a 21-foot- (6.4 m) deep manhole. Twenty seconds later he started gasping for air and fell from the chain seat face down into the accumulated water at the bottom of the manhole. An autopsy determined that oxygen deficiency was the cause of death. Air testing and a respirator could have saved his life.

The Bottom Line: You can't see a breathing hazard. Make sure the air is tested and you have the right respirator for the job.

Source: OSHA

Step 4 If the face piece caves in slightly for a few seconds, a good face seal has been obtained.

A respirator must be clean, in good condition, and all of its parts must be in place for it to give proper protection. Respirators must be cleaned every day. Failure to do so will limit their effectiveness and offer little or no protection. For example, suppose the respirator was worn for two weeks and not cleaned. The bacteria from breathing into the respirator, plus the airborne contaminants that managed to enter the face piece, have made the inside of the respirator very unsanitary. Continued use may cause more harm than good. Remember, only a clean and complete respirator will provide the necessary protection. Follow these general guidelines:

- Inspect the condition of a respirator before and after each use.
- To get a proper face seal, do not wear a respirator if the face piece is distorted or if it is worn and cracked.
- Do not wear respirators if any part is missing. Replace worn straps or missing parts before use.
- Do not expose respirators to excessive heat or cold, chemicals, or sunlight.
- Clean and wash a respirator each day. Remove the cartridge and filter, hand wash the respirator using mild soap and a soft brush, and let it air dry overnight.
- Sanitize a respirator each week. Remove the cartridge and filter, and then soak the respirator in a sanitizing solution for at least two minutes. Thoroughly rinse with warm water and let it air dry overnight. Store the clean and sanitized respirator in its resealable plastic bag. Do not store the respirator face down. This will cause distortion of the face piece.

2.12.2 Hearing Protection

Some pneumatic tools, like jackhammers, can be very noisy. Wear ear protection when using these tools. OSHA requires hearing protection when workers are exposed to sound levels that average at or above 85 **decibels (dBA)** over an eight-hour period (known as an eight-hour time-weighted average or TWA). Actually, .85 dBA is not very loud. As *Table 4* shows, 85 dBA is about as loud as the alarm on a typical digital alarm clock. Even at this level, long-term exposure can pose a risk to a person's hearing. A jackhammer's sound intensity is typically about 100 dBA; when working with or near a jackhammer, always wear hearing protection.

The competent person on site, who is trained in fitting hearing protectors, must help workers choose the correct size and type of protection for their working environment. The protector selected should be comfortable to wear and reduce noise levels sufficiently to prevent hearing loss.

Table 4 Decibel Levels of Familiar Sounds

Noise Source	Decibel Level
Busy traffic	75dBA
Noisy restaurant	80dBA
Digital alarm clock	85dBA
Average factory	88dBA
Screaming child	90dBA
Subway train	100dBA
Diesel truck	100dBA
Helicopter	105dBA
Power mower	105dBA
Power saw	110dBA
Needle gun	112dBA
Abrasive blasting	115dBA
Football stadium	117dBA
600 cfm compressor	120dBA
Rock concert	120dBA
Car horn	120dBA
Jackhammer	130dBA
Air raid siren	130dBA
Pain starts	**140dBA**
Gunshots	140dBA
Jet engine	140dBA
Rocket launching	180dBa

201T04.EPS

Hearing protectors must adequately reduce the noise level for each employee's work environment. Generally, each type and model of hearing protection (earplug or earmuff) has a noise reduction rating (NRR) for how many decibels it will reduce the noise. Once the NRR number for the hearing protection is known, subtract 7 dBA. Doing this takes into account the background noise of the work site. The leftover NRR number is how low it will make the noise that enters your ears. Thus, if the sound of a needle gun is measured at 103 dBA and the earplugs have a NRR number of 30, subtract 7 from 30 to give a protection of 23. Subtracting 23 from 103 dBA results in only 80 dBA of exposure. Workers must be shown how to use and care for the protectors and must be supervised on the job to make sure that they know how and when to wear them.

A sound's intensity level depends on the distance from the source. The farther from the sound's source, the quieter it will seem. As a rule of thumb, the decibel level may be above the legal limit if you have to raise your voice to be heard one foot away. Other signs of high noise levels are temporary hearing loss or ringing in the ears.

Some workers may experience hearing loss even if noise is below the legal limit. It's best to avoid noise exposure whenever possible. If there's any reason to think the noise level may be too high, the company can have the level measured with instruments. This is called noise monitoring.

For more guidance on when to wear ear protection, what type of ear protection to wear, and how to wear it correctly, contact a supervisor or safety technician.

3.0.0 Work-Zone Safety

> **WARNING**
>
> There are specific requirements for working on or near railroads, highways, power plants, and other facilities. Always acquire and follow the safety plan for the specific work zone. Failure to do so can result in death, serious injury, and legal liabilities.

Coating work is often done in or near public areas. Creating a clear work zone is an important part of working safely. Barricades, fencing, caution tape, signs, and cones are used to mark a construction work zone. To ensure everyone's safety, keep people and vehicles away from the work area.

Signs, tags, and color codes are used in the workplace to protect employees from hazardous conditions and to assist them in responding to emergencies (*Figure 9*). For signs, tags, and color codes to be effective, all workers must understand what they mean and know what action they are required to take. This helps to avoid confusion and ensures their effectiveness.

Figure 9 Communication tags and signs.

Signals such as alarms, bells, buzzers, whistles, and horns can also be used to communicate hazards to workers. For example, back-up alarms are used on forklifts, construction equipment, and trucks. Fire alarms are used to clear work areas. Conveyer belt lines have buzzers, bells, and/or whistles to let workers know they are about to be started.

Barricades are another way to warn of potential danger. They are used on construction sites to keep out unauthorized personnel and control traffic (*Figure 10*). Barricades are also used to control pedestrian traffic outside or walking traffic in rooms and hallways that have ongoing coating operations.

It's important to recognize all of the signs and signals related to a job site and make sure they are properly placed and working correctly. Doing this can save lives.

3.1.0 Signs

All work sites have specific markings and signs to identify hazards and provide emergency information (*Figure 11*). Learn to recognize these types of signs:

- *Danger signs* – These signs warn workers of immediate hazards.
- *Caution signs* – These signs warn workers of potential hazards.
- *Informational signs* – These signs provide workers with general information not related to safety.
- *Safety signs* – These signs provide workers with general information related to safety issues.

When working at an unfamiliar location, always follow the posted signs.

3.2.0 Safety Tags

Accident-prevention tags are used as a temporary way of warning workers about immediate and potential hazards (*Figure 12*). They are similar to signs; however, they are not designed to be used in place of signs or as a permanent means of protection. For example, an Out of Order tag may be used on damaged equipment until it can be disposed of or repaired. A Do Not Start tag may be placed on machinery during lockout procedures. Tags can be an effective means of protecting workers and property. Tags and the devices used to attach them must meet specific physical requirements to ensure their durability and effectiveness.

Figure 10 Typical uses of barricades.

3.2.1 Do Not Start Tags

Do Not Start tags are used to prevent the unexpected energizing of equipment that could result in injury, equipment damage, or both. The tags must be placed in a conspicuous location that effectively blocks the starting mechanism. The tag is white. The square panel is red and the letters may be white, gray, or visibly etched on the tag.

3.2.2 Danger Tags

Danger tags are used where an immediate hazard exists. They tell workers that specific precautions must be observed. Danger tags are white. White letters appear on a red oval in a black square. A danger tag may read DANGER – UNSAFE – DO NOT USE.

3.2.3 Caution Tags

Caution tags are used to warn workers of potential hazards and to caution workers against unsafe practices. These tags inform workers that proper precautions should be taken. For example, a tag may read CAUTION – Stop Machinery to Clean, Oil, or Repair. Caution tags are yellow. They contain a black square panel with yellow letters.

3.2.4 Out of Order Tags

Out of Order tags should only be used to notify workers that a particular machine or tool is out of order and that using it may present a hazard. Out of Order tags are white. They contain a black square with white letters.

3.2.5 Radiation Tags

Radiation tags notify workers of actual or potential radiation hazards. Radiation tags are yellow with a black or reddish-purple panel. Any letters used against the yellow background are black. The radiation symbol should also appear on the tag.

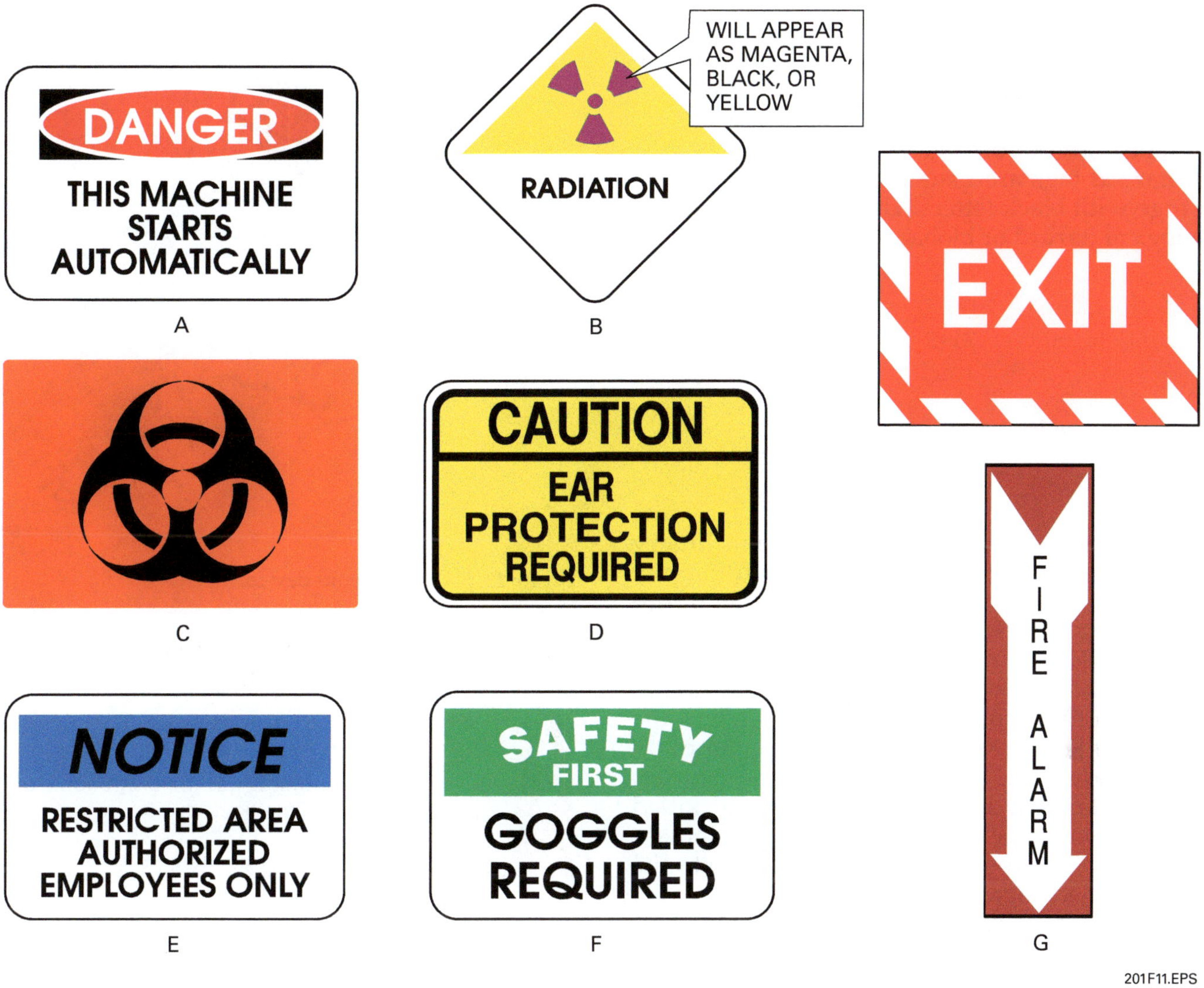

Figure 11 Signs.

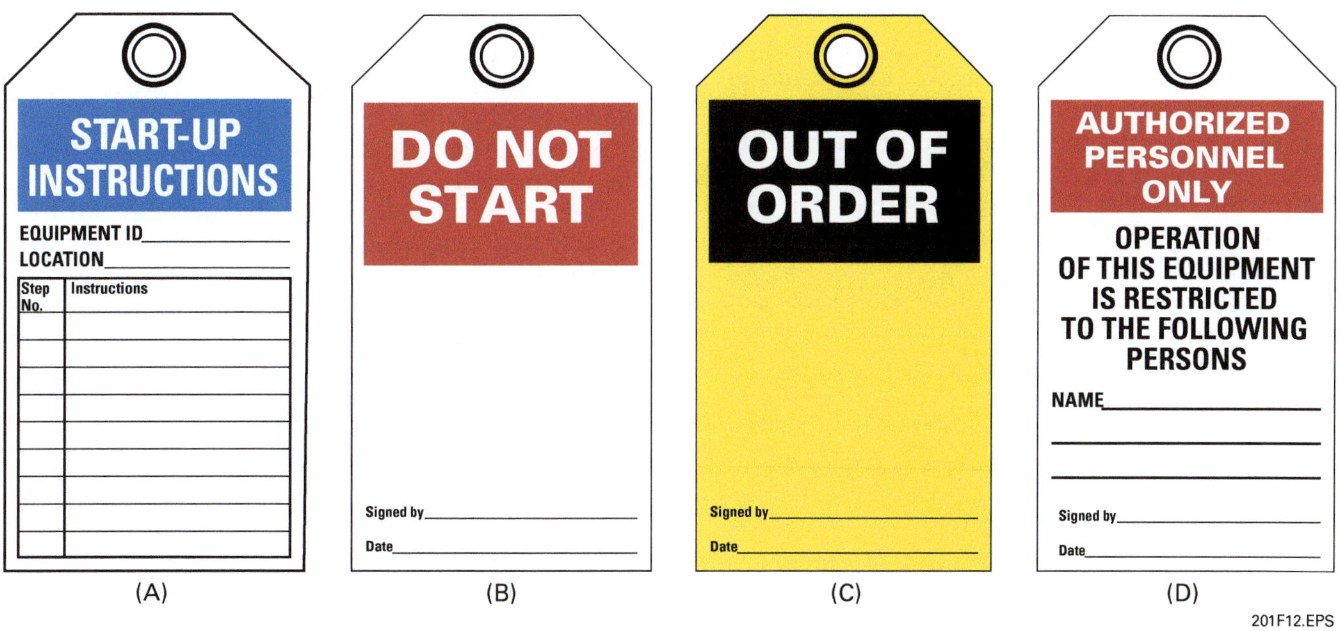

Figure 12 Examples of safety tags.

3.2.6 Biological Hazard Tags

Biological hazard tags signify the presence of an actual or potential biological hazard. Biological hazards can include any infectious agents that present a risk or potential risk to workers. Biological hazard tags are usually white or fluorescent orange with black letters. The tag also shows the biological hazard or biohazard symbol.

3.3.0 Lockout/Tagout

> **WARNING**
>
> Lockout/tagout varies by job site.

Many accidents on a job site involve machinery and equipment. These accidents often happen because there is an uncontrolled release of energy that can cause machinery to start up while being serviced or maintained. Failure to lockout and tag machinery before working on it is a major cause of injury and death on job sites. Some of those injuries include electrocution, amputation of body parts, and severe crushing injuries. In one instance, a maintenance worker was killed when he was knocked off balance by equipment that moved unexpectedly, knocking him into the spinning blade of a 36-inch (91 cm) saw.

The proper implementation of an effective lockout/tagout system can eliminate this type of hazard. It safeguards workers from unexpected releases from various **energy sources**. An energy source is any source of electrical, mechanical, hydraulic, pneumatic, chemical, thermal, or other energy. Energy can be stored in a system or machine even after the power is shut off. Stored energy in a system must be released or **grounded** to prevent unexpected operation of the machine while it is being serviced. For example, although the power has been shut off, residual air pressure remains in a compressor, which can activate **pneumatic tools**.

There is an additional danger if the machine contains acids, chemicals, flammable liquids, or high-temperature liquids or gases. Stored energy may cause the explosive release of these hazardous materials, resulting in serious injuries.

When anyone is working on or around any of these hazards, mechanical and other systems must be shut down, drained, or de-energized. **Energy-isolating devices**, such as switches, circuit breakers, valves, or other components, are used with tags and locks (*Figure 13*) to make sure that motors are not started, valves are not opened or closed, and any other changes that would endanger workers cannot be made.

Figure 13 Lockout/tagout device.

> **WARNING**
>
> Electrical equipment is not the only source of lockout/tagout accidents.

Some disconnect devices are equipped with keyed interlocks for protection during operation. These locks are called kirklocks and are relied upon to ensure proper sequence of operation only. They are not to be used for the purpose of locking out a circuit or system. Where disconnects are installed for use in isolation, they should never be opened under load. When opening a disconnect manually, it should be done quickly using a positive force. Again, lockouts should be used when the disconnects are open.

Generally, each lock has its own key, and the person who puts the lock on keeps the key. That person is the only one who can remove the lock. Tags typically have the word DANGER on them (*Figure 14*).

3.3.1 Lockout/Tagout Devices

Lockout and tagout protects workers from all possible sources of energy. In a lockout, an energy-isolating device such as a disconnect switch or circuit breaker is placed in the OFF position and locked. *Figure 15* shows some examples of the types of **lockout devices** that can be used with key or combination locks. Multiple lockout

Figure 14 Typical safety tags for lockout.

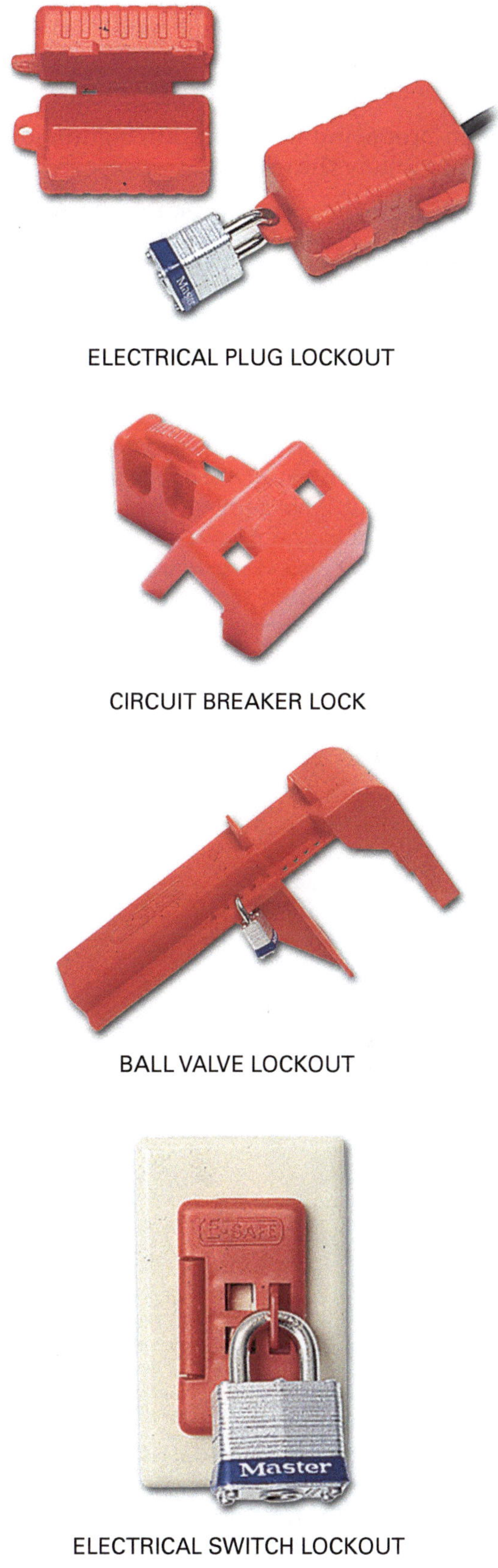

Figure 15 Lockout devices.

devices (*Figure 16*) are used when more than one person has access to the equipment. In some instances, as with work involving valves, a chain and lock can be used to hold the valve in place and keep it from being turned.

In a tagout, components that power up equipment and machinery, such as switches, are set in a safe position and a written warning or **tagout device** is attached to them (*Figure 17*).

An effective lockout/tagout program should include the following:

- An inspection of equipment by a trained individual who is thoroughly familiar with the equipment operation and associated hazards.
- The purchase of locks, tags, and blocks.
- Identification and labeling of lockout devices.
- A standard written operating procedure that is followed by all employees.

3.3.2 Lockout/Tagout Procedures

The exact procedures for lockout/tagout may vary at different companies and job sites. Ask the supervisor to explain the lockout/tagout procedure on the job site. Know this procedure and follow it. This is for everyone's safety. If unsure about the lockout/tagout procedures, ask the supervisor. When working at another facility, a company's authorized person must communicate with the onsite facility regarding lockout/tagout to ensure that the process, machine, or area has been completely de-energized.

The following is an example of a typical lockout/tagout procedure. It is made up of these four components:

- Preparation for lockout/tagout
- Sequence for lockout/tagout
- Restoration of energy
- Emergency removal authorization

When preparing for a lockout/tagout, it is important to follow these steps.

Step 1 Check the procedures to ensure that no changes have been made since last using a lockout/tagout.

Step 2 Identify all authorized and affected employees involved with the pending lockout/tagout.

NOTE

There is a difference between turning off a machine and actually disengaging a piece of equipment. When turning off a control switch, a circuit is opening; however, there is still electrical energy at the switch. A short in the switch or someone turning on the machine may start it running again.

To perform a lockout/tagout, follow these steps.

Step 1 Notify all authorized and affected personnel that a lockout/tagout is to be used and explain why it is necessary.

Step 2 Shut down the equipment or system using the normal OFF or STOP procedures.

Step 3 Lock out all energy sources and test disconnects to be sure they cannot be moved to the ON position. Next, open the control cutout switch. If there is no cutout switch, block the magnet in the switch open position before working on electrically op-

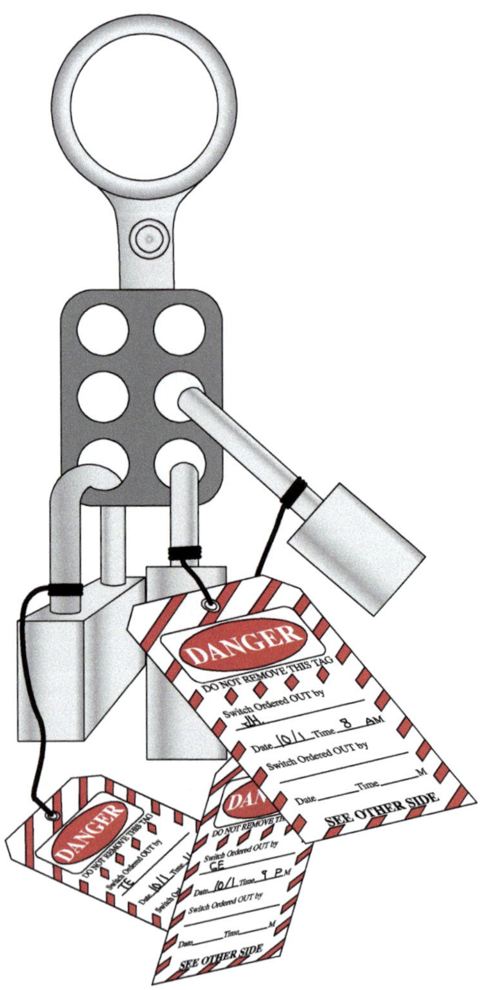

201F16.EPS

Figure 16 Multiple lockout/tagout device.

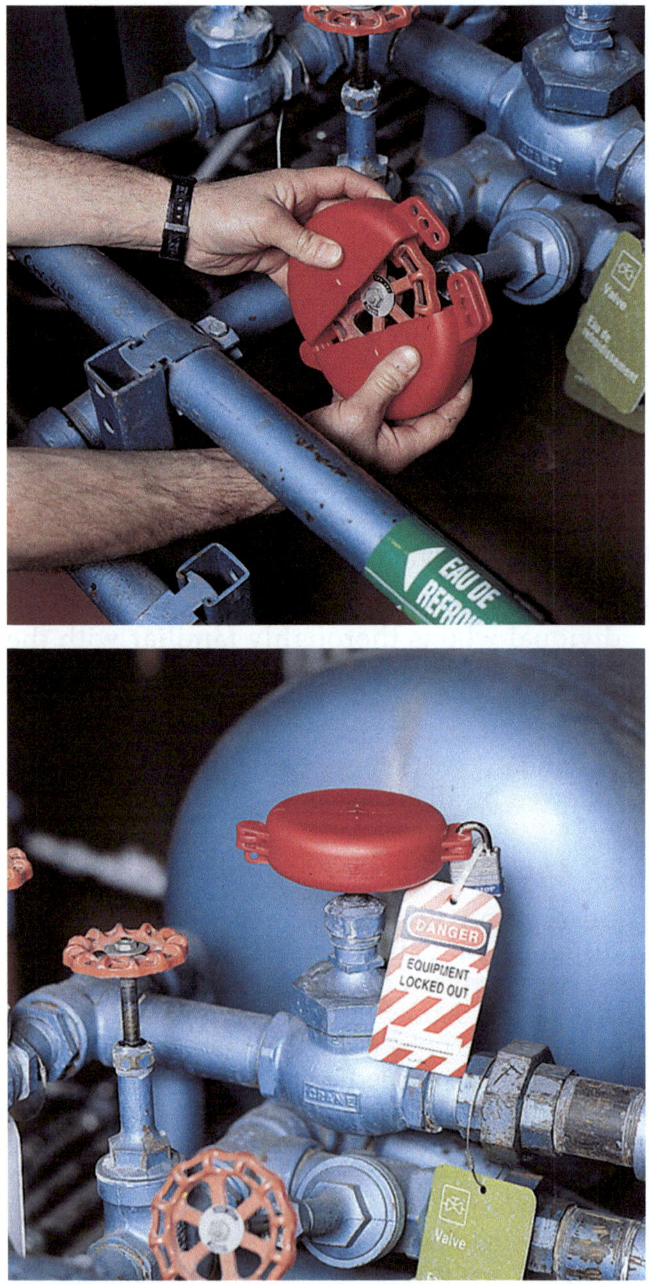

201F17.EPS

Figure 17 Placing a lockout/tagout device.

erated equipment or apparatus such as motors or relays. Remove the control wire.

Step 4 Lock and tag the required switches or valves in the open position. Each authorized employee must affix a separate lock and tag, if necessary.

Step 5 Dissipate any stored energy by attaching the equipment or system to ground.

Step 6 Using a known power source, verify that the test equipment is functional.

Step 7 Confirm that all switches are in the open position and use test equipment to verify that all components are de-energized.

Step 8 If it is necessary to temporarily leave the area, retest upon returning to ensure that the equipment or system is still de-energized.

After work is done on the machinery or equipment, use these steps to restore energy.

Step 1 Completely reassemble and secure the equipment or system.

Step 2 Confirm that all equipment and tools are accounted for and removed from the equipment or system.

Step 3 Replace and reactivate all of the safety controls.

Step 4 Remove the locks and tags from the isolation switches. Each employee must remove his or her own lock and tag.

Step 5 Notify all affected personnel that the lockout/tagout has ended and the equipment or system will be re-energized.

Step 6 Operate or close the isolation switches to restore energy.

On-Site

P.R.O.P.E.R. Lockout/ Tagout Procedures

Using P.R.O.P.E.R. lockout/tagout procedures is the best way to prevent accidents, injury, and death.

P – Process shut down.
R – Recognize energy sources.
O – Off (shut energy sources off).
P – Place locks and tags.
E – Release stored energy.
R – Return controls to neutral.

Think About It

Lockout Method

Which lockout method would you rather use if your life was at stake?

(A) TAPE

(B) CIRCUIT BREAKER LOCK

201SA01.EPS

There are times when emergency removal of a lockout/tagout device is required. The device can only be removed by an authorized person. If the authorized employee is absent or the key is lost, they must be informed that the lockout/tagout device has been removed. Written verification of the action taken, including informing the authorized employee of the removal, must be recorded in the job journal by the person who has taken such action.

3.3.3 Lockout/Tagout Safeguards

It is important to know how to protect workers on the job. Everyone must be aware of what activities are being done on the site and understand how to perform them safely. The following are the most common safeguards that should be used to keep the job site safe:

- Never operate any device, valve, switch, or piece of equipment that has a lock or a tag attached to it.
- Use only tags that have been approved for the job site.
- If a device, valve, switch, or piece of equipment is locked out, make sure the proper tag is attached.
- Lock out and tag all electrical systems.
- Lock out and tag pipelines containing acids, explosive fluids, and high-pressure steam.
- Lock and tag motorized vehicles and equipment when they require repair or are being replaced. Also, disconnect or disable any starting devices.

3.4.0 Barricades and Barriers

Any opening in a wall, floor, or the ground is a safety hazard. There are two types of protection for these openings: they can be covered or they can be guarded. Cover any hole whenever possible. When it is not practical to cover a hole, use barricades. If the bottom edge of a wall opening is less than three feet above the floor and would allow someone to fall four feet or more, place guards around the opening. There are several different guard methods:

- Railings are used across wall openings or as barriers around floor openings to prevent falls (*Figure 18A*).
- Warning barricades alert workers to hazards but provide no real protection (*Figure 18B*). Typical warning barricades are made of plastic tape or rope strung from wire or between posts. The tape or rope is color-coded:

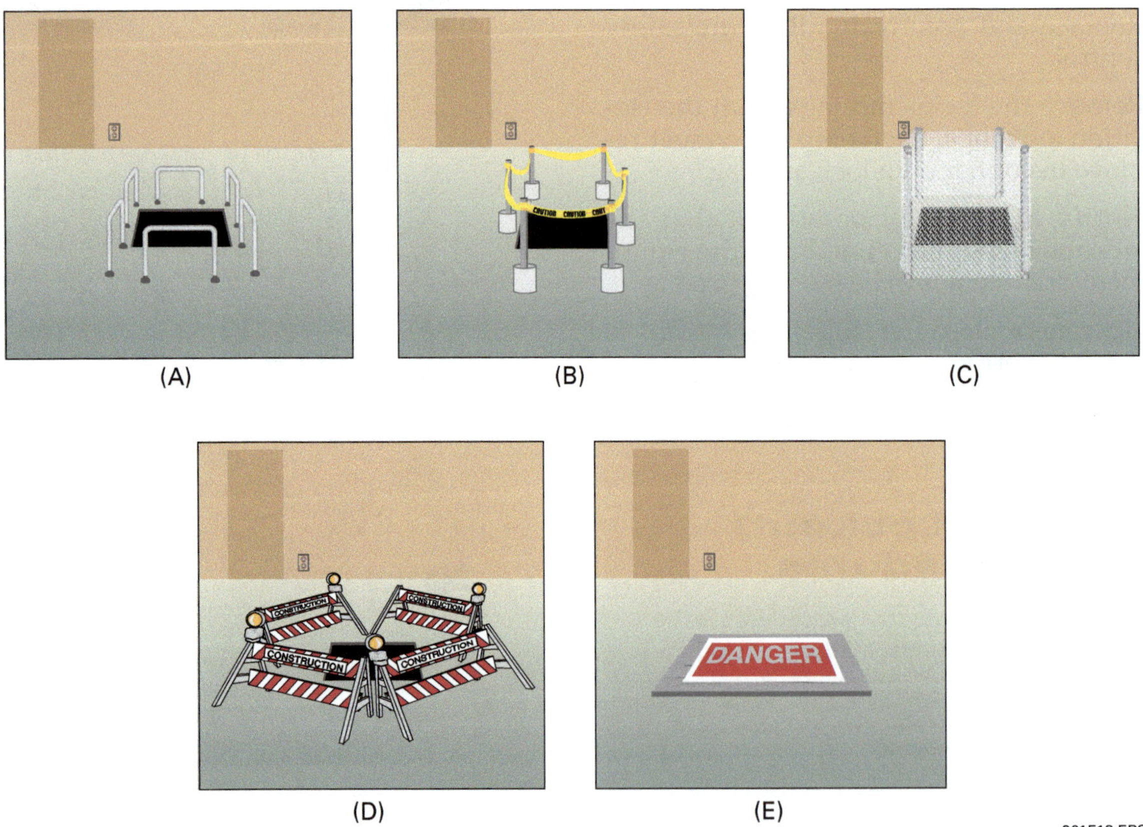

201F18.EPS

Figure 18 Common types of barricades and barriers for a hole in the floor.

- Red means danger. No one can enter an area with a red warning barricade. A red barricade is used when there is danger from falling objects or when a load is suspended over an area.
- Yellow means caution. You can enter an area with a yellow barricade, but you must know what the hazard is, and be careful. Yellow barricades are used around wet areas or areas containing loose dust. Yellow with black lettering warns of physical hazards such as bumping into something, stumbling, or falling.
- Yellow and purple indicates a radiation warning. No one may pass a yellow and purple barricade. These barricades are often used where piping welds are being X-rayed.

- Protective barricades provide both a visual warning and protection from injury (*Figure 18C*). They can be wooden posts and rails, posts and chain, or steel cable. People are not able to get past protective barricades.
- Blinking lights are placed on barricades so they can be seen at night (*Figure 18D*).
- Hole covers are used to cover open holes in the floor or ground (*Figure 18E*). They must be secured and labeled. They must be strong enough to support twice the weight of anything that may be placed on top of them.

The types of barriers and barricades will vary between job sites. There may also be different procedures for when and how barricades are put up. Learn and follow the policies at each job site.

4.0.0 Aboveground Work

Much of industrial coatings work occurs above ground, so ladders and **scaffolding** are a common sight on coating jobs. They are also a frequent source of fatal accidents. The most common accidents associated with ladders and scaffolding are falling, being struck by falling objects, and electrocution. In one instance, a worker died when he slipped from a fixed ladder attached to a water tower and fell 40 feet to the ground. He died because he was not using the proper safety equipment. Always use fall protection equipment with 100 percent tie-off.

Safety is a top priority on a job. It is each worker's responsibility to learn how to set up, use, and maintain equipment. This will keep everyone safe.

4.1.0 Ladders

Ladders are some of the most important tools on a job site. Different types of ladders should be used in different situations (*Figure 19*). Aluminum ladders are corrosion-resistant and can be used in situations where they might be exposed to the elements. They are also lightweight and can be used where they need to be frequently lifted and moved. Wooden ladders, which are heavier and sturdier than fiberglass or aluminum ladders, can be used when heavy loads must be moved up and down. Fiberglass ladders are nonconductive and also very durable, so they are useful in situations involving electrical work or where some amount of rough treatment is unavoidable. Both fiberglass and aluminum are easier to clean than wood.

Selecting the right ladder for the job at hand is important to completing a job as safely and efficiently as possible. When selecting a ladder, consider its features and how it meets the needs of the job. Always consider the highest **duty rating** and weight limit needed, as well as the height requirements. A ladder that is too long or too short will not allow the work surface to be reached easily, safely, or comfortably.

When selecting a ladder for a job, it is important to choose one that will extend at least 36 inches (91 cm) above the landing surface you are trying to reach. Always place the base of the ladder so that the distance between the base and the wall is one-quarter of the ladder length from the base elevation to the point where the ladder touches the wall (*Figure 20*).

Keep the following precautions in mind when setting up and using any type of ladder:

- Do not use a metal ladder when performing any type of electrical work or whenever there is a possibility of coming into contact with electrical conductors.
- Place the ladder in a stable manner.
- Place the ladder so that it leaves at least 6 inches (15 cm) of clearance in back of the ladder.
- Place the ladder so that it leans against a solid and immovable surface. Never place a ladder against a window, door, doorway, sash, loose or movable wall, or box.
- Face the ladder when climbing up or down.
- Climb or descend the ladder one rung at a time. Never run up or slide down a ladder.
- Do not use ladders during high winds. If using a ladder in windy conditions, make sure to lash the ladder securely in order to prevent slippage.
- Check to make sure work boot soles are free of oil, mud, and grease.
- Keep both hands free to hold the ladder securely while climbing. Use a rope to raise and lower any tools and materials.

Figure 19 Types of ladders.

- Never rest any tools or materials on the top of a ladder.
- Move the ladder in line with the work to be done. Never lean sideways away from the ladder in order to reach the work area.
- Never stand on the top two rungs of a ladder.
- Use ladders only for short periods of elevated work. If working from a ladder for extended periods, use a **lifeline** fastened to a safety belt.
- Never use makeshift substitutes for ladders.

- Lay the ladder on the ground when finished using it, unless it is anchored securely at the top and bottom where it is being used.
- Never use stepladders for straight-ladder work.

Each ladder is designed for a specific purpose and climbing conditions, and they generally fall into four categories: straight, extension, step, and fixed.

4.1.1 Straight Ladders

Straight ladders consist of two rails, rungs between the rails, and safety feet on the bottom of the rails (*Figure 21*). The straight ladders used in construction are generally made of wood or fiberglass. Metal ladders conduct electricity and should not be used around electrical equipment.

Ladders should always be inspected before use. Be sure to follow these guidelines when inspecting a ladder:

- Check the rails and rungs for cracks or other damage, including loose rungs. If any damage is found, do not use the ladder.
- Check the entire ladder for loose nails, screws, brackets, or other hardware. If any hardware problems are found, tighten the loose parts or have the ladder repaired before using it.
- Make sure the feet are securely attached, that there is no damage, and that they are not worn down. Do not use a ladder if its safety feet are not in good working order. If necessary for uneven surfaces, attach and use a stabilizer leg (*Figure 22*).

Setup is the next step after inspecting a ladder. Use these guidelines to make sure the ladder is set up safely:

201F20.EPS

Figure 20 Proper ladder positioning.

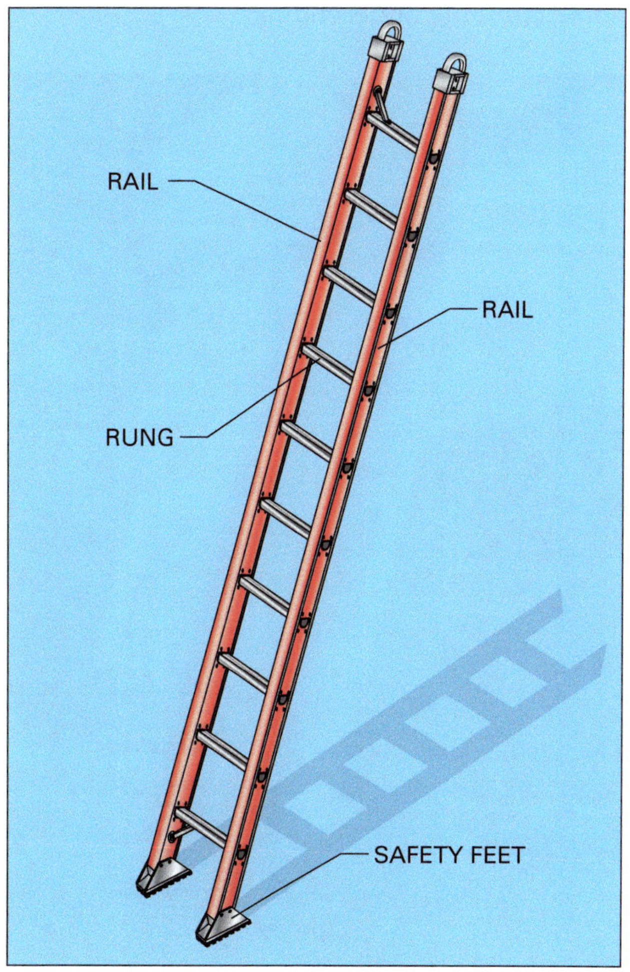

201F21.EPS

Figure 21 Straight ladder.

- Place the straight ladder at the proper angle before using it. A ladder placed at an improper angle will be unstable and could cause a fall.
- Straight ladders should be used only on stable and level surfaces unless they are secured at both the bottom and the top to prevent any accidental movement (*Figure 23*).
- The distance between the foot of a ladder and the base of the structure it is leaning against must be one-quarter the distance between the ground and the point where the ladder touches the structure.

Once the ladder has been inspected and set up, work can begin. Here are some safeguards to use while working on a ladder:

- Never try to move a ladder while someone is on it. If a ladder must be placed in front of a door that opens toward the ladder, the door must be locked or blocked open so that it cannot strike the ladder.
- Never use a ladder as a work platform by placing it horizontally.
- Make sure the ladder is properly secured before climbing or decending.
- Make sure the ladder's feet are solidly positioned on firm, level ground.
- Keep both hands on the rails when climbing a straight ladder.

201F22.EPS

Figure 22 Ladder stabilizer leg.

201F23.EPS

Figure 23 Securing a ladder.

- Always check to ensure that the top of the ladder is firmly positioned and in no danger of shifting to the right or left once climbing begins.
- Always keep body weight in the center of the ladder between the rails.
- Never go up or down a ladder while facing away from it. Face the ladder at all times.
- Don't carry tools while climbing a ladder. Instead use a hand line or tagline and pull tools up once you've reached the place you will be working.

WARNING

Remember that the addition of a person's weight will affect the ladder's steadiness while climbing. Test the ladder first by applying some weight to it without actually beginning to climb. This ensures that the ladder will remain steady.

4.1.2 Extension Ladders

An extension ladder is actually two straight ladders. They are connected so the overlap between them can be adjusted and the length of the ladder can be changed as needed (*Figure 24*).

Extension ladders are positioned and secured following the same rules as straight ladders. There are, however, some safety rules that are unique to extension ladders:

- When adjusting the length of an extension ladder, always reposition the movable section from the bottom, not the top, to make sure the rung locks are properly engaged after making the adjustment.

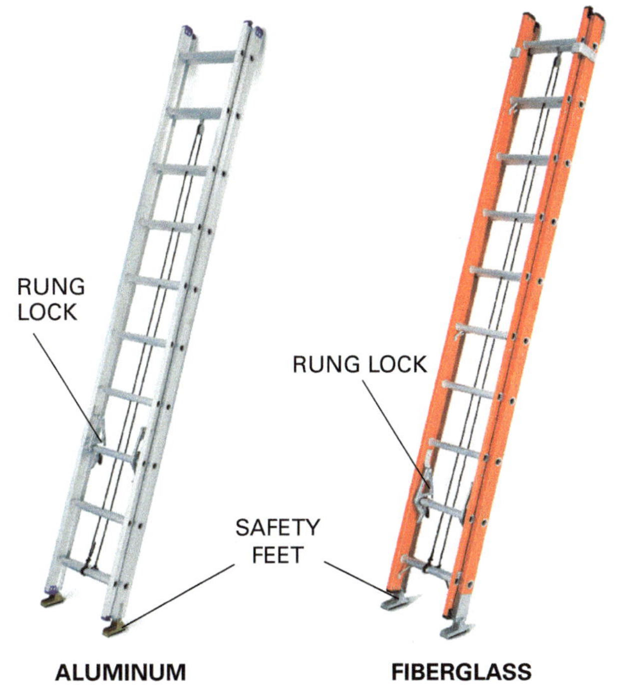

Figure 24 Extension ladders.

- Make sure the section locking mechanism is fully hooked over the desired rung.
- Make sure that all ropes used for raising and lowering the extension are clear and untangled.
- Make sure the extension ladder overlaps between the two sections (*Figure 25*). For ladders up to 36 feet (11 meters) long, the overlap must be at least 3 feet (91 cm). For ladders 36 to 48 feet (11 to 15 meters) long, the overlap must be at least 4 feet (1.2 meters). For ladders 48 to 60

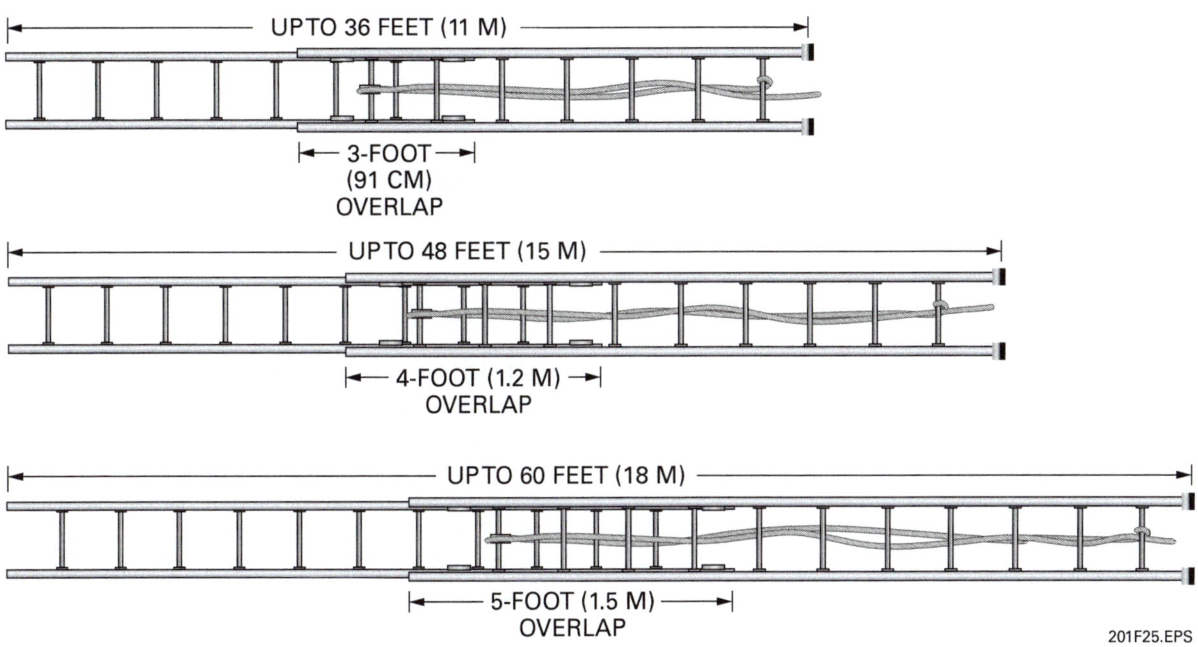

Figure 25 Overlap lengths for extension ladders.

feet (15 to 18 meters) long, the overlap must be at least 5 feet (1.5 meters).

- Never stand above the highest safe standing level on a ladder. On an extension ladder, this is the fourth rung from the top. Standing on a higher rung may result in a loss of balance or even a fall. Some ladders have colored rungs to show the highest safe standing level.

4.1.3 Stepladders

Stepladders are self-supporting ladders made of two sections hinged at the top (*Figure 26*). The section of a stepladder used for climbing consists of rails and rungs like those on straight ladders. The other section consists of rails and braces. Spreaders are the hinged arms between the sections that keep the ladder stable and prevent it from folding while in use. A stepladder may have a pail shelf to hold paint or tools.

Inspect stepladders in the same manner as straight and extension ladders. For stepladders, though, pay special attention to the hinges and spreaders to be sure they are in good repair. Also, be sure the rungs are clean. A stepladder's rungs are usually flat, so oil, grease, or dirt can easily build up on them and make them slippery.

Follow these rules when using stepladders:

- Be sure that all four feet are on a hard, even surface when positioning a stepladder. If they're not, the ladder can rock from side to side or corner to corner when climbing it.
- Never stand on the top step or the top of a stepladder. Putting any weight this high will make the ladder unstable. The top of the ladder is made to support the hinges, not to be used as a step.

201F26.EPS

Figure 26 Typical stepladder.

- Make sure the spreaders are locked in the fully open position when the ladder is in position.
- Never use the braces for climbing even though they may look like rungs. They are not designed to support weight.

NOTE

OSHA requires that all ladders used for construction be rated for heavy duty.

Figure 27 shows some common ladder safety precautions.

4.1.4 Fixed Ladders

Fixed or stationary ladders (*Figure 28*) are ladders that cannot be readily moved or carried because they are a permanent part of a building or structure. Fixed ladders must be capable of supporting at least two loads of 250 pounds (114 kilograms) each. Each step or rung must be capable of supporting a single concentrated load of at least 250 pounds (114 kilograms) applied in the middle of the step or rung. For ladders built after March 15, 1991, the rungs and steps must be corrugated, knurled, dimpled, coated with skid-resistant material, or otherwise treated to minimize slipping.

The angle of a fixed ladder must be no greater than 90 degrees from the horizontal, as measured at the back of the ladder. The ladder must be equipped with cages, wells, ladder safety devices, or **self-retracting lifelines** when the climb is less than 24 feet (7 meters), but the top of the ladder is a distance greater than 24 feet (7 meters) above lower levels.

Ladders where the total length of the climb equals or exceeds 24 feet (7 meters) must be equipped with one of the following:

- Ladder safety devices
- Self-retracting lifelines and rest platforms at intervals not to exceed 150 feet (46 meters)
- A cage or well and multiple ladder sections

4.2.0 Scaffolding

Scaffolding consists of elevated working **platforms** that support workers and materials. They are a very common sight on many construction jobs. Typical scaffolding is shown in *Figure 29*. The main part of the scaffolding is the working platform. A working platform should have a guardrail system that includes a top rail, midrail, toeboard, and screening. To be safe and effective, the top rail should be approximately 42 inches (1.07 m) high, the midrail should be located halfway between

Figure 27 Ladder safety.

the toeboard and the top rail, and the toeboard should be a minimum of 4 inches (10 cm) high.

If people will be passing or working under the scaffolding, the area between the top rail and the toeboard must be screened. Finally, the platform planks must be laid closely together. For safety purposes, the ends of the planks must overlap at least 6 inches (15 cm) and no more than 12 inches (30 cm).

There are many types of scaffolding used in residential and commercial construction. The three most common are supported scaffolding, swing scaffolding, and beam-suspended scaffolding.

201F28.EPS

Figure 28 Typical fixed ladder.

4.2.1 Scaffolding Hazards

Improper or careless use of scaffolding can result in accidents, injury, and/or death. Those who work on scaffolding can minimize their risks by being aware of the hazards involved and following the proper safety procedures and guidelines to minimize those hazards.

201F29.EPS

Figure 29 Typical scaffolding.

The main hazards involved with the use of scaffolding are falls, workers being struck by falling objects, and electric shock.

Falls can happen because fall protection has not been provided, is not used, or is installed or used improperly. Poorly planked scaffolding causes many falls. Working on scaffolding when conditions are dangerous, such as in high winds, ice, rain, and lightning, also leads to accidents. Likewise, falls happen when scaffolding collapses because of improper construction.

Fall protection is required on any scaffolding 10 feet (3 meters) or more above a lower level. Fall-protection devices consist of guardrail systems, **personal fall-arrest systems**, and/or safety nets. A guardrail system normally serves as adequate fall protection for most scaffolding.

Guardrail systems must extend around all open sides of the scaffolding. The side facing the work surface does not need to have a guardrail if it is located less than 14 inches (35 cm) away from the work surface. Any opening on a scaffolding platform must be protected by a guardrail system, including the access opening(s) and platforms that do not extend across the entire width of the scaffolding.

People who work or pass under scaffolding may be hit by falling objects. Tools, materials, debris, and scaffolding parts may fall to the surface below. Those working on scaffolding may also be injured if there are others working above them, or if the structure or workpiece extends above the work level of the scaffolding. Any worker who is exposed to the danger of falling objects is required to wear a hard hat. Depending on the situation, additional protection may be needed such as debris nets, screens or mesh, canopy structures, and toeboards. Barricades that prevent access under the scaffolding can also be used to protect workers and others.

> **WARNING**
>
> If the scaffolding shifts or begins to collapse, stop working and exit the scaffolding immediately.

Because most scaffolding is made of metal, electric shock is always a hazard. Never assume that you can work around high-voltage wires just by avoiding contact. High voltages can **arc** through the air and cause electrocution without direct contact. When scaffolding must be erected close to power lines, the utility company must be called in to de-energize, move, and/or cover the lines with insulating protective barriers.

Always refer to the competent person on site if you have any questions about the safety of scaffolding.

4.2.2 General Scaffolding Safety Guidelines

Safety begins by getting training in the proper use of scaffolding. It is equally important to always use the right safety equipment, including a hard hat and personal fall-protection systems.

A competent person must instruct the erection crew and all users of the scaffold must be trained before they can work on it. This competent person must be present whenever the scaffold is moved, altered, or taken down. In addition, the competent person must inspect the scaffold at the beginning of each shift, prior to work. This ensures that a knowledgeable person has verified the structure for safety. Do not climb a scaffold until it has been inspected.

Never work on scaffolding, in a cable lift work cage, or on a cable lift modular platform if you have any of the following conditions:

- Are subject to seizures
- Become dizzy or lightheaded when working at an elevation
- Take medication that might affect your stability and/or performance
- Are under the influence of drugs and/or alcohol

When working on scaffolding, always follow these guidelines:

- Never interchange parts of a scaffolding system made by different manufacturers.
- Erect and use scaffolding according to the manufacturer's instructions. Scaffolding must also be erected and used in accordance with all local, state, and federal/OSHA requirements.
- Attach a green, red, or yellow tag (*Figure 30*) as needed to any scaffolding that is assembled and erected to alert users of its current mechanical and/or safety status. However, do not rely solely on these tags. Inspect all parts of scaffolding before each use.
- If the scaffolding shifts, exit the scaffolding immediately.

4.2.3 Safety Guidelines for Supported Scaffolding

Supported scaffolding (*Figure 31*) is built from the ground up at a job site. Always inspect the erected scaffolding before use. Use the following guidelines for erecting and using tubular supported scaffolding:

- Inspect all scaffolding parts before assembly.
- Never use parts that are broken, damaged, or deteriorated. Be cautious of rusted materials.
- Follow the manufacturer's recommendations for the proper methods of erecting and using scaffolding.
- Do not interchange parts from different manufacturers.
- Do not force braces or other parts to fit. Adjust the level of the scaffolding until the connections can be made easily.
- Provide adequate sills or underpinnings for all scaffolding built on filled or soft ground. Compensate for uneven ground by using adjusting screws or leveling jacks.

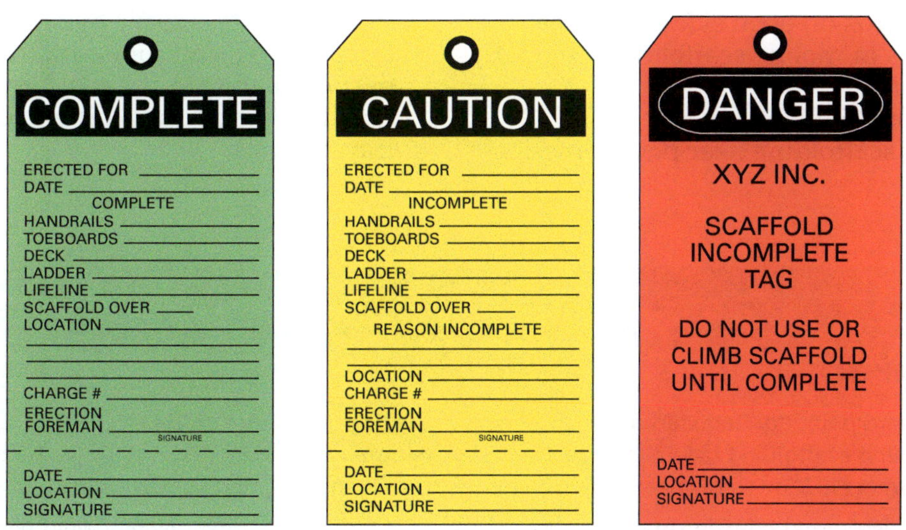

201F30.EPS

Figure 30 Typical scaffolding tags.

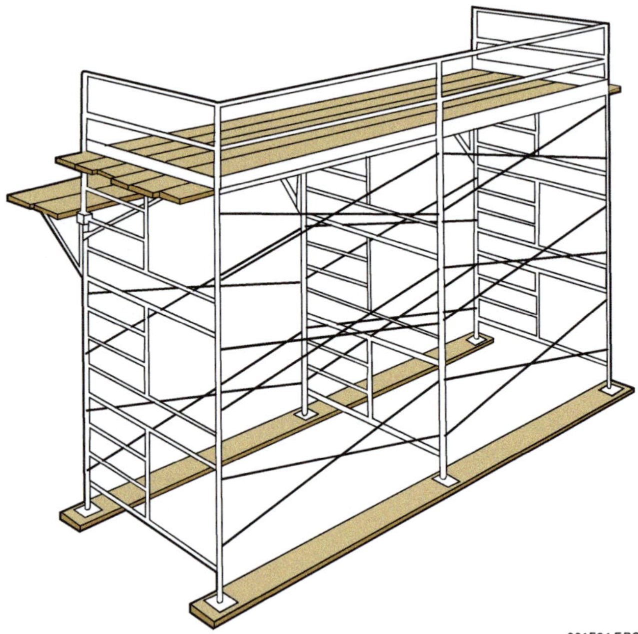

Figure 31 Supported scaffolding.

- Do not use boxes, concrete blocks, bricks, or other similar objects to support scaffolding.
- Keep scaffolding free of clutter and slippery material.
- Be sure scaffolding is plumb and level at all times. Follow the prescribed spacing and positioning requirements for the parts of the scaffolding. Anchor or tie-in scaffolding to the building at prescribed intervals.
- Use ladders rather than cross braces to climb the scaffolding. Position ladders with caution to prevent the scaffolding tower from tipping.
- Do not work on scaffolding that is more than 6 feet (1.8 meters) high without guardrails, midrails, and toeboards on open sides and ends.
- Lock the casters of mobile scaffolding when it is positioned for use.
- Do not ride on mobile scaffolding.
- Avoid building scaffolding near power lines.

4.2.4 Safety Guidelines for Swing and Other Suspended Scaffolding

Swing and modular scaffolding (*Figure 32*) is suspended by ropes or cables in a manner that allows it to be raised or lowered as needed. Another type of suspended scaffolding is a work cable lift cage (*Figure 33*). A work cage is typically suspended with rigging devices that attach to I-beams with various sizes of clamps and rollers. It is important to follow these guidelines when erecting and using swing and other suspended scaffolding:

- Follow the manufacturer's recommendations for installation, use, and maintenance of the equipment. Before installation, inspect all parts of a structure to which rigging and tieback lines will be secured to ensure that they can support the load.
- Be sure rigging devices are of the proper size and design to support the scaffolding and that they are installed properly. If counterweights are used to secure the inner end of outrigger beams, they should be fastened to the outrigger. Roofing materials and sand bags are not appropriate counterweights.
- Check that vertical lifelines for each person are installed perpendicular to the face of the structure and are secured to a solid support.
- Check for power lines or electric service wires on the job site. If they pose a hazard, contact the utility company to have them temporarily deactivated.
- Observe the scaffolding's load capacity; never overload the equipment.
- Inspect all scaffolding equipment each day. Check ropes and cables thoroughly for wear, fraying, corrosion, brittleness, damage, or other conditions that may weaken them. Have them replaced as necessary by qualified personnel.
- Keep suspension ropes and cables straight and perpendicular to the platform during use. Do not affix them to anything to change the line of travel.

Case History

Be Prepared

Two employees were painting the exterior of a three-story building when one of the two outriggers on their two-point suspension scaffolding failed. One painter safely climbed back onto the roof while the other fell approximately 35 feet (10.7 m) to his death. Neither painter was wearing an approved safety harness.

The Bottom Line: Be prepared for the unexpected. When working on scaffolding, always wear approved PPE.
Source: OSHA

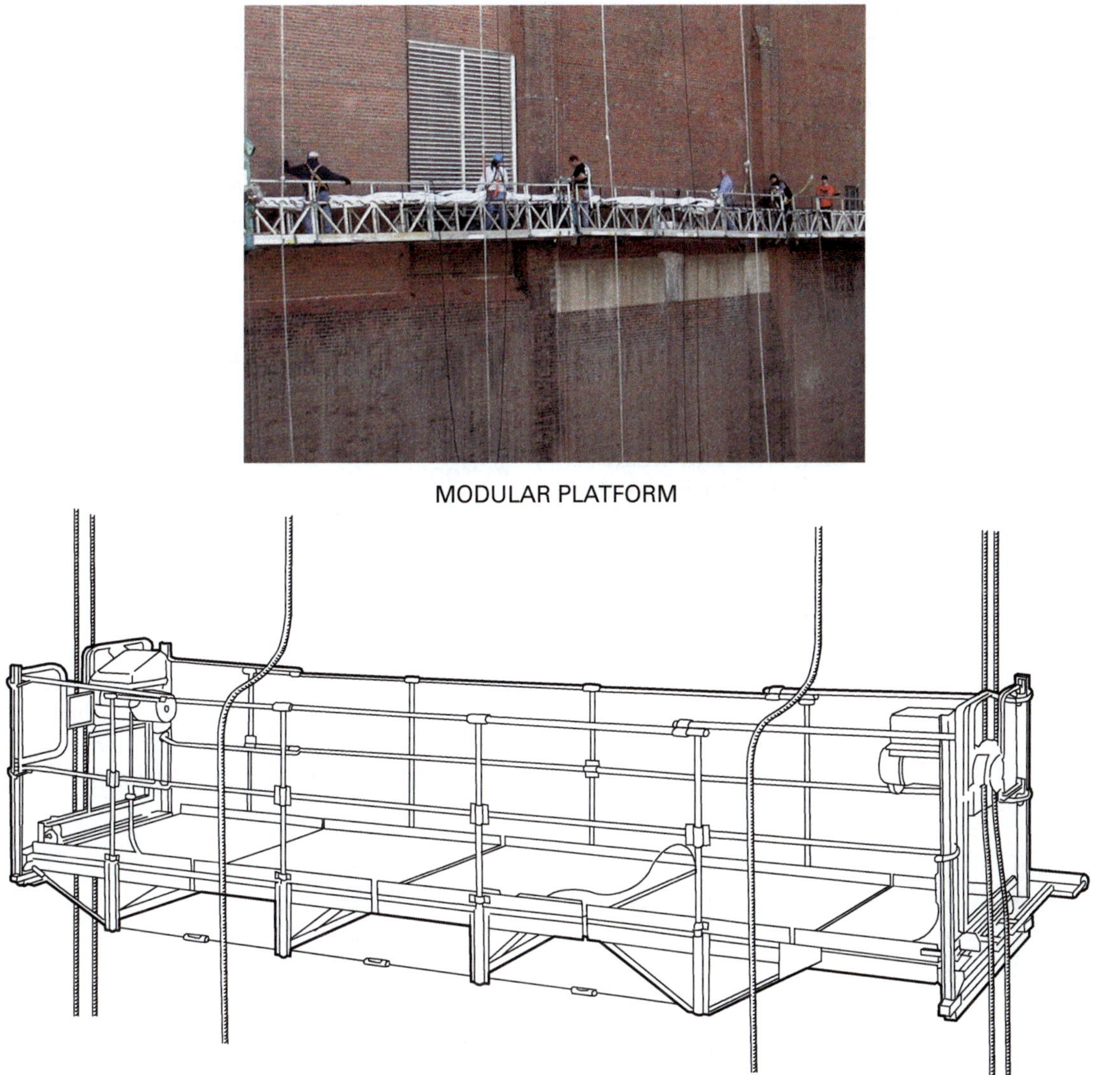

Figure 32 Swing and modular scaffolding.

- Lash the scaffolding to the building or structure to prevent it from swaying.
- Stay off scaffolding during storms or high winds; watch for icy or slippery platforms.
- Guarantee safe access to the swing stage at all times.
- Use two-way radios for communication between workers on the scaffolding and on the ground.
- Do not combine two or more **two-point swing scaffolding** units to form one unit.
- Raise and lash the scaffolding in a safe position when not in use.

Figure 33 Cable lift work cage.

Case History

Work Independently

Two workers were sandblasting a 110-foot (34-meter) high water tank while working on a two-point suspension scaffolding 60 to 70 feet (18 to 21 meters) above the ground. The scaffolding attachment point failed, releasing the scaffolding cables, and the scaffolding fell to the ground. Both workers died instantly because they were not tied off independently, nor was the scaffolding equipped with an independent attachment system.

The Bottom Line: Make sure that all safety equipment is in place before using any scaffolding.
Source: OSHA

4.3.0 Falls

Falls are the leading cause of industrial deaths. In fact, more than one third of all deaths in the construction industry are the result of a fall. Fall protection is required when workers are exposed to falls from work areas with elevations that are 6 feet (1.8 meters) or higher. The following types of work areas put the worker at risk of falling:

- Scaffolding
- Ladders
- Leading edges
- Ramps or runways
- Wall or floor openings
- Roofs
- Concrete forms
- Unprotected sides and edges

Falls happen because of the inappropriate use or lack of fall-protection systems (*Figure 34*). They also happen because of worker carelessness. It is your responsibility to learn how to set up, use, and maintain fall-protection equipment. Not only will this keep you alive and uninjured, it could save the lives of your co-workers.

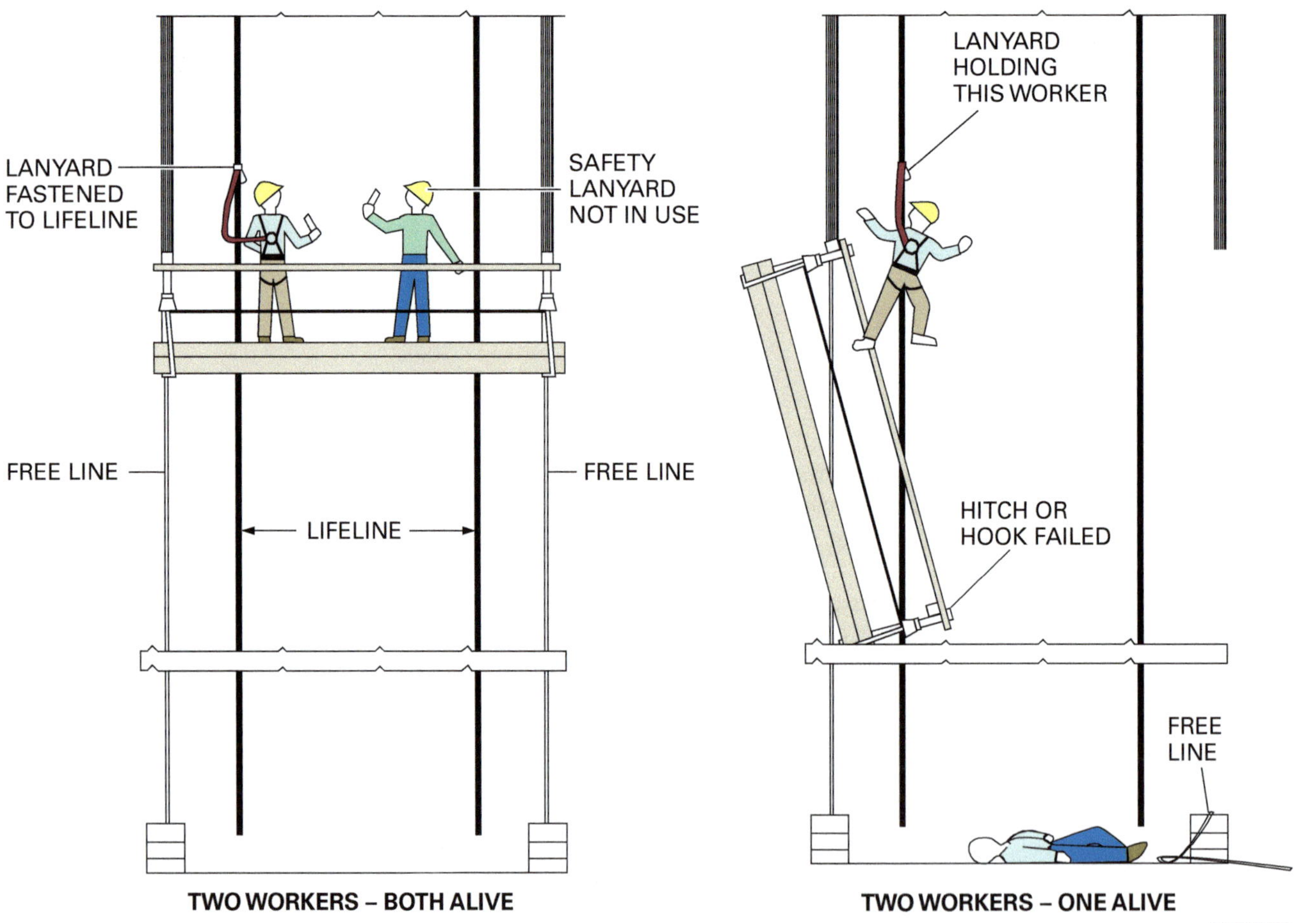

Figure 34 Proper and improper safety harness use.

4.3.1 Falling Hazards and Safeguards

Falls are classified into two groups: falls from an elevation and falls on the same level. Falls from an elevation can happen when someone is doing work from scaffolding, work platforms, decking, concrete forms, ladders, or excavations. Falls from elevations are almost always fatal. This is not to say that falls on the same level aren't also extremely dangerous. When a worker falls on the same level, usually from tripping or slipping, head injuries often occur. Sharp edges and pointed objects, such as exposed rebar, can cut or stab the worker.

The following safe practices can help prevent slips and falls:

- Wear safe, strong work boots that are in good repair.
- Watch where you step. Be sure your footing is secure.
- Do not get into an awkward position. Stay in control of movements at all times.
- Maintain clean, smooth walking and working surfaces. Fill holes, ruts, and cracks. Clean up slippery material and litter.
- Install cables, extension cords, and hoses so that they will not become tripping hazards.
- Do not run on scaffolding, work platforms, decking, roofs, or other elevated work areas.

The best way to survive a fall from an elevation is to use fall-protection equipment. The three most common types of fall-protection equipment are guardrails, personal fall-arrest systems, and safety nets.

4.3.2 Personal Fall-Arrest Systems

Personal fall-arrest systems catch workers after they have fallen. They are designed and rigged to prevent a worker from **free falling** a distance of more than 6 feet (1.8 meters) and hitting the ground or a lower work area. When describing personal fall-arrest systems, these terms must be understood:

- *Free-fall distance* – The vertical distance a worker moves after a fall before a deceleration device is activated.
- *Deceleration device* – A device such as a shock-absorbing **lanyard** or self-retracting lifeline that brings a falling person to a stop without injury.
- *Deceleration distance* – The distance it takes before a person comes to a stop when falling. The required deceleration distance for a fall-arrest system is a maximum of 3½ feet (1 meter).
- *Arresting force* – The force needed to stop a person from falling. The greater the free-fall distance, the more force is needed to stop or arrest the fall.

Personal fall-arrest systems use specialized equipment:

- **Body harnesses**
- Lanyards
- Deceleration devices
- Lifelines
- Anchoring devices and equipment **connectors**

Full-body harnesses with sliding back D-rings (*Figure 35*) are used in personal fall-arrest systems. They are made of straps that are worn securely around the user's body. This allows the arresting force to be distributed throughout the body, including the shoulders, legs, torso, and buttocks. This distribution decreases the chance of injury. When a fall occurs, the sliding D-ring moves to the nape of the neck. This keeps the worker in an upright position and helps to distribute the arresting force. The worker then stays in a relatively comfortable position while waiting for rescue.

Selecting the right full-body harness depends on a combination of job requirements and per-

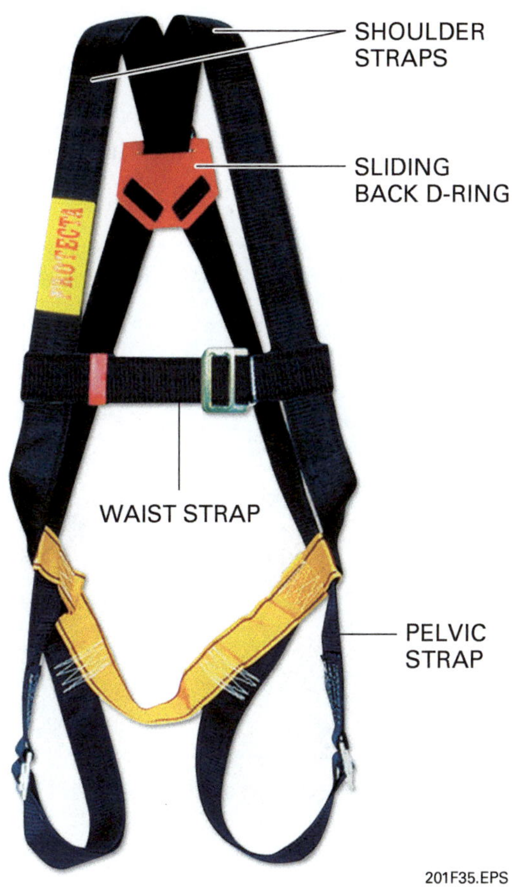

Figure 35 Full-body harness with sliding back D-ring.

sonal preference. Harness manufacturers normally provide selection guidelines in their product literature. Some types of full-body harnesses can be equipped with front chest D-rings, side D-rings, or shoulder D-rings. Harnesses with front chest D-rings are typically used in ladder climbing and personal-positioning systems (*Figure 36*). Those with side D-rings are also used in personal-positioning systems. Personal-positioning systems allow workers to hold themselves in place, keeping their hands free to accomplish a task.

A personal-positioning system should not allow a worker to free fall more than 2 feet (61 cm). The **anchorage** that it's attached to should be able to support at least twice the impact load of a worker's fall or 3,000 pounds (1,361 kilograms), whichever is greater.

Sliding back D-rings are the most common type of D-ring. The procedure for putting on a sliding back D-ring full-body harness is as follows:

Step 1 Hold the harness by the back D-ring.

Step 2 Shake the harness, allowing all the straps to fall into place.

Step 3 Unbuckle and release the waist and/or leg straps.

Step 4 Slip the straps over the shoulders so that the D-ring is located in the middle of the back (*Figure 37*).

Step 5 Fasten the waist strap. It should be tight but not binding.

201F36.EPS

Figure 36 Harness with front chest D-ring.

201F37.EPS

Figure 37 Properly positioning the D-ring.

Step 6 At each leg, pull the straps between the legs and buckle.

Step 7 After all the straps have been buckled, tighten all friction buckles so that the harness fits snugly but allows full range of movement.

Step 8 Pull the chest strap around the shoulder straps and fasten it in the mid-chest area. Tighten it enough to pull the shoulder straps taut.

Lanyards are short, flexible straps with connectors on each end (*Figure 38*). They are used to connect a body harness to a lifeline, deceleration device, or anchorage point. There are many kinds of lanyards made for different uses and climbing situations. All must have a minimum breaking strength of 5,000 pounds (2,268 kilograms). They come in both fixed and adjustable lengths and are made out of steel, rope, or nylon webbing. Some have a shock absorber (*Figure 39*) that absorbs up to 80 percent of the arresting force when a fall is being stopped. When choosing a lanyard for a particular job, always follow the manufacturer's recommendations.

NOTE

In the past, body belts were often used instead of a full-body harness as part of a fall-arrest system. As of January 1, 1998, however, they were banned from such use. This is because body belts concentrate all of the arresting force in the abdominal area and often resulted in broken backs when the worker fell. In addition, they cause the worker to hang in an uncomfortable and potentially dangerous position while awaiting rescue.

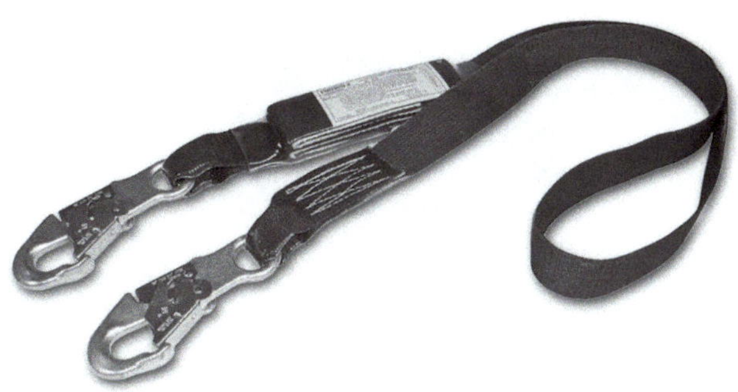

201F38.EPS

Figure 38 Typical lanyard.

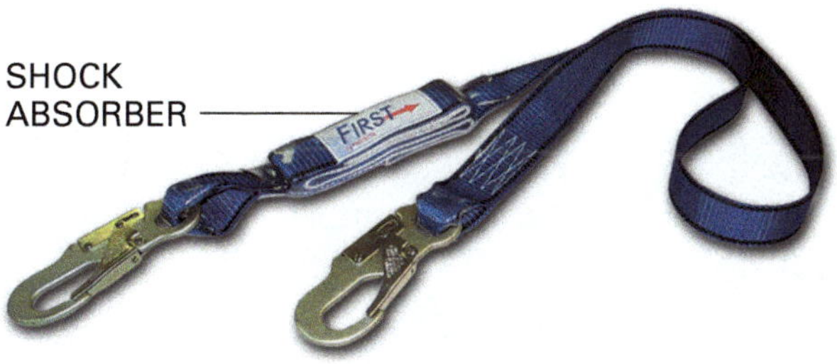

201F39.EPS

Figure 39 Lanyard with a shock absorber.

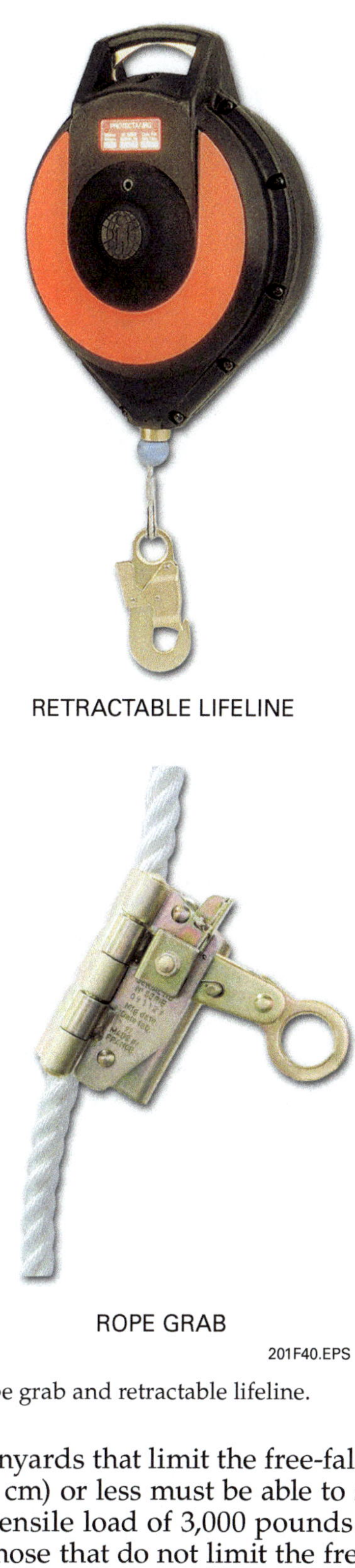

201F40.EPS

Figure 40 Rope grab and retractable lifeline.

CAUTION

When activated during the fall-arresting process, a shock-absorbing lanyard stretches in order to reduce the arresting force. This potential increase in length must always be taken into consideration when determining the total free-fall distance from an anchor point.

Deceleration devices limit the arresting force to which a worker is subjected when the fall is stopped suddenly. Rope grabs and self-retracting lifelines are two common deceleration devices (*Figure 40*). A rope grab connects to a lanyard and attaches to a lifeline. In the event of a fall, the rope grab is pulled down by the attached lanyard, causing it to grip the lifeline and lock in place. Some rope grabs have a mechanism that allows the worker to unlock the device and slowly descend down the lifeline to the ground or surface below.

Self-retracting lifelines provide unrestricted movement and fall protection while climbing and descending ladders and similar equipment or when working on multiple levels. Typically, they have a 25- to 100-foot (8- to 30-meter) galvanized-steel cable that automatically takes up the slack in the attached lanyard, keeping the lanyard out of the worker's way. In the event of a fall, a centrifugal braking mechanism engages to limit the worker's free-fall distance. Self-retracting lifelines and lanyards that limit the free-fall distance to 2 feet (61 cm) or less must be able to support a minimum tensile load of 3,000 pounds (1,361 kilograms). Those that do not limit the free fall to 2 feet (61 cm) or less must be able to hold a tensile load of at least 5,000 pounds (2,268 kilograms).

Lifelines are ropes or flexible steel cables that are attached to an anchorage. They provide a means for tying off personal fall-protection equipment. Vertical lifelines (*Figure 41*) are suspended vertically from a fixed anchorage. A fall-

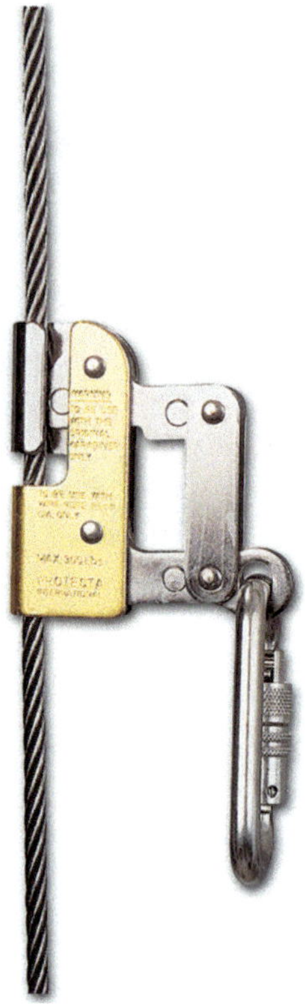

Figure 41 Vertical lifeline.

arrest device such as a rope grab is attached to the lifeline. Vertical lifelines must have a minimum breaking strength of 5,000 pounds (2,268 kilograms). Each worker must use his or her own line. This is because if one worker falls, the movement of the lifeline during the fall arrest may also cause the other workers to fall. A vertical lifeline must be connected in a way that will keep the worker from moving past its end, or it must extend to the ground or the next lower working level.

Horizontal lifelines (*Figure 42*) are connected horizontally between two fixed anchorages and are highly engineered systems that are normally permanently installed. Horizontal lifelines must be designed, installed, and used under the supervision of a qualified person. The more workers tied off to a single horizontal line, the stronger the line and anchors must be.

Anchoring devices (*Figure 43*), commonly called tie-off points, support the entire weight of the fall-arrest system. The anchorage must be capable of supporting 5,000 pounds (2,268 kilograms) for each worker attached. Overhead beams are considered anchorage points.

The D-rings, buckles, carabiners, and snaphooks (*Figure 44*) that fasten and/or connect the parts of a personal fall-arrest system are called connectors. There are regulations that specify how these are to be made, and that require D-rings and snaphooks to have a minimum tensile strength of 5,000 pounds (2,268 kilograms).

Figure 42 Horizontal lifeline and tieoff.

> **NOTE**
> Only locking-type snaphooks are permitted for use in personal fall-arrest systems.

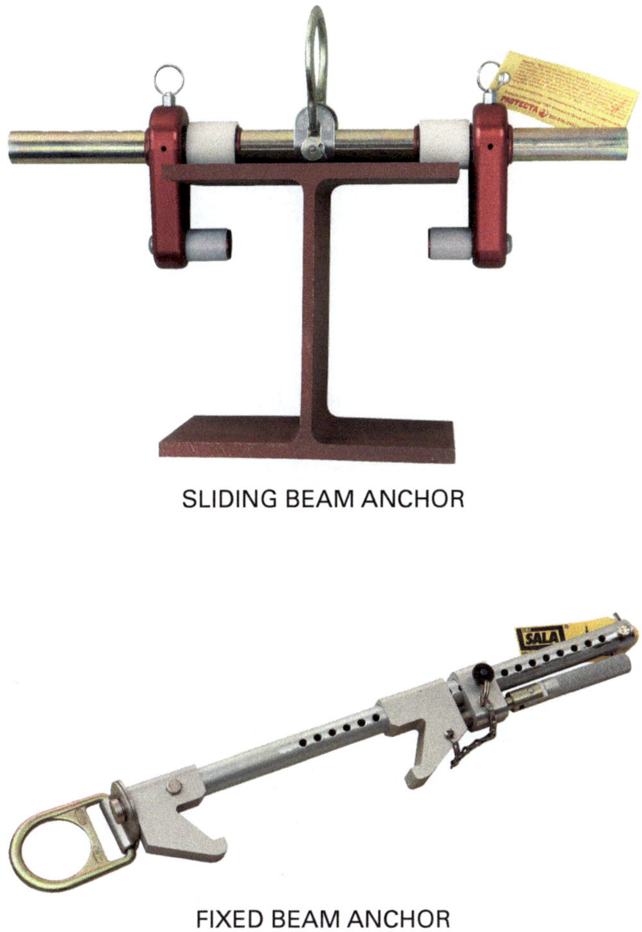

Figure 43 Lifeline beam anchorage devices.

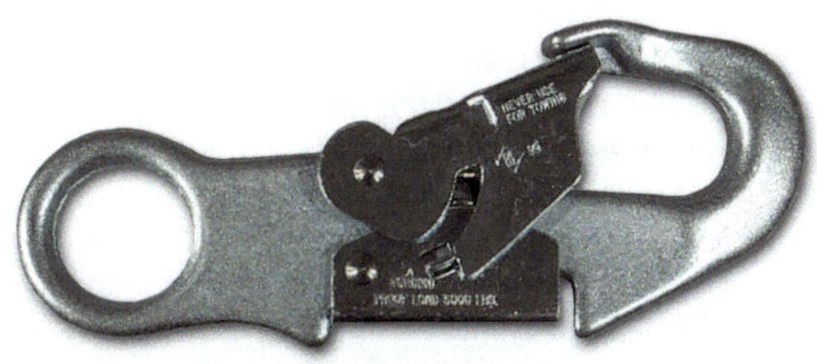

Figure 44 Double locking snaphook.

Connecting the body harness either directly or indirectly to a secure anchor point is called tying off. Tying off is always done before getting into a position from which a fall can occur. Follow the manufacturer's instructions for selecting a suitable anchorage point as well as the best tie-off methods for the equipment.

In general, an anchorage point should be:

- Directly above the worker
- Easily accessible
- Damage-free and capable of supporting 5,000 pounds (2,268 kilograms) per worker
- High enough so that no lower level is struck should a fall occur
- Separate from work basket tie-offs

Be sure to check the manufacturer's equipment labels and allow for any equipment stretch and deceleration distance.

Before using fall-protection equipment on the job, know the basics of fall protection and the proper use of the equipment. All equipment supplied by an employer must meet established standards for strength. Always read the instructions and warnings on any fall-protection equipment. Before each use, inspect the equipment using the following guidelines:

- Examine harnesses and lanyards for mildew, wear, damage, and deterioration.
- Make sure no straps are cut, broken, torn, or scraped.
- Check for damage from fire, chemicals, or corrosives.
- Make sure all hardware is free of cracks, sharp edges, and burrs.
- Check that snaphooks close and lock tightly and that buckles work properly.
- Check ropes for wear, broken fibers, pulled stitches, and discoloration.
- Make sure lifeline anchors and mountings are not loose or damaged.

Do not mix equipment from different manufacturers. All substitutions must be approved by a supervisor. All damaged or defective parts must be taken out of service immediately and tagged as unusable or destroyed. If the equipment was used in a previous fall, remove it from service until it can be inspected by a qualified person.

4.3.3 Safety Net Systems

Safety nets are used for fall protection on bridges and similar projects. They must be installed as close as possible, not more than 30 feet (9 meters), beneath the work area. There must be enough clearance under a safety net to prevent a worker who falls into it from hitting the surface below. There must also be no obstruction between the work area and the net.

Depending on the actual vertical distance between the net and the work area, the net must extend 8 to 13 feet (2.4 to 4 meters) beyond the edge of the work area. Mesh openings in the net must be limited to 36 square inches (91 square cm) and 6 inches (15 cm) on the side. The border rope must have a 5,000-pound (2,268-kilogram) minimum

On-Site

Drop Testing Safety Nets

Safety nets should be drop-tested at the job site after the initial installation, whenever relocated, after a repair, and at least every six months if left in one place. The drop test consists of dropping a 400-pound (181-kilogram) bag of sand of 29 to 31 inches (74 to 78 cm) in diameter into the net from at least 42 inches (1 meter) above the highest walking/working surface at which workers are exposed to fall hazards. If the net is still intact after the bag of sand is dropped, it passed the test.

breaking strength, and connections between net panels must be as strong as the nets themselves. Safety nets must be inspected at least once a week and after any event that might have damaged or weakened them. Worn or damaged nets must be removed from service.

4.3.4 Rescue After a Fall

Every elevated job site should have an established rescue and retrieval plan. Planning is especially important in remote areas that are not readily accessible to a phone service. Before beginning work, be aware of what an employer's rescue plan calls for the event of a fall. Find out what rescue equipment is available and where it is located. Learn how to use the equipment for self-rescue and the rescue of others.

If a fall occurs, any employee hanging from the fall-arrest system must be rescued safely and quickly. An employer should have previously determined the method of rescue for fall victims, which may include equipment that lets victims rescue themselves, a system of rescue by co-workers, or a way to alert a trained rescue squad. If a rescue depends on calling for outside help such as the fire department or rescue squad, all the needed phone numbers must be posted in plain view at the work site. In the event a co-worker falls, follow the employer's rescue plan. Call any special rescue service needed. Communicate with the victim and monitor him or her constantly during the rescue.

5.0.0 Roadway Work

Roadway work presents complex work situations in which workers face multiple injury risks under conditions that may change without warning. The most common sources of injury in highway work zones are the passing motor vehicles that enter the work space. Eighty percent of roadwork injuries are caused by out-of-control vehicles in the work space.

It is important to know how to stay safe in a consistently unsafe environment. The following guidelines can help save lives:

- Ensure that all the appropriate traffic control devices, such as signs (*Figure 45*), barriers, barricades (*Figure 46*), signals, and pavement markings, are in place before beginning work in a lane or road shoulder.
- NEVER work in an open lane of traffic.
- Always wear high-visibility, retro-reflective safety apparel (*Figure 47*) and PPE appropriate for the roadway involved. This must conform to the *ANSI/ISEA Standard 107-2004.*
- Only those personnel trained to be flaggers may control traffic.
- Be constantly aware of the surroundings and the equipment and traffic in the area.
- Make sure that communication is available between all roadway workers and operators of all equipment and vehicles in the work space.
- Understand where work is permitted and where it is prohibited for safety. Access to only the designated work space in the temporary traffic control (TTC) area is permitted for all workers, equipment, vehicles, and materials.

201F45.EPS

Figure 45 Typical highway construction signs.

Figure 46 Barricades and concrete (Jersey) barriers used in road construction.

5.1.0 Temporary Traffic Control (TTC)

Supervisors and/or dedicated personnel trained in TTC are responsible for planning and supervising the installation and maintenance of TTC zones and devices. Since 2000 and the advent of the *Millennium Edition of the National Manual on Uniform Traffic Control Devices (MUTCD)*, only personnel trained in TTC techniques are permitted to perform this work as part of the *MUTCD* and the *Code of Federal Regulations (CFR)*. Allowing untrained workers to perform this function can result in injury or death, and the supervisor(s) can be subject to fine and imprisonment for criminal negligence.

There are numerous TTC devices used at work sites, such as signs, signals, barricades, barriers, pavement markings, lights, etc., all of which must be installed in accordance with the TTC plan produced by an engineer trained in TTC techniques.

5.1.1 Signs

Traffic control signs provide general and specific warnings and instructions to road users, both alerting and warning them to the work and workers present and the actions that must be taken to properly and safely traverse the work area. These directions provide the information to enable the road user to avoid conflict with the workers and equipment in the work zone. It can be as simple as a single road worker symbol sign on a straight low volume road with a very short work zone, or as complex as a several-miles long series of warnings and directions on a freeway.

High-speed highways and freeways are as dangerous to road users as they are to workers due to the large differences in speeds and the size and location of vehicles, equipment, and materials.

5.1.2 Work Zone Traffic Control

Traffic control through a work zone is based upon several major items: the size of the work area, approach speeds, volume of traffic, and proximity of workers to the passing traffic. In a painting operation there is less heavy equipment moving in the work zone compared with other road construction activities, unless the painting operation is part of a highway or bridge rehabilitation project. These are mixed use operations areas, which become more dangerous without an internal traffic control plan (ITCP). An ITCP is a plan by the contractor to separate as much as possible the pedestrian workers from the motorized workers and their large equipment and vehicles. These plans frequently provide the use of traffic controllers and good communication equipment to keep all personnel in contact with one another.

On the roadside of the work area, work space cones and barricades are used to physically delineate and, along with portable barriers, separate workers from the traffic traversing through the TTC zone. Buffer spaces should always be provided to give errant vehicles room to recover before entering the work space. Additionally, on high speed arterials and freeways shadow or protection vehicles with mounted impact attenuators should be used to further protect the workers and road users in the work area.

CLASS	DESCRIPTION	WIDTH OF RETROREFLECTIVITY	SPEED	RECOMMENDED USERS
3	Greatest visibility & highest conspicuity	≥ 50mm ½ inch ¼ mile visibility	> 50 MPH	Roadway construction workers Vehicle operators Utility workers Survey crews Emergency responders Railway workers, Accident site investigators
2	Superior visibility & more conspicuous than Class 1	≥ 35mm 1⅜ inches 1,000 feet visibility	> 25 to 50 MPH	School crossing guards Parking and toll gate personnel Airport ground crews Railway workers Law enforcement personnel directing traffic
1	Away from traffic; conspicuous	≥ 25mm 1 inch 500 feet visibility	≥ 25 MPH	Parking service attendants Warehouse workers in equipment traffic Shopping cart retrievers and sidewalk maintainers

NOTE:
- Apparel background (outer) material color shall be either fluorescent orange-red or fluorescent yellow-green as per the standard.
- Retroflective material shall be either orange, yellow, white, silver, yellow-green, or a fluorescent version of these colors, visible at 1,000 ft minimum, and designed to identify the wearer as a person.

201F47.EPS

Figure 47 High-visibility safety apparel chart.

5.2.0 Flaggers

The *MUTCD* indicates that a person who directs traffic is a flagger. It further indicates that because flaggers are responsible for public safety and make the greatest number of contacts with the public, flaggers should be trained in safe traffic control practices and public contact techniques. Many states now require that the flaggers have special training and certification.

Flaggers' qualifications include the following:

- Sense of responsibility for safety of the public and the workers
- Adequate training in safe TTC techniques
- Average intelligence
- Good physical condition (sight, hearing, and mobility)
- Mental alertness

- Ability to react rationally and calmly in an emergency
- Courteous but firm manner
- Neat appearance

The *MUTCD* states that flaggers, day and night, shall wear safety apparel meeting the requirements of International Safety Equipment Association (ISEA) *American National Standard for High-Visibility Apparel* and labeled as meeting the *ANSI Standard 107-2004* performance for Class 2 risk exposure. For nighttime activity, safety apparel labeled as meeting the performance for Class 3 risk exposure should be considered for flagger wear.

The STOP/SLOW paddle should be the primary and preferred hand-signaling device, because it gives road users more positive guidance than red flags. Red flags should be limited to emergency situations when the paddle is not available.

Flaggers should be positioned far enough in advance of the work space to provide adequate warning to workers if a vehicle is likely to enter the work space.

6.0.0 Confined Spaces

Spaces on a job site are considered confined when their size and shape restrict the movement of anyone who must enter, work in, and exit the space. Confined spaces often have poor ventilation and are difficult to enter and exit. For example, employees who work in process vessels generally must squeeze in and out through narrow openings and perform their tasks while in a cramped or awkward position. In some cases, confinement itself creates a hazard.

In a confined space, hazards such as poor air quality, toxins, explosions, fire, and moving machinery parts tend to be far more deadly. Over an 8-year period in the 1990s an average of 38 workers per year were killed in confined spaces.

Most confined spaces have restricted entrances and exits. Workers are often injured as they enter or exit through small doors and hatches. It can also be difficult to move around in a confined space, and workers can be struck by moving equipment. Escapes and rescues are much more difficult in confined spaces than they are elsewhere.

Confined spaces are entered for inspection, equipment testing, repair, cleaning, or emergencies. They should only be entered for short periods of time.

Some of the confined spaces employees may work in include the following:

- Boilers
- Tunnels
- Sewers
- Underground utility vaults
- Pipelines
- Pits
- Air ducts
- Process vessels (*Figure 48*)

A written confined-space entry program can protect everyone. It will identify the hazards and specify the equipment or support that is needed to avoid injury. All industrial, and some construction sites, have written confined-space entry programs. Know and follow company policy.

6.1.0 Confined-Space Classification

Confined spaces must be inspected before work can begin. This helps to identify possible hazards. A space is considered confined if it can be described as follows:

- Is large enough and so configured that an employee can bodily enter and perform assigned work
- Has a limited or restricted means of entry or exit (such as tanks, vessels, silos, storage bins, hoppers, vaults, and pits)
- Is not designed for continuous employee occupancy

Figure 48 Entry into a process vessel.

After an inspection by a company-authorized person, the confined space is classified based on any hazards that are present. The two classifications are **non-permit required confined space** and **permit required confined space** (*Figure 49*).

> **WARNING**
> No one may enter a confined space unless they have been properly trained and have a valid entry permit.

6.1.1 Non-Permit Required Confined Space

A non-permit required confined space is a work space free of any mechanical, physical, electrical, and **atmospheric hazards** that can cause death or injury. After a space has been classified as a non-permit required space, workers can enter using the appropriate PPE for the type of work to be performed. Always check with a supervisor if it is unclear what PPE is required. Also be aware that coatings work often introduces additional hazards into the space. These include mechanical hazards from abrasive blasting and pressurized airless spray hoses; atmospheric hazards from abrasive dust; and vapor release from the coatings themselves.

6.1.2 Permit Required Confined Space

A **permit required confined space** is a confined space that has real or possible hazards. These hazards can be atmospheric, physical, electrical, or mechanical. *OSHA CFR 1910.146* defines a permit required confined space as a confined space with the following characteristics:

- Contains or has the potential to contain a hazardous atmosphere
- Contains a material that has the potential for engulfing an entrant
- Has an internal configuration such that an entrant could be trapped or asphyxiated by inwardly converging walls or by a floor that slopes downward and tapers to a small cross-section
- Contains any other recognized serious safety or health hazard

> **WARNING**
> All permit required confined spaces must be identified, and the associated permit must be posted.

An entry permit must be issued and signed by the job-site supervisor before the confined space is entered. No one is allowed to enter a confined

Figure 49 Non-permit required and permit required confined spaces.

space unless there is a valid entry permit. The permit is to be kept at the confined space while work is being performed. Always check with a supervisor if it is unclear whether or not a permit is needed to enter a confined space.

6.2.0 Entry Permits

Confined spaces can be extremely dangerous. Entry into the space begins when any part of the body passes the entrance or opening of a confined space. An entry permit is required before entering these spaces (*Figure 50*).

An entry permit is a job checklist that verifies that the space has been inspected. It also lets everyone on the site know about the hazards of the job. All entry permits must be filled out and signed by the supervisor before anyone enters the space. The permit must also be posted at the entrance to the site and be available for workers to review. Entry permits must include the following information:

- A description of the space and the type of work that will be done
- The date the permit is valid and how long it lasts
- Test results for all atmospheric testing including oxygen, toxin, and flammable material levels
- The name and signature of the person who did the tests
- The name and signature of the entry supervisor
- A list of all workers, including supervisors, who are authorized to enter the site
- The means by which workers and supervisors will communicate with each other
- Special equipment and procedures that are to be used during the job
- Other permits needed for work done in the space, such as welding
- The contact information for the emergency response rescue team

> **NOTE**
> Some confined spaces require both a permit to enter and a permit to work in the space.

6.3.0 Atmospheric Hazards

Atmospheric hazards are the most common hazards in a confined space. In a hazardous atmosphere, the air can have either too little or too much oxygen, be explosive or flammable, or contain toxic gases. Special meters are used to detect these atmospheric hazards (*Figure 51*).

6.3.1 Oxygen-Deficient or Oxygen-Enriched Atmospheres

A confined space that does not have enough oxygen is called an **oxygen-deficient** atmosphere; a confined space that has too much oxygen is called an **oxygen-enriched** atmosphere. For safe working conditions, the oxygen level in a confined space must range between 19.5 percent and 23.5 percent by volume, with 21 percent being considered the normal level. Oxygen concentrations below 19.5 percent by volume are considered oxygen-deficient; those above 23.5 percent by volume are considered oxygen-enriched.

> **WARNING**
> Never enter a confined space if unsure how safe it is.

Many of the following processes that occur in a confined space use oxygen and may reduce the percentage of oxygen to an unsafe level:

- Burning
- Rusting of metal
- Breaking down of plants or garbage
- Oxygen mixing with other gases

> **WARNING**
> Normal breathing in a confined space can also create an oxygen deficiency.

When the oxygen in a confined space is reduced, it becomes harder to breathe. The symptoms of insufficient oxygen happen in this order:

1. Fast breathing and heartbeat
2. Impaired mental judgment
3. Extreme emotional reaction
4. Unusual fatigue
5. Nausea and vomiting
6. Inability to move your body freely
7. Loss of consciousness
8. Death

6.3.2 Combustible Atmospheres

Air in a confined space becomes combustible when chemicals, gases, or paint dust reach a certain concentration. Flammable gases and vapors can be trapped in confined spaces. Dust and work byproducts from spray painting or surface preparation work can also form a combustible atmosphere.

Attachment 16 – 2
Confined-Space Entry Permit
Master Card / Safe Work Ticket No.________________

1. Work Description: ____
Equip. Name / Number & Location or Area ______________________________
Purpose of Entry____
Valid Start Date______________________ Duration Time________ to ________

2. Hazardous Materials:
What did the equipment last contain? ______________________________
Will the work generate a hazardous atmosphere? ☐ Yes ☐ No If yes, specify hazards and controls.
____

3. Rescue Requirements:
☐ External, by attendant ☐ Complex Rescue, by rescue team at point of entry
☐ Non-IDLH and/or Simple Rescue, by rescue team on-site ☐ IDLH, by rescue team at point of entry
Has the rescue team been notified of the entry? ☐ Yes ☐ N/A Time of notification______________
How will the rescue team be summonsed for an emergency? ☐ Radio Channel: ____ ☐ Other: ________

4. Gas Test Requirements:
LEL/O_2 - Instrument Mfg./No. ______/________ Bump Check Time/Gas Tester - _____/_____
Toxicity - Instrument Mfg./No. ______/________ Bump Check Time/Gas Tester - _____/_____
Frequency of Testing ☐ Continuous* ☐ Other - Specify - ______________________
* Continuous monitoring results must be recorded every three hours

Acceptable Levels	Results								
Oxygen: 19.5% - 23.5%									
Combustible Gas: %LEL - <10%									
Other _________ < PEL* ______									
Other _________ < PEL* ______									
Other _________ < PEL* ______									

* Entry in excess of the PEL will require appropriate PPE.

5. Ventilation / Exhaust Equipment:
☐ None required, natural ventilation adequate ☐ Forced air ventilation ☐ Exhaust ventilation
Equipment Type: ☐ Air powered horn ☐ Electric blower Volume Required - _______ cfm

6. Personal Protection:
☐ Gloves (type) ______________ ☐ Respirator (type) ____________________
☐ Goggles or face shield ☐ Self Contained Breathing Equipment
☐ Lifelines Attached to Harness ☐ Other, specify: ____________________
☐ Chemical Resistant Suit, Specify Type______________________________

7. Fire Protection: ☐ None required ☐ Portable Fire Extinguisher – type and size:__________
☐ Fire Watch ☐ Other, specify: ______________________________

201F50A.EPS

Figure 50 Entry permit (1 of 2).

8. Condition of Area and Equipment:

Required Yes	N/A	THESE KEY POINTS MUST BE CHECKED
		a. Equipment locked and tagged out?
		b. Piping is disconnected, capped or plugged and/or blinded?
		c. Equipment emptied, washed, purged & ventilated?
		d. Low voltage or GFCI-protected equipment provided? e. Explosion-proof electrical equipment provided?
		f. Provisions are made to barricade or post signs at entry points when attendant is not on duty.

Other Requirements:

9. Special Instructions: ☐ None ☐ Check with issuer before starting work

10. Approval	Permit			Permit Acceptance		
	Supt. / Area Supv.	Date	Time	Maint. Supv. / Engineer / Contractor Supv.	Date	
Issued by						
Endorsed by						
Endorsed by						

11. Individual Review / Entrant Roster: I have been instructed in the proper Work Permit, Confined Space Entry, Lockout/Tagout Procedures, associated physical and atmospheric:hazards and have reviewed the gas testing results

Entrants	**Date**	**Time In / Out**		**Time In / Out**		**Time In / Out**		**Time In / Out**		**Time In / Out**	

I have been informed of the duties and responsibilities for an attendant, the associated physical and atmospheric hazards, and have reviewed the gas testing results.

Attendants	**Date**	**Time On / Off**		**Time On / Off**		**Time On / Off**	

12. Job Completion:

☐ Yes ☐ N/A Has the rescue team been notified ?
☐ Yes ☐ No Is the work on equipment complete & the confined space ready to return to service?
☐ Yes ☐ No Has the worksite been cleaned and made safe?
Workers answering above questions: ______________________________

13. Post Job Review: Were any hazards encountered or created during entry operations?
☐ Yes ☐ No If yes, describe: ______________________________
Possible solutions: ______________________________

Forward to job file within 7 days of job completion.

201F50B.EPS

Figure 50 Entry permit (2 of 2).

Figure 51 Detection meter.

Some flammable gases are lighter than air and have a higher concentration at the top of a confined space. Vapors from fuels are generally heavier than air and will form a greater concentration at the bottom of the space.

A spark, flame, or surface temperatures in excess of the flash point of chemicals or gases will cause an explosion in a combustible atmosphere.

WARNING

Explosion-proof lights, motors, exhaust fans, and other equipment must be used to prevent fires and explosions in any combustible area. Unless intrinsically safe, do not use cell phones in combustible areas.

Intrinsically safe/explosion-proof equipment must be used in hazardous combustible areas. They are used in tanks, containments, and other confined spaces. The equipment is required to be rated and marked in accordance with the *National Electrical Code®* (*NEC®*). The ratings inform users under what conditions the equipment will not cause a fire or an explosion. Ratings are defined by Class, Division, and Group. For example, Class I, Division 1 or 2, Group C and D (or IIB and IIA) ratings pertain to most of the volatile solvents. The Division 1 rating pertains to locations where the solvent would normally be present. The Division 2 rating pertains to locations where the solvent would not be expected to be present unless an abnormal condition occurs.

Class II, Group E ratings pertain to combustible metallic powder such as aluminum, magnesium, or other metal dusts that by particle size, abrasiveness, or conductivity present an equivalent hazard. Class II, Group F ratings pertain to carbon, charcoal, or coke dusts. Class II, Group G ratings pertain to plastic or chemical (dry paint) dust.

Because the ignition temperatures for various solvents and other materials vary widely, the operating surface temperature of intrinsically safe/explosion-proof equipment is also required by the *NEC®* to be marked on the equipment directly or by means of a *NEC®* code number. Always make sure that the equipment operating temperature will always be below the flash point listed on the MSDS for the solvent or material being used.

A variety of intrinsically safe/explosion-proof equipment is available. Examples of some equipment include the following:

- Portable lighting, including tank lights, telescoping-stand lights, and fluorescent lights with up to 50,000 ft^2 (4,645 m^2) of coverage
- Communications equipment such as radios and cell phones
- Portable power distribution equipment with **ground fault circuit interrupter (GFCI)** circuit breakers including extension cords with on/off switches
- Test equipment and tools such as infrared thermometers and non-sparking tool sets
- Powered clamp-on paint mixers

6.3.3 Toxic Atmospheres

Toxic gases and vapors come from many sources. They can be deadly when they are inhaled or absorbed through the skin above certain concentration levels. In spaces with no ventilation, high concentrations can gather and quickly become toxic. Even in lower doses, some chemicals can seriously affect breathing and brain functions.

The harmful effects of toxic gases and vapors vary. Many toxic gases, such as carbon monoxide, cannot be detected by sight or smell. Some toxic gases have harmful effects that may not show up until years after contact. Others, such as nitric oxide, can kill quickly.

NOTE

The MSDS attached to the entry permit will have information about any toxins that may be encountered.

6.3.4 Monitoring the Atmosphere in Confined Spaces

The air inside a permit required confined space must be tested before anyone is allowed to enter. Atmospheric tests must be done in this order:

1. Oxygen content
2. Flammable gases and vapors
3. Potential toxic contaminants

The test for oxygen must be performed first because most combustible gas meters are oxygen dependent and will not provide reliable readings in an oxygen-deficient atmosphere. The test for combustible gases is performed next because the threat of fire or explosion is usually more urgent and more life-threatening than exposure to toxic gases and vapors.

Various instruments are used to test and monitor confined-space atmospheres. Portable, battery-operated, gas-detection meters can measure oxygen levels by changing the sensor in the detection meter. Gas-detection meters must be calibrated and operated according to the manufacturer's instructions. The meter must be able to detect oxygen and combustible gases at the levels specified in *OSHA 29 CFR 1910.146*. Pre-entry gas testing should be done for each four feet of vertical rise in a confined space to take into account the stratification of gas and vapor that may be in the space. A CSP, CIH, registered safety engineer, or marine chemist should review the findings prior to entry per *OSHA 29 CFR 1910.146, Appendix B*.

6.4.0 Additional Hazards

In addition to atmospheric hazards, there are several other physical and environmental hazards in confined spaces:

- Electric shock
- Purging
- Falling objects
- Engulfment
- Extreme temperatures
- Noise
- Slick or wet surfaces
- Moving parts

6.4.1 Electric Shock

Electric shock can occur when **power tools** and line cords are used in an area where there are wet floors or surfaces. Tools and equipment should be grounded or a ground fault circuit interrupter should be used when working in a confined space.

6.4.2 Purging

Purging happens when toxic, corrosive, or natural gases enter and mix with the air in a confined space. Purging is most often experienced when working in pipes. Purging gases are used to clean pipelines. These gases can create an oxygen imbalance in the space that will suffocate workers almost immediately. Once purging is complete, appropriate ventilation must be established to render the atmosphere safe. Air monitoring is necessary to verify air quality.

6.4.3 Falling Objects

Materials or equipment can fall into confined spaces and strike workers. This usually happens when another worker enters or exits a confined space with a top-side opening, such as a manhole. Vibrations can also cause materials or tools to fall and strike workers.

6.4.4 Engulfment

Engulfment occurs when a worker is buried alive by a liquid or material that enters a confined space. Small, loose material stored in bins and hoppers, such as grain, sand, or coal, can engulf and suffocate a worker. Heavier materials or liquids that enter a confined space can crush or strangle a worker.

6.4.5 Extreme Temperatures

Confined spaces that are too hot or too cold can be hazardous to workers. Spaces that are too hot can cause heat stroke or heat exhaustion, while spaces that are too cold can cause **hypothermia**. Another temperature-related hazard in a confined space is steam. Steam is extremely hot and can cause serious or deadly burns.

6.4.6 Noise

Noise in a confined space can be very loud. This happens because the size of the space is small and sound bounces off the walls. Too much noise, or noise that is too loud, can permanently damage hearing. It can also prevent workers from communicating. If this happens, an evacuation warning could be missed.

6.4.7 Slick or Wet Surfaces

Workers can get seriously hurt by slips and falls on slick or wet surfaces. Wet surfaces also add to the chance of electrocution from electrical circuits, equipment, and power tools.

6.4.8 Moving Parts

Workers can get struck or trapped by moving parts such as augers or belts. This usually happens when a worker slips or falls. It also can happen if the operator of the moving machinery is wearing loose clothing or jewelry.

6.5.0 Responsibilities and Duties

Everyone involved in confined space work must have special training. This includes entrants doing work in a confined space, attendants at the opening of the space, entry supervisors, and rescue workers.

6.5.1 Entrants

Entrants are people who enter a confined space to do the work. They must be aware of the dangers of the job and know how to protect themselves. An entrant must do the following:

- Obtain a valid entry permit.
- Know the atmospheric, fire, and toxic contamination hazards.
- Use the specified PPE, including face and eye protection, gloves, aprons, and coveralls as required.
- Stay in contact with the attendant to make sure he or she is being monitored and can be told to evacuate, if needed.
- Alert the attendant when warning signs or symptoms of exposure exist.
- Know how to escape when necessary.
- Exit the space immediately upon hearing an evacuation alarm or if an uncontrolled hazard is detected within the space.

> **NOTE**
> Retrieval methods should be part of an emergency procedure. However, exercise caution when using retrieval devices in situations where the device may cause worker entanglement.

6.5.2 Attendants

An attendant stays outside of the confined space and communicates directly with the entrant. The attendant has constant contact with the entrant through telephone, radio, visual, or other approved means. An attendant's basic responsibility is to protect the entrant. To do this, the attendant must do the following:

- Set up a station at the exit of the confined space.
- Keep count of the personnel inside.
- Know what work is allowed or disallowed in the space.
- Know how to monitor for safety.
- Test the atmosphere remotely using a probe (*Figure 52*) and recognize safe and unsafe levels.
- Maintain contact with entrants and be able to recognize symptoms of physical distress.
- Order evacuation when problems occur.
- Know how to call rescuers and use the alarm system.
- Refuse entry to unauthorized personnel.
- Never enter the space for rescue attempts.

Figure 52 Attendant checking the atmosphere.

6.5.3 Supervisors

The entry supervisor is the person responsible for safe confined-space entry operations. This means that he or she is responsible for the lives of all workers involved in the operation. The entry supervisor must do the following:

- Authorize and oversee entry.
- Be well trained in entry procedures.
- Know the hazards of the confined space.
- Make sure entry permits are correct and complete.
- Make sure that proper lockout/tagout procedures are performed.
- Make sure that all equipment required for entry is available and understand how it works.
- Make sure that workers understand the job.
- Know rescue plans and rescue workers.
- Be responsible for canceling entry authorization and terminating entry when unacceptable conditions are present or when the job is completed.
- Evaluate problems and recognize when re-entry is safe.

- Ensure that explosion-proof equipment is used.
- Notify all workers in the area or facility that work is being done in a confined space.

6.5.4 Rescue Workers

There are two types of rescues: non-entry rescues and entry rescues.

A non-entry rescue can be done by anyone outside of the confined space. It is usually done by the attendant. In this type of rescue, winches and tripods can be used to pull the entrant out of the space without endangering other workers (*Figure 53*).

Rescue workers have a responsibility to do the following:

- Know the hazards of the confined space.
- Know the proper use of personal protective and rescue equipment, including respirators.
- Practice rescue operations and know how to perform rescues.
- Be trained in first aid and CPR.
- Know the company's confined-space rescue program.

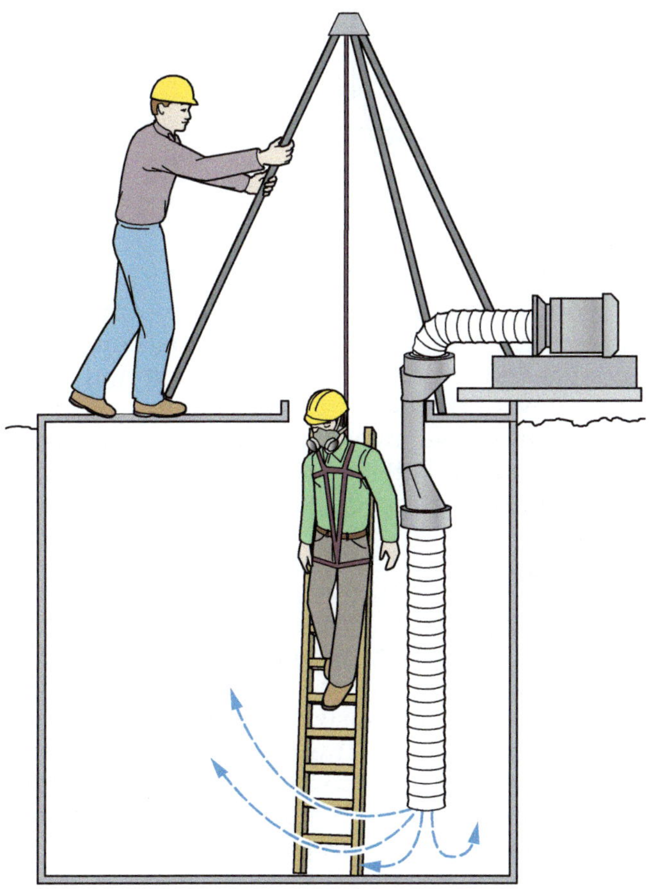

Figure 53 Non-entry rescue.

Entry rescues can be extremely hazardous. Only trained rescue workers are allowed to enter a confined space during an emergency. More rescue workers are killed in confined-space accidents than the workers they are trying to rescue.

6.6.0 Safeguards

It is important to understand how to protect yourself and your co-workers when working in a confined space. In order to do this, everyone must be aware of what is happening on the site and understand how to work safely. These are the most common safeguards that should be used during confined space operations:

- Monitoring and testing
- Ventilation
- PPE
- Communications
- Training

6.6.1 Monitoring and Testing

The air in a confined space must be tested before any workers enter it. Testing must be done by a properly trained, qualified attendant. The air is tested for oxygen content, explosive gases or vapors, and toxic chemicals. This can be done by inserting a wand attached to a gas meter into the confined space.

NOTE

Entrants and attendants have the right to witness or review the gas testing results and may request additional tests if deemed necessary.

The atmosphere in a confined space may need to be monitored during the entire job. This is done by attaching monitors to entrants or by using outside devices. When the atmosphere is monitored, workers can be assured that the air quality is good and that they will immediately know about changes in the atmosphere that would require them to leave.

6.6.2 Ventilation

If the air in a confined space is hazardous or has the potential of becoming dangerous, the space must be ventilated immediately to remove toxic gases or vapors and replace lost oxygen. Ventilators blow clean air into the space and must stay on as long as workers are in the space (*Figure 54*).

Figure 54 Proper positive ventilation.

It's important to remember that just because air is being blown into or removed from the space, it doesn't mean that the air is being ventilated. Toxic gases can hide in confined spaces. Make sure the attendant has carefully tested the entire space before entering it.

6.6.3 Personal Protective Equipment

Every job requires some type of PPE. Standard PPE includes hard hats, safety goggles and glasses, boots, and gloves. On a confined-space job site, the following items may be needed in addition to the standard equipment:

- Full body harness
- Lifelines
- Air-purifying respirator
- Air-supplying respirator

> **NOTE**
> Always check the entry permit to make sure you have the PPE needed to enter a confined space.

6.6.4 Communication

All workers on a site must be able to communicate with one another. It is especially important for attendants and entrants to be able to communicate. This allows attendants to warn entrants about dangers and order an evacuation when necessary. Communication between workers is another way to monitor the confined space.

6.6.5 Training

Entering a confined space requires specialized training. In fact, no one is allowed to enter a confined space unless they have been properly trained and authorized by the site supervisor to enter the space. Training gives workers the knowledge needed to complete their jobs safely and efficiently. Do not enter any confined space work area without the proper training.

7.0.0 Safety Guidelines For Industrial Coating

Industrial coating work can be dangerous to people and the environment. Some of the dangers will be obvious, such as the use of hand and power tools that can inflict serious injuries if defective or used improperly; the use of chemicals that are flammable, may irritate the skin, or be harmful when inhaled; flying debris when working with high pressure waterjetting and abrasive blasting equipment; and exposure to high amounts of irritating dust. Some dangers will not be obvious, such as inhalation of invisible and dangerous vapors; carrying toxic dust home on your clothing; and being exposed to any number of toxic substances in old coatings. Follow all safety rules and wear the proper PPE.

7.1.0 Fire Hazards

For a fire to start, three things are needed in the same place at the same time: fuel, heat, and oxygen. If one of these three is missing, a fire will not start.

Fuel is anything that will combine with oxygen and heat to burn. When pure oxygen is present, such as near a leaking oxygen hose or fitting, material that would not normally be considered fuel (including some metals) will burn.

Heat is anything that will raise a fuel's temperature to the flash point. The flash point is the temperature at which a fuel gives off enough gases (vapors) to burn. The flash points of many fuels are quite low: room temperature or less. When the burning gases raise the temperature of a fuel to the point at which it ignites, the fuel itself will burn and keep burning even if the original source of heat is removed.

Oxygen is always present in the air. What is needed for a fire to start can be shown as a fire triangle (*Figure 55*). If any one element of the triangle is missing, a fire cannot start. If a fire has started, removing any one element from the triangle will put it out.

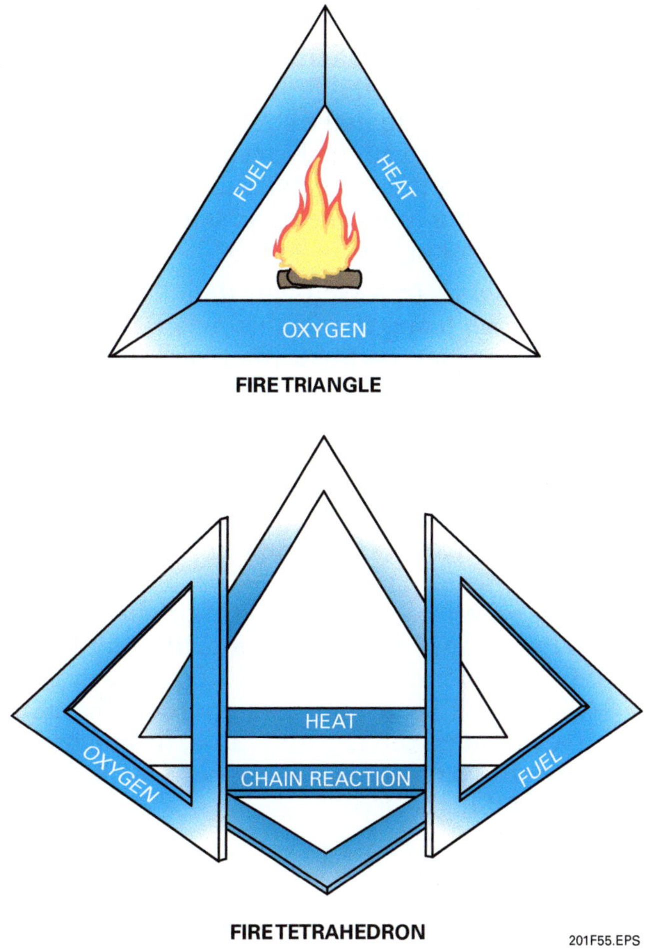

Figure 55 Fire triangle and tetrahedron.

Some fire prevention experts add a fourth side to the fire triangle concept, resulting in the development of a new model called the Fire Tetrahedron. The fourth element involved in the combustion process is referred to as the chemical chain reaction. Specific chemical chain reactions between fuel and oxygen molecules are essential to sustaining a fire once it has begun.

Think About It

Safety

1. A coating team has been assigned to coat a vacant four-story building. There's a large parking lot surrounding the building, but it's not used, and the whole area is securely fenced and gated, so there is no danger of vehicular or pedestrian traffic. Will flaggers and barricades be needed? Can ladders be safely used? Or should scaffolding be used? Will fall protection devices be needed?
2. What if the building is located on a busy street with no surrounding parking lot or fence?
3. A coating team needs to perform surface preparation and coating work on the exterior portion of a bridge that is 40 feet (12.2 m) above ground. What safety devices will be needed?

7.1.1 Fire Prevention

The best way to provide fire safety is to prevent a fire from starting in the first place. The best way to prevent a fire is to make sure that the three elements needed for fire (fuel, oxygen, and heat) are never present in the same place at the same time.

The following are some basic safety guidelines for fire prevention:

- Always work in a well-ventilated area, especially when using flammable materials such as shellac, lacquer, paint stripper, or construction adhesives.
- Never smoke or light matches when working with flammable materials.
- Keep oily rags in approved, self-closing metal containers.
- Store combustible materials only in approved containers.
- Know where to find fire extinguishers, what kind of extinguisher to use for different kinds of fires, and how to use the extinguishers.
- Make sure all extinguishers are fully charged. Never remove the tag from an extinguisher; it shows the date the extinguisher was last serviced and inspected.

7.1.2 Fire Fighting

Always be aware of the potential for fire on any job site. Know the location of fire-fighting equipment and which equipment to use on different types of fires. Only qualified personnel are authorized to fight fires.

Most companies tell new employees where fire extinguishers are kept. If you do not know, be sure to ask. Also ask how to report fires. The telephone number of the nearest fire department should be clearly posted in the work area. If the company has a company fire brigade, know how to contact them. Learn the company's fire-safety procedures.

7.2.0 Hand and Power Tools

Tools are a common part of our lives, so it may be easy to forget that they can pose serious safety risks. All tools are built with safety in mind, but serious tool-related accidents still occur. Hand and power tools can cause many different types of injuries:

- Burns
- Cuts
- Sprains
- Electrical shock
- Eye injuries
- Hearing loss
- Broken bones

Fire is a significant hazard when working with hand and power tools. Fires occur when sparks come in contact with flammable or combustible materials. Sparks from power tools as well as striking **impact tools** can ignite fumes, rags, paper, or solvents being used in the area. To minimize this risk, use the right tools for the job, be aware of the environment, and know what to do if there is a fire.

7.3.0 Hand Tool Safety

Hand tools are non-powered tools such as scrapers, chipping hammers, and brushes (*Figure 56*). Hand tools are dangerous if they are misused or improperly maintained. The following are gen-

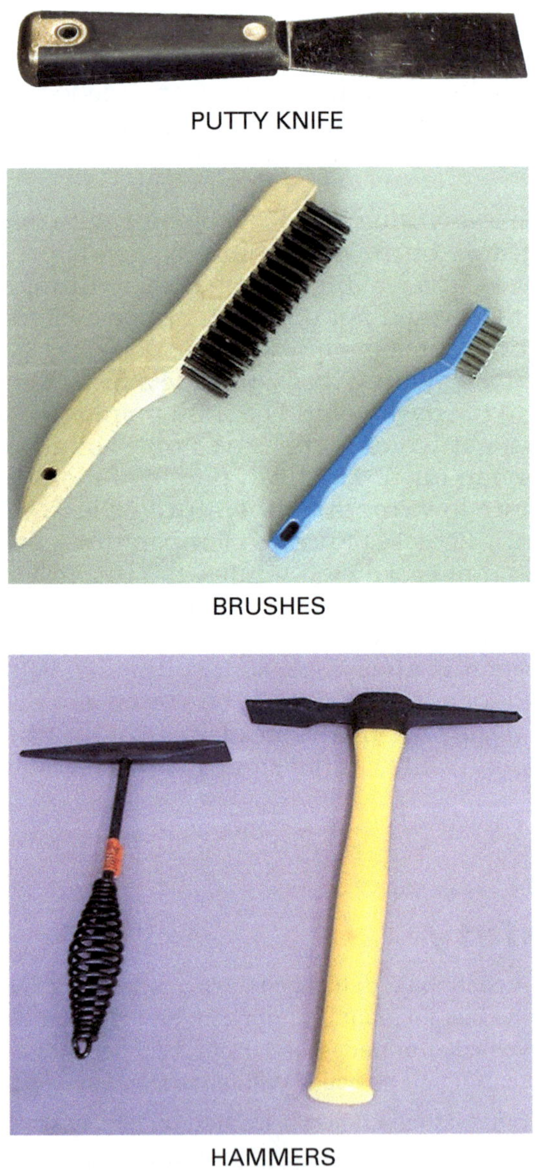

201F56.EPS

Figure 56 Hand tools.

eral guidelines for the use and maintenance of hand tools:

- Using the wrong tool can be dangerous. Use tools only for their designated purpose.
- Always maintain your tools properly. If the wooden handle on a tool such as a hammer is loose, splintered, or cracked, the head of the tool may fly off and strike the user or another person.
- Repair or replace damaged or worn tools.
- Impact tools such as chisels, wedges, and drift pins are unsafe if they have mushroomed heads. The heads might shatter when struck, sending sharp fragments flying.

In general, risks are greatly reduced by inspecting and maintaining tools regularly and always wearing appropriate PPE such as safety goggles, hard hats, and filtering masks. Also, keep floors dry and clean to prevent accidental slips that can result in injuries caused by the tools being used.

7.3.1 Impact Tools

Impact tools such as chisels are meant to be struck by a hammer. The two most common types of injuries associated with impact tools are hammer strikes and eye injuries from flying debris. Getting struck by a hammer is usually the result of carelessness or not paying attention. Stay focused on the task when using a hammer.

Injuries caused by flying tool fragments from damaged tools can be prevented by inspecting the tool before using it. Look for mushroomed heads, cracks, chips, or other signs of damage or weakness. Eye injuries can be prevented by wearing proper eye protection while using impact tools.

7.3.2 Dust and Suspended Particles

Some hand tools create a lot of airborne dust and particles, especially during surface preparation operations. Short-term exposure to these particles may cause minor irritation of the eyes, nose, or sinuses. In some individuals, exposure may trigger a more serious reaction such as an asthma attack. Long-term exposure over the course of a career can impact a person's lungs, possibly causing chronic bronchitis, pneumonia, or other pulmonary disorders. To help prevent these conditions, filtering masks and eye protection should be worn when using a tool that creates dust.

7.4.0 Power Tools

Power tools (*Figure 57*) can be hazardous when used improperly or not well maintained. Most of the risks associated with hand tools are also risks when using power tools. When adding a power source to a tool, however, the risk factors increase.

Power tools are powered by different sources:

- Electricity
- Pneumatics (air pressure)
- Hydraulics (fluid pressure)

Know the safety rules and proper operating procedures for each tool used. Specific operating procedures and safety rules for using a tool are provided in the operator's/user's manual supplied by the manufacturer. Before operating any power tool for the first time, always read the manual to become familiar with the tool. If the manual is missing, contact the manufacturer for a replacement.

7.4.1 Safety Guidelines for Power Tools

Before explaining more about power tools, it is important to understand the general safety rules that apply when using all power tools, regardless of type.

201F57.EPS

Figure 57 Power tools.

Follow these general guidelines in order to prevent accidents and injury:

- Never use a sparking power tool in a combustible atmosphere. Always use an intrinsically safe/explosion-proof tool.
- Never carry a tool by the cord or hose.
- Keep cords and hoses away from heat, oil, and sharp edges.
- Do not attempt to operate any power tool unless properly trained and checked out by an instructor.
- Always wear eye protection, a hard hat, and any other required PPE when operating power tools.
- Wear face and hearing protection when required.
- Wear proper respiratory equipment when necessary.
- Wear the appropriate clothing for the job being done. Wear close-fitting clothing that cannot become caught in moving tools. Roll up or button long sleeves, tuck in shirttails, and tie back long hair. Do not wear any jewelry, including watches or rings.
- Do not distract others or let anyone distract you while operating a power tool.
- Do not engage in horseplay.
- Do not run or throw objects.
- Consider the safety of others, as well as yourself. Observers should be kept at a safe distance away from the work area.
- Never leave a power tool running while it is unattended.
- Assume a safe and comfortable position before using a power tool. Be sure to maintain good footing and balance.
- Secure work with clamps or a vise, freeing both hands to safely operate the tool.
- To avoid accidental starting, never carry a tool with your finger on the switch.
- Be sure that a power tool is properly grounded and connected to a GFCI before using it.
- Ensure that power tools are disconnected before performing maintenance or changing accessories.
- Do not use dull or broken tools or accessories.
- Use a power tool only for its intended use.
- Keep feet, fingers, and hair away from the blade and/or other moving parts of a power tool.
- Never use a power tool with guards or safety devices removed or disabled.
- Never operate a power tool if hands or feet are wet.
- Keep the work area clean at all times.
- Become familiar with the correct operation and adjustments of a power tool before attempting to use it.
- Keep tools sharp and clean for the best performance.
- Follow the instructions in the user's manual for lubricating and changing accessories.
- Keep a firm grip on the power tool at all times.
- Select electric extension cords of sufficient size to service the particular power tool being used.
- Report unsafe conditions to an instructor or supervisor.
- Remove damaged tools from use and tag out with Do Not Use tags.

7.4.2 Electrically Powered Tools

The most serious danger in using electrically powered tools is that of electrocution. Electricity can cause burns, shocks, explosions, electrocution, and fires. Electrical shocks can be minor and uncomfortable or they can be severe, causing burns or death. Even a small amount of current can cause the heart to stop pumping in rhythm, which will result in death unless immediate medical attention is provided. Electrical shock can also cause a loss of balance, muscle control, or consciousness, which could then cause the victim to fall and/or drop a tool. A fall from a ladder or scaffolding can obviously be quite serious.

To prevent electrical shock, tools must provide at least one of the following types of protection:

- *Double insulated* – Double insulation is more convenient than three-wire cords. The user and tools are protected in two ways: by normal insulation on the wires inside, and by a housing that cannot conduct electricity to the operator in the event of a malfunction.
- *Powered by a low-voltage isolation transformer* – If the electrically powered tools do not have either a ground plug or double insulation, check with a supervisor to make sure that you are protected by a low-voltage isolation transformer.
- *Grounded with a three-wire cord* – Three-wire cords have two current-carrying conductors and one grounding conductor. You are probably familiar with the three-prong plug common on electrically powered tools (*Figure 58*). When there is a three-prong cord that means the tool is powered by a grounded three-wire cord and should only be plugged into a three-prong, grounded receptacle (*Figure 59*). Any time an adapter is used to accommodate a two-hole receptacle, the adapter wire must be attached to a known ground. Never remove the third prong (grounding conductor) from a plug. When using a three-prong extension cord, make sure that it is properly grounded at its source.

Here are some general tips to help stay safe when using electrically powered tools:

- Electrically powered tools should be used on jobs for which they are intended.
- Wear all appropriate PPE, such as gloves and safety footwear.
- When not in use, store power tools in a dry place.
- Electrically powered tools should not be used in damp or wet places.
- The work area should be well lit.

Because malfunctioning electrically powered tools can cause sparks, these tools can also cause fires and explosions. Know about any fire hazards in the work area and avoid using electrically powered tools around flammable materials, fumes, and gases.

Static electricity can also produce sparks that can ignite flammable materials. **Bonding** and grounding are used to control static electricity. Bonding connects conductive objects using a copper wire. This equalizes the charge between the objects. Grounding connects the bonded objects to a ground rod or other grounding system in order to safely discharge static electricity or other fault currents.

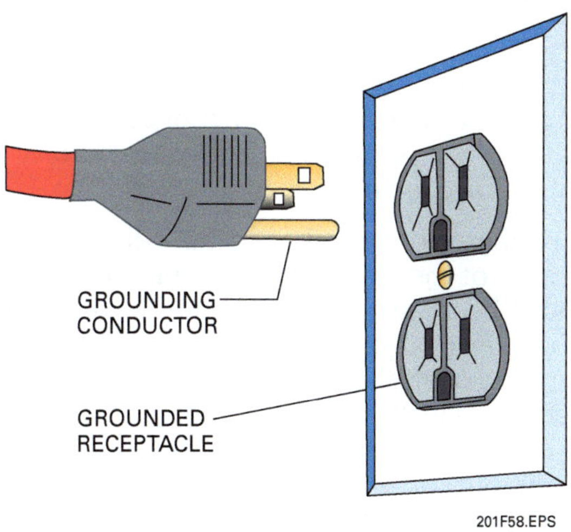

Figure 58 Grounded three-prong plug.

Finally, electrical cords and extension cords pose a tripping hazard. Avoid running cables and cords across elevated work areas and scaffolding. Extension cords should be brightly colored to make them more visible. Cords and cables should be run somewhere other than walkways, or at least along a wall, rather than in the middle of a walkway or across it. Occasionally, it may be necessary to run a cord or cable across a walkway. If this cannot be avoided, either tape the cord down and put a carpet over it, or place it in a cord runner designed to minimize the tripping hazard.

7.4.3 Pneumatic Tools

Pneumatic tools such as chippers and sanders are powered by compressed air. The most serious risk in using these tools is the danger of getting hit by one of the tool's attachments.

The use of pneumatic tools often creates flying debris. Wear proper eye protection when using a pneumatic tool. Full-face protection is highly recommended. Screens must be set up to protect nearby workers from being struck by flying fragments around chippers, riveting guns, staplers, or air drills.

Airless spray guns that atomize paints and fluids at high pressures (1,000 psi or more) must be equipped with automatic or visual safety devices that will prevent the trigger from being pulled until the safety device is manually released.

Pneumatic tools are powered by an air supply delivered through air hoses (*Figure 60*). When using pneumatic tools, always check to see that they are fastened securely to the air hose to prevent them from being disconnected. Most pneumatic tools have a short wire or a positive locking device attaching the air hose to the tool. These ensure that the tool is attached correctly.

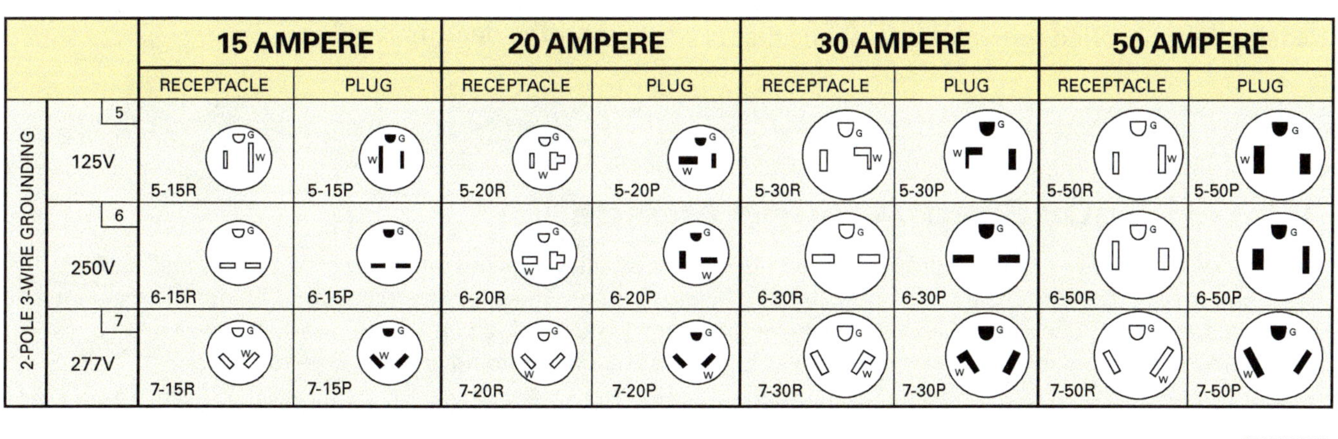

Figure 59 Three-prong receptacle configurations.

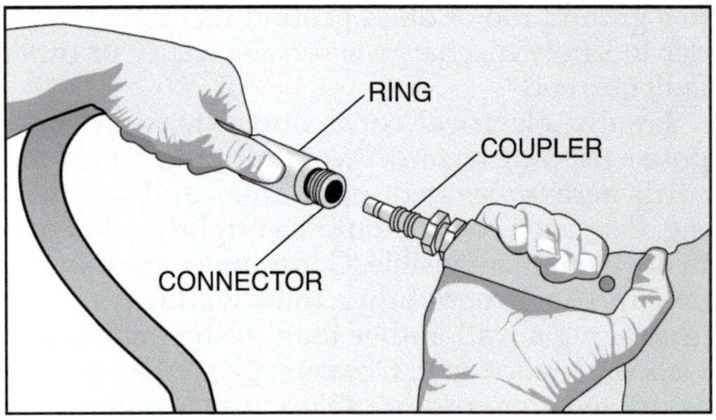

201F60.EPS

Figure 60 Pneumatic air hose connection.

If the air hose is more than ½ inch in diameter, a safety excess flow valve must be installed at the source of the air supply to shut off the air automatically in case the hose breaks. In general, the same safety precautions should be taken with an air hose that are recommended for electric cords. This is because the same types of accidents and injuries can happen with both, including tripping.

7.5.0 Abrasive Blasting

Abrasive blasting operations can be particularly dangerous. Abrasives and compressed air are shot from the equipment under great pressure, and improper handling of the blast gun can seriously injure or kill a person. This pressure can spray abrasives long distances, so it is important that people and equipment in the area of the surface preparation operations are protected.

For the safety of the blast-cleaning operator, quality work clothing and PPE must be used. As shown in *Figure 61*, these include the following:

- Safety boots with steel insert toe cap
- Strong leather gloves with long cuffs
- Heavy protective clothing (possibly coveralls)
- Air-fed blasting helmets or hoods with a replaceable visor and leather cape or respirators, adequately supplied with clean, fresh air that is monitored and alarmed to alert the worker of contaminated air
- Eye and hearing protection

201F61.EPS

Figure 61 Abrasive blasting PPE.

Because abrasive blast cleaning at high pressure is a dangerous operation, it is essential to use good safety practices to protect operators, spectators, and other personnel. Important safety considerations are as follows:

- Only the operator should be allowed within the area of the blasting operation, and he or she should be aware of the location of other workers involved in the operation.
- Warning notices and barricades should be posted to alert those not involved in the blasting operation of the hazards.
- A look-out (or pot-man) should be on the alert for hazardous situations that may develop.
- All equipment should be inspected and tested for proper operation before use, including all hoses and connectors, electrical connections, and the deadman valve.

Case History

Watch Your Step Around Hoses

A worker tripped over a trailing compressed-air hose left in an alleyway and fell hard, breaking his arm and injuring his back. The air tube line had not been put away after use. The alleyway was poorly lit and the incident happened at night. Following an investigation, the company installed floodlighting in poorly lit areas and relocated the air line and hose. Safety awareness and housekeeping training is now given to all staff.

The Bottom Line: Good housekeeping is an important part of safety. Store all tools and hoses after use.
Source: The National Institute for Occupational Safety and Health (NIOSH)

- Workers should never remove PPE during the operation so as to prevent injury.
- All blasting equipment and the structure being cleaned should be grounded to prevent the buildup of static electricity and electrical shocks.

7.6.0 Waterjetting

Waterjetting is the use of a high-pressure water stream to clean surfaces. Waterjetting respiratory requirements may be less stringent than these requirements for other methods of surface preparation, but protective clothing is essential. Even under low pressure, water is powerful enough to break human skin, and water under high pressure can be lethal. Water under pressure of 10,000 psi (69 Mpa) can slice through concrete, and your skin presents much less resistance.

In addition to the standard equipment, all workers in the area of a waterjetting operation must wear the following PPE (*Figure 62*):

- Waterproof suits
- Head protection
- Protective goggles or a full-face shield

Shotgun operators must wear the following PPE:

- Protective toe guards
- Shin guards
- Gloves
- Full-face shield
- In some cases, body armor

Lance operators are not required to wear shin and toe guards, only rubber boots. Hearing protectors or earplugs are required for those working near noisy equipment, but all workers should consider using hearing protection. In addition, when there is a possibility that toxic substances are being removed from the surface being cleaned, workers (especially the operator) may need breathing equipment or masks.

Waterjetting uses only water to clean, but the water is under pressure (as high as 72,000 psi [496 MPa]), so the cleaning power is remarkable. When holding a waterjet nozzle, it is like holding a loaded gun. Use the following guidelines during waterjetting operations:

- Always be aware of where the lance is pointed. Avoid having to look away from the lance during operation.
- Always be aware of potential hazards.
- Never point the nozzle at your body. The pressure of the water from the nozzle can peel away body armor.

201F62.EPS

Figure 62 Waterjetting PPE.

- Never use water out of the nozzle to clean shoes or face shields.
- Never point the nozzle of the lance at anybody—even someone far away from the operation. The pressure is so great that it is capable of traveling long distances and may still have enough force to knock a person down.
- Test the dump valve at the start of the operation. If the water does not cut off when the dump valve is used, stop work at once. Never try to repair the system while it's pressurized. Shut off pressure at the pump, drain the system, and have the valve repaired or replaced. This applies to leak repair as well.
- Shut down all equipment when it is unattended. Use lockout devices as needed.
- When working in a confined space, be certain to vent exhaust properly. Watch for symptoms of carbon monoxide poisoning (headache, dizziness, and nausea, building to confusion and weakness, followed by unconsciousness and a red complexion). For early symptoms, get into fresh air. For advanced symptoms, medical personnel must be notified and oxygen should be administered. Do not resume work until the source of the carbon monoxide is found and corrected.

All blast accidents must be reported immediately. Even when the blast does not visibly penetrate the skin, the pressure can force water, air bubbles, and toxins under the skin. This is true even if there is no blood showing. Small amounts of air bubbles entering the circulatory system are unlikely to cause a problem, but when amounts approach the volume of a fluid ounce, they can adversely affect the cardiac system. Larger amounts can be fatal. In addition, bacteria can be forced under the skin and cause an infection. Accident victims need to go to a doctor, hospital, or emergency room immediately. If the incident takes place in a remote area, the wound should be covered. The worker should remain quiet until medical assistance arrives. This is important because activities can increase circulation and spread air and bacteria throughout the body.

7.7.0 Chemical Stripping

When working with chemical strippers, study the MSDS for the product(s) being used. At the minimum, wear eye protection to keep splashes out of the eyes (consider face protection, too). Wear protective gloves and protective clothing when working with substances that can eat away fabric.

Depending on the product being used, some type of respirator may be needed. When using an air purifier type, be certain to use the correct cartridge for the product (Refer to *Table 3*). Information about the type of respirator cartridge needed can be found on the product MSDS.

During stripping operations, many of the products used will be flammable, so be certain that no ignition sources are present. This includes open flames, welding torches, unsealed electrical switches and connections, and sparks generated by grinding on or striking metal. Always use sealed explosion-proof lighting systems in the stripping operation area. If unsure whether existing lighting is explosion proof, don't use it. Instead use portable explosion-proof lighting. Monitoring devices may be needed to detect levels of volatile vapors in the air (*Figure 63*). Remember to store rags used to wipe off strippers in approved containers for disposal.

7.8.0 Mixing and Thinning

During mixing and thinning operations, it is important to read and follow the instructions for the product being used. Never substitute one chemical for another without first determining whether it is suitable. Check the product data sheet or contact the manufacturer to confirm this information. In addition, check the MSDS for every product being used to determine how to safely use the product.

Post signs and erect barricades to identify the area for personnel not involved in coating operations. Reroute foot and vehicular traffic around the area whenever necessary. Don't assume that people will know to stay out of the area.

Practice good housekeeping procedures during mixing and thinning. Always cap containers of solvents and thinners after use. Ensure that all containers are labeled to indicate the contents.

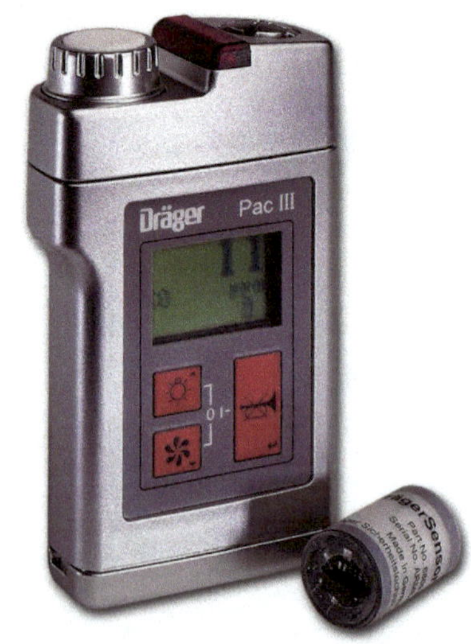

201F63.EPS

Figure 63 VOC monitoring device.

Wipe up all spills immediately. Label and sample mixed or thinned coatings in accordance with the employer's or client's policy.

It is always a good idea to wear eye or face protection during mixing and thinning operations, even when it is not required. Using full eye goggles will help prevent eye injuries from splashes. There are few substances that will not cause some amount of irritation if splashed in the eyes.

Be certain that the area being used for mixing and thinning operations is properly ventilated. When necessary, use intrinsically safe/explosion-proof lighting and powered mixing tools.

7.9.0 Coating Application

During coating operations, post signs and erect barricades to keep personnel not involved in coating operations out of the area. Place barricades beyond the immediate area to prevent stray drops and sprays of coating from reaching people, automobiles, and vegetation. Whenever possible, reroute foot and vehicular traffic far from the coating operation.

7.10.0 Debris Management

In addition to chemicals, abrasives, and water hazards, debris generated during surface preparation may contain heavy metals such as lead, arsenic, and cadmium. These metals are very toxic. Do not inhale or swallow the dust or allow the dust to contact skin for long periods.

During concrete cleaning, **silica** may be released into the atmosphere. Silica can be irritating to the skin, eyes, and lungs. Repeated long-term exposure to free silica (which means it's not bonded with another element) can cause a serious lung disease called silicosis.

The content of an old coating cannot be determined just by looking at it. The coating must be sampled and tested in a laboratory to establish its content. Once the content is determined, the type of PPE can also be determined, as well as the steps to take to protect people and the environment.

Wearing the appropriate protective equipment is essential. Vacuum devices will help to collect the debris for disposal and keep the dust down in the work area. There may be monitoring devices in the area to assess the levels of potentially harmful materials. OSHA limits employee exposure to many dangerous elements averaged over an eight-hour work day. For lead, the concentration is 50 micrograms per cubic meter of air; for cadmium, it is 200 micrograms per cubic meter of air; and for arsenic, it is 10 milligrams per cubic meter of air. A CSP or CIH should be consulted for the correct monitoring procedure and interpretation of test results.

To protect the environment, all surface preparation debris such as water, dry or wet abrasives, solvents, rags, and paint chips, must be disposed of in an approved manner.

SUMMARY

Industrial coating work can be dangerous to people as well as the environment. Some of the dangers are immediately obvious, such as using chemicals, working with high-pressure waterjetting and abrasive blasting equipment, and being exposed to high concentrations of irritating dust. Some dangers will not be obvious, such as inhaling invisible and dangerous vapors, carrying toxic dust home on clothing, and being exposed to any number of toxic substances in old coatings. That is why it is important to know as much as possible about the materials being used as well as any existing coatings, and then take the necessary steps to protect yourself and others as well as the environment.

To get information about the products being used, check the manufacturer's instructions about how to use them. Review the MSDSs for information about required PPE. The MSDS provides complete information about the hazards of a product. When working with old coatings and structures, they must be tested to determine the PPE and containments needed.

There is no substitute for safety. One accident can have permanent consequences. Know and follow all safety precautions, and use common sense to keep everyone safe.

Review Questions

1. Which of the following organizations was created to protect the environment?
   a. EPA
   b. OSHA
   c. NACE
   d. ASTM

2. OSHA requires hearing protection for sounds that average _____ dBA over an eight-hour period.
   a. 65
   b. 85
   c. 105
   d. 145

For Questions 3 through 7, match the type of sign or tag with the corresponding description.

3. _____ Danger sign
4. _____ Do Not Start tag
5. _____ Informational sign
6. _____ Safety sign
7. _____ Accident prevention tag

   a. Provides general information not related to safety.
   b. Provides information related to safety issues.
   c. Informs workers that an immediate hazard exists.
   d. A temporary way of warning workers about immediate and potential hazards.
   e. Prevents the unexpected energizing of equipment.

8. An unintentional release of energy can be prevented by using _____.
   a. a GFCI
   b. lockout/tagout
   c. switches
   d. valves

9. Lockout/tagout systems are used only to protect workers from electrical hazards.
   a. True
   b. False

10. A typical lockout/tagout procedure is made up of _____.
    a. three components
    b. four components
    c. five components
    d. six components

11. The first step of a lockout/tagout procedure is _____.
    a. shutting down the equipment
    b. dissipating any stored energy
    c. verifying that the test equipment is operational
    d. notifying all authorized and affected personnel

12. A yellow and purple barricade tape is used for a _____.
    a. falling object hazard
    b. radiation hazard
    c. wet area hazard
    d. toxic hazard

For Questions 13 through 15, match the type of ladder to the corresponding description.

13. _____ Aluminum ladder
14. _____ Fiberglass ladder
15. _____ Wooden ladder

    a. Lightweight and corrosion-resistant
    b. Heavy and sturdy
    c. Durable and nonconductive

16. Fixed ladders must be able to support at least two loads of _____.
    a. 150 lbs (69 kilograms)
    b. 200 lbs (91 kilograms)
    c. 250 lbs (114 kilograms)
    d. 300 lbs (136 kilograms)

17. Fall protection is required on any scaffolding that is _____ feet or more above a lower level.
    a. 6
    b. 10
    c. 12
    d. 14

For Questions 18 through 20, match the correct term to the definition.

18. _____ Deceleration device

19. _____ Deceleration distance

20. _____ Free-fall distance

a. The distance it takes before a person comes to a stop when falling
b. A shock-absorbing lanyard or self-retracting lifeline that brings a falling person to a stop without injury
c. The vertical distance a worker moves after a fall before a deceleration device is activated
d. Falling force distance
e. Equipment that supports the entire weight of the fall-arrest system

21. Short, flexible straps with connectors on each end are called _____.

   a. anchorages
   b. deceleration devices
   c. lanyards
   d. lifelines

22. Safety nets should be placed no more than _____ feet beneath the work area.

   a. 10
   b. 20
   c. 30
   d. 35

23. Entry into a confined space begins when _____.

   a. the permit is issued
   b. all of the entrants have been assigned a job
   c. any part of the body passes through the entrance
   d. the first entrant enters the confined space

24. An explosion in a combustible atmosphere can be caused by _____.

   a. flammable clothing or materials
   b. a flame or spark
   c. toxic vapors at the top of the confined space
   d. vibrations from tools being used

25. Common injuries associated with impact tools are due to _____.

   a. flying debris
   b. electric shock
   c. toxic dust
   d. excessive force

Trade Terms Introduced in This Module

Accident: An unplanned event preceded by an unsafe act, unsafe condition, or both, which disrupts activities and often results in personal injury or property damage.

Acute effect: An immediate reaction to chemical exposure, such as nausea or death.

Acute exposure: A short-term, usually intense, exposure to a chemical.

Anchorage: A secure point of attachment for lifelines, lanyards, or deceleration devices.

Apparatus: An assembly of machines used together to perform a particular job.

Arc: The flow of electrical current from one point to another through a gas such as air.

Atmospheric hazard: A potential danger in the air or a condition of poor air quality.

Best practices plan: A safety program that includes the best available practices and equipment to ensure worker safety.

Body harness: Straps that may be secured about the worker in a manner that will distribute the fall-arrest forces over at least the thighs, pelvis, waist, chest, and shoulders, with means for attaching it to other components of a personal fall-arrest system.

Bonding: Connecting conductive objects to establish electrical continuity.

Chronic effect: A delayed or long-term reaction to chemical exposure. Chronic effects can result in serious diseases, such as cancer.

Chronic exposure: A long-term, usually low-dose, exposure to a chemical.

Combustible: Air or materials that can explode and cause a fire.

Competent person: A company-appointed person who can identify working conditions or surroundings that are unsanitary, hazardous, or dangerous to employees and who has authorization to correct or eliminate these conditions promptly.

Compliance: A safety plan that meets the minimum OSHA safety standards.

Confined space: An area that is large enough for a person to work in, but with limited ways of entering and exiting, and that is not meant for continuous human occupancy.

Connector: A device that is used to couple (connect) parts of the personal fall-arrest system and positioning device systems together. It may be an independent component of the system, such as a carabiner. It may also be an integral component of part of the system, such as a buckle or D-ring sewn into a body belt or body harness, or a snaphook spliced or sewn to a lanyard.

Decibels (dBA): A measure of sound intensity or loudness. The higher the decibel level, the louder and more potentially damaging the sound is.

Duty rating: American National Standards Institute (ANSI) rating assigned to ladders. It indicates the type of use the ladder is designed for (industrial, commercial, or household) and the maximum working load limit (weight capacity) of the ladder. The working load limit is the maximum combined weight of the user, tools, and any materials bearing down on the rungs of a ladder.

Electrically powered tools: Tools that use electrical current to operate.

Energy source: Any source of electrical, mechanical, hydraulic, pneumatic, chemical, thermal, or other energy.

Energy-isolating device: Any mechanical device that physically prevents the transmission or release of energy. These include, but are not limited to, manually operated electrical circuit breakers, disconnect switches, line valves, and blocks.

Free fall: The act of falling before a personal fall-arrest system begins to apply force to arrest the fall.

Ground fault circuit interrupter: A protective device that functions to de-energize a circuit or portion thereof within an established period of time when a current to ground exceeds some predetermined value. This value is less than that required to operate the overcurrent protective device of the supply circuit.

Grounded: Connected to ground or to a conductive body that extends the ground connection.

Guards: Devices that protect tool operators from dangerous moving parts such as blades, gears, and pulleys.

Hazard Communication (HazCom) Standard: A federal OSHA regulation requiring employers to educate and inform workers about chemical hazards on the job site (*29 CFR 1910.1200*).

Hypothermia: A life-threatening condition caused by exposure to very cold temperatures.

Impact tools: Tools that must strike or be struck to accomplish their task. Impact tools include hammers, chisels, and taps.

Lanyard: A flexible line of rope, wire rope, or strap that generally has a connector at each end for connecting the body belt or body harness to a deceleration device, lifeline, or anchorage.

Leading edge: The edge of a floor, roof, or formwork for a floor or other walking/working surface (such as the deck) which changes location as additional floor, roof, decking, or formwork sections are placed, formed, or constructed. A leading edge is considered to be an unprotected side and edge during periods when it is not actively and continuously under construction.

Lifeline: A component consisting of a flexible line connected vertically to an anchorage at one end (vertical lifeline), or connected horizontally to an anchorage at both ends (horizontal lifeline), and which serves as a means for connecting other components of a personal fall-arrest system to the anchorage.

Lockout: The placement of a lockout device on an energy-isolating device, in accordance with an established procedure, ensuring that the energy-isolating device and the equipment being controlled cannot be operated until the lockout device is removed.

Lockout device: Any device that uses positive means such as a lock to hold an energy-isolating device in a safe position, thereby preventing the energizing of machinery or equipment.

Material safety data sheet (MSDS): A document that must accompany any hazardous material. The MSDS identifies the substance and gives the exposure limits, the physical and chemical characteristics, the kind of hazard it presents, precautions for safe handling and use, and specific control measures.

Non-permit required confined space: A work space free of any atmospheric, physical, electrical, and mechanical hazards that can cause death or injury.

Oxygen-deficient: An atmosphere in which there is not enough oxygen. This is usually considered less than 19.5 percent oxygen by volume.

Oxygen-enriched: An atmosphere in which there is too much oxygen. This is usually considered more than 23.5 percent oxygen by volume.

Permit required confined space: A confined space that has real or possible hazards. These hazards can be atmospheric, physical, electrical, or mechanical.

Personal fall-arrest system: A system used to stop an employee in a fall from a working level. It consists of an anchorage, connectors, and a body harness, and may include a lanyard, deceleration device, lifeline, or suitable combinations of these. As of January 1, 1998, using a body belt for fall arrest is prohibited.

Platform: A work surface elevated above lower levels. Platforms can be constructed using individual wood planks, fabricated planks, fabricated decks, and fabricated platforms.

Pneumatic tools: Tools that use pressurized air to operate.

Power tools: Tools that require a power source, such as electricity, hydraulics, or pneumatics, to operate.

Scaffolding: A temporary supported framework or suspended platform or work area designed to support workers, materials, and equipment at elevated or otherwise inaccessible job sites.

Self-retracting lifeline: A deceleration device containing a drum-wound line that can be slowly extracted from, or retracted onto, the drum under slight tension during normal employee movement, and which, after the onset of a fall, automatically locks the drum and arrests the fall.

Silica: A white or colorless crystalline compound used to manufacture a wide variety of materials, especially glass and concrete.

Tagout: The placement of a tagout device on an energy-isolating device, in accordance with an established procedure, to indicate that the energy-isolating device and the equipment being controlled may not be operated until the tagout device is removed.

Tagout device: Any prominent warning device, such as a tag and a means of attachment, that can be securely fastened to an energy-isolating device in accordance with an established procedure. The tag indicates that the machine or equipment to which it is attached is not to be operated until the tagout device is removed in accordance with the energy-control procedure.

Two-point swing scaffolding: A manual or power-operated platform supported by hangers at two points suspended from overhead supports in a way that allows it to be raised or lowered to the working position.

Additional Resources

This module presents thorough resources for task training. The following resource material is suggested for further study.

ASTM International; www.astm.org

Environmental Protection Agency; www.epa.gov

Field Safety Participant Guide. NCCER. Upper Saddle River, NJ: Prentice Hall.

Hydroblasting Trainee Guide. NCCER. Upper Saddle River, NJ: Prentice Hall.

NACE International; www.nace.org

Occupational Safety and Health Administration; www.osha.gov

OSHA 29 CFR 1926, Safety and Health Standards for Construction

Safety Orientation Pocket Guide. NCCER. Upper Saddle River, NJ: Prentice Hall.

Safety Technology Participant Guide. NCCER. Upper Saddle River, NJ: Prentice Hall.

Figure Credits

Photo courtesy of Bullard, Module opener, 201F07 (hard hat), 201F61

Sherwin-Williams Co., 201F04

Bacou-Dalloz, 201F07 (safety goggles and glasses)

North Safety Products USA, 201F07C (work gloves), 201F08 (full facepiece respirator, self-contained breathing apparatus, and supplied-air respirator), 201F15, 201F17, 201SA01 (circuit breaker lock)

Topaz Publications, Inc., 201F07 (ear plugs and ear muffs), 201F10, 201F13, 201F45, 201F46, 201F56, 201F57, 201F62

Timberland, 201F07 (safety shoes)

Scott Health and Safety, 201F08 (half-mask respirator and air-purifying respirator)

Mike Powers, 201SA01 (tape)

Louisville Ladder Group, 201F19

Ridge Tool Company (RIDGID®), 201F22, 201F24, 201F26

Werner Co., 201F27

Access Safety Group, 201F28

Spider, a division of SafeWorks, LLC, 201F32 (modular platform), 201F33

Photo courtesy of Capital Safety (DBI-SALA and PROTECTA), 201F35, 201F36, 201F38–201F44

Champion America, 201F47 (vest)

NACE International, 201F47 (chart)

RAE Systems, Inc., 201F51

Draeger Safety, Inc., 201F63

Firefree Coatings, Inc., Appendix

Material Safety Data Sheet

Practice Exercise

An MSDS for a fire-resistant coating is provided in this Appendix. Use it to complete the table below. Refer to Section 2.2.0 for interpretation of the flammability codes.

Firefree Coatings, Inc. **FireFree88®**

Hazard	Fire-Resistant Coating
Is it flammable?	
Is the waste hazardous?	
How should the waste be disposed of?	
What PPE is needed?	
What is the first aid for skin contact?	
How do you clean up a spill?	

201A01.EPS

Material Safety Data Sheet

FIREFREE Coatings, Inc.
580 Irwin Street, No. 1
San Rafael, CA 94901

Product: FireFree 88®
MSDS Revision No. 4
November 2007

Emergency and Information:

(888) 990-3388

Section I Product Identification

Product: **FireFree 88®**
Product Class: **Latex Intumescent Fire Retardant, Resistant Paint**

National Paint and Coatings Association, Hazardous Material Identification System:

Health Hazard		See Section VI and IX for more information.
Flammability Hazard	0 -	not considered flammable.
Reactivity Hazard	0 -	minimal
Disposal Procedures	2 -	Glasses, gloves and dust respirator if paint dust is anticipated in elevated concentrations. **Application of wet paint should not present a health hazard.**

Section II Hazardous Ingredients

NONE

1 Confidential

201A01A.EPS

Material Safety Data Sheet

FIREFREE Coatings, Inc.
580 Irwin Street, No. 1
San Rafael, CA 94901

Product: Fire Free 88
MSDS Revision No. 3

Page 2

This product contains pigments and other solids necessary to formulate an intumescent fire retardant paint. These materials may be dust hazards if; excessive quantities of cured paint is released into the air. Due to the anticipated surface coating thickness, it is unlikely that excessive exposures would occur. Work practices designed to subdue dust during paint removal, such as wet methods, would be prudent. Application of wet paint would not present such a hazard under normal application conditions. Spray application in confined areas may require special controls and use of personal protective equipment. An in can preservative has been added at levels not considered hazardous.

Section III Physical Data

Boiling Range:	**212°F**	**Vapor Density:**	**Neutral**
Evaporation Rate:	**Slower than ether**	**Percent volatile:**	**48 to 50%**
Density:	**10.4-lbs/gallon**	**Water solubility:**	**miscible with water**
VOC:	**33 grams/litre**		

Section IV Fire and Explosion Data

Flammability Classification: Not regulated
Lower Explosive Limit: N/A
Hazards of decomposition products: None

Flash Point: >250°F
Extinguishing Media: N/A
Special Fire Protective Procedures:
None

Unusual Fire and Explosion Hazard:
Closed containers may explode due to internal pressure when exposed to extreme heat

2 Confidential

201A01B.EPS

Material Safety Data Sheet

FIREFREE Coatings, Inc.
580 Irwin Street, No. 1
San Rafael, CA 94901

Product: Fire Free 88
MSDS Revision No. 3

Page 3

Section V Reactivity Data

Stability:	Stable
Hazardous Polymerization:	Will not occur
Hazardous Decomposition Products:	May produce hazardous gases (carbon monoxide) if heated to very high temperatures

Section VI Health Hazard Information

Summary of risks: NONE

Section VII Spill, Leak and Disposal Procedures

Spill/Leak Procedures:

Ventilate area. Soak liquid into an absorbent and dispose of according to local, State and Federal regulations. If the paint cures and removal is attempted, dust may be released. Clean-up personnel should wear approved respiratory protection, gloves, and goggles to prevent irritation from contact and/or inhalation.

Waste management/Disposal:

This product does not exhibit any characteristics of a hazardous waste. Follow all local, state and Federal regulations for proper disposal.

3 Confidential

201A01C.EPS

Material Safety Data Sheet

FIREFREE Coatings, Inc.
580 Irwin Street, No. 1
San Rafael, CA 94901

Product: Fire Free 88
MSDS Revision No. 3

Section VIII Special Information

Under normal application conditions, no special precautions or needs for personal protective equipment are anticipated. Exposure to this material should be limited to protect against irritation and inhalation. The use of personal protective equipment including goggles, gloves, respirators, long sleeve clothing and head covering may be recommended.

Respiratory Protection: Provide workers with NIOSH/MSHA-approved respirators in accordance with requirements.

Respiratory Type

Optional disposal respirator. Half-mask air purifying respirator equipped with high efficiency particulate air (HEPA) filter cartridge. Full face-piece air purifying respirator equipped with high efficiency particulate air (HEPA) filter cartridges or powered air purifying respirator with HEPA filter cartridges. Any supplied-air respirator operated in positive pressure mode.

SARA Title III information.

This product does not contain reportable quantities of substances listed under SARA Title III sections 302, 304, and 313.

DOT Class.

Not regulated.

201A01D.EPS

Material Safety Data Sheet

FIREFREE Coatings, Inc.
580 Irwin Street, No. 1
San Rafael, CA 94901

Product: Fire Free 88
MSDS Revision No. 3

Section IX Acronyms and Definitions Used in This MSDS

DOT:	Department of Transportation
MSHA:	Mine Safety and Health Administration
NIOSH:	National Institute for Occupational Safety and Health
OSHA:	Occupational Safety and Health Administration
SARA:	Superfund Amendments and Reauthorization Act

Reasonable care has been taken in the preparation of the information contained in this document. FIREFREE Coatings, Inc., intends no warranties, makes no representation and assumes no responsibility as to the accuracy or suitability of such information for application to purchaser's intended purposes or for consequences of its use.

201A01E.EPS

NCCER CURRICULA — USER UPDATE

NCCER makes every effort to keep its textbooks up-to-date and free of technical errors. We appreciate your help in this process. If you find an error, a typographical mistake, or an inaccuracy in NCCER's curricula, please fill out this form (or a photocopy), or complete the online form at **www.nccer.org/olf**. Be sure to include the exact module ID number, page number, a detailed description, and your recommended correction. Your input will be brought to the attention of the Authoring Team. Thank you for your assistance.

Instructors – If you have an idea for improving this textbook, or have found that additional materials were necessary to teach this module effectively, please let us know so that we may present your suggestions to the Authoring Team.

NCCER Product Development and Revision

13614 Progress Blvd., Alachua, FL 32615

Email: curriculum@nccer.org

Online: www.nccer.org/olf

❑ Trainee Guide ❑ Lesson Plans ❑ Exam ❑ PowerPoints Other ____________

Craft / Level: Copyright Date:

Module ID Number / Title:

Section Number(s):

Description:

Recommended Correction:

Your Name:

Address:

Email: Phone:

Corrosion Protection

69202-10

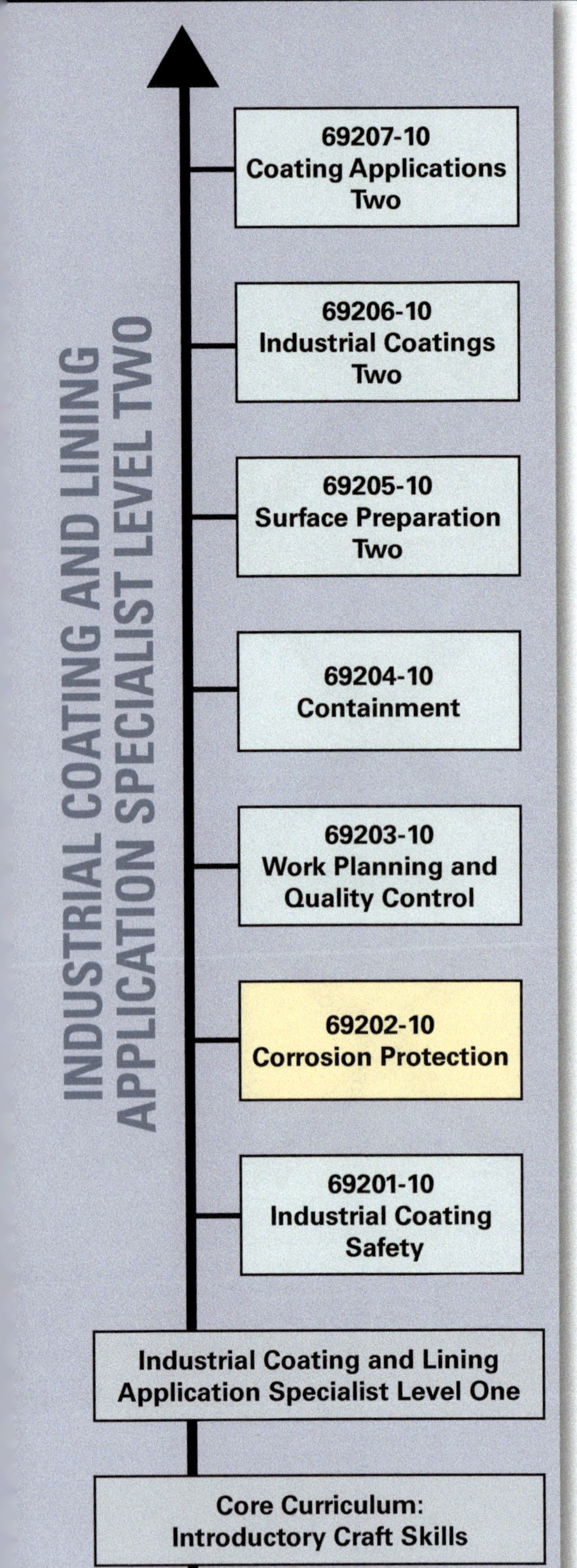

This course map shows all of the modules in *Industrial Coating and Lining Application Specialist Level Two.* The suggested training order begins at the bottom and proceeds up. Skill levels increase as you advance on the course map. The local Training Program Sponsor may adjust the training order.

Objectives

When you have completed this module, you will be able to do the following:

1. Identify the components of a corrosion cell.
2. Describe dissimilar metal corrosion.
3. Describe how concrete corrodes.
4. Describe how concrete rebar corrodes and the consequences of that corrosion.
5. List the types and classifications of corrosion.

Trade Terms

Aggregate
Electrolyte
Ions
Noble
Sacrificial coating

Prerequisites

Before you begin this module, it is recommended that you successfully complete *Industrial Coating and Lining Application Specialist Level One* and *Industrial Coating and Lining Application Specialist Level Two*, Module 69201-10.

Contents

Topics to be presented in this module include:

Figures and Tables

1.0.0 Introduction

Corrosion is a costly structural problem. A study funded by the Federal Highway Administration revealed that corrosion costs about $276 billion a year in the United States alone. Corrosion affects the appearance and safety of a structure and can also be detrimental to the environment when pipelines, railcars, tankers, trains, or storage tanks corrode.

NACE defines corrosion as "the deterioration of a material, usually a metal, that results from a chemical or electrochemical reaction with its environment." Materials corrode over time, and some corrode faster than others. Industrial Coating Specialists apply coatings that help stop or slow down corrosion. Understanding how to stop corrosion first requires an understanding of how corrosion occurs. This module explains how metals and concrete corrode. It also covers some of the methods used to control corrosion.

Some metals, such as copper and aluminum, are said to oxidize. This means that they chemically react with the oxygen and moisture in the air. Think of a shiny new penny. As the penny ages, its copper surface combines with the oxygen in the air and turns into copper oxide, which causes it to appear dull. Over time, copper carbonate is formed as the copper is exposed to oxygen, water, and carbon dioxide. The copper carbonate causes the penny's surface to turn a characteristic blue-green color, and along with copper oxide, can protect the metal from corrosion in some environments.

2.0.0 Corrosion Cell

The corrosion of metals is caused by electrochemical activity, which is sometimes called galvanic action. The electro portion of the word electrochemical refers to the electron flow between anodic and cathodic areas within the metal. The chemical portion involves the exchange, or flow, of **ions** between the anodic and cathodic areas through an **electrolyte** on the surface of the metal. Both flows must be present for corrosion to occur. The four components needed to support current flow and cause corrosion are called a corrosion cell (*Figure 1*). These components are described as follows:

- *Anodic areas* – Anodic areas are the electron donors for a metallic pathway. These areas are the parts of steel that corrode.
- *Cathodic areas* – Cathodic areas are electron receptors. They are the surface areas of steel that receive the electrons through the metallic pathway.
- *Metallic pathway* – This is the conductor that completes the path for electron flow. It is the interior of the steel substrate.
- *Electrolyte* – An electrolyte is a liquid or corrosive medium that conducts electricity and is usually water-based. Pure water is a poor conductor. However, the presence of chemical salts, such as sodium or magnesium chloride, or the sulfate compound contamination of water allows it to readily conduct electricity by ion exchange.

Steel is a mixture of many different elements. It is not a perfectly uniform or homogeneous metal. Because of these traits, steel plate and structures can have many tiny anodic and cathodic areas on their surfaces. These areas can differ slightly from each other in their electrical potential. As a result, steel has three of the four elements necessary to create a corrosion cell. The same conditions exist in most other metals.

The corrosion process can be demonstrated by explaining how a dry-cell battery circuit uses galvanic corrosion to generate electrical power (*Figure 2*). Note that the four corrosion cell elements are present as follows:

- An electrolyte comprised of a moist ammonium chloride and zinc chloride paste inside the battery case. In the case of steel, the electrolyte is rain or dew that has been contaminated by chemical salts or compounds deposited on the steel's surface. Sometimes water used for cleaning may be contaminated with the same materials.
- A negative electrode (zinc case) that corresponds to the anode in a corrosion cell.
- A positive electrode (carbon rod) that corresponds to the cathode in a corrosion cell.

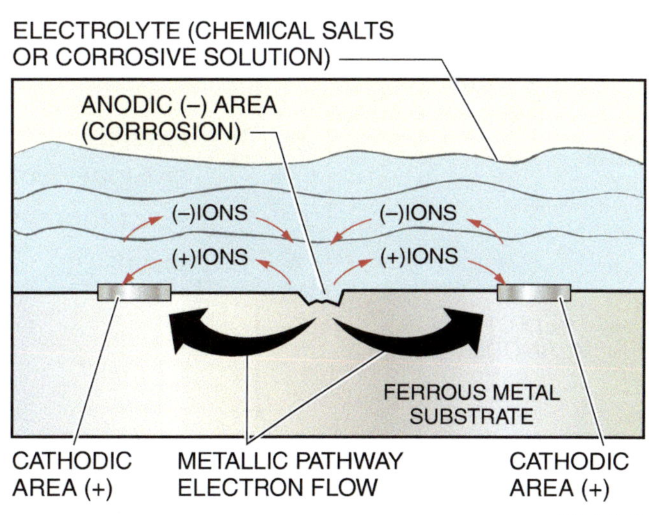

Figure 1 A corrosion cell on a steel surface.

- A conductive wire that corresponds to the metallic pathway in a corrosion cell.

The anode is the part of the metal that corrodes (dissolves) in the electrolyte when electron current flows in a corrosion cell. The anode dissolves in the form of positively charged ions. The electrons left behind cause the immediate area of the anode to become negatively charged. In the battery example, the surplus electrons at the anode flow through a wire, a bulb, and a closed switch to the cathode, which lights the bulb. The same process takes place in a corrosion cell through the metallic pathway provided by the steel substrate. The cathode, which corresponds to the carbon rod of the battery, is the less active area (+) on the metal's surface. This area consumes the electrons. On steel, the reaction generally ionizes the liquid electrolyte on the surface to form hydrogen gas and negatively charged (–) hydroxyl ions. The negatively charged (–) hydroxyl ions combine with the dissolved positively charged (+) metal ions at the anode to complete the circuit and form compounds, such as ferrous hydroxide. With exposure to oxygen, the ferrous hydroxide further reacts to become iron oxide, or rust (*Figure 3*).

Some metals, such as gold and platinum, are resistant to corrosion, but their high cost makes their widespread use impractical. The list shown in *Figure 4* contains common metals in order of their susceptibility to corrosion, from the most to the least corrosive. Note that steel and cast iron, two commonly used metals, are in the middle of the list.

Figure 3 Corroded steel.

1. MAGNESIUM
2. ZINC
3. ALUMINUM
4. CARBON STEEL
5. CAST IRON
6. COPPER
7. STAINLESS STEEL
8. SILVER
9. GOLD
10. PLATINUM

202F04.EPS

Figure 4 Metals listed in order of their susceptibility to corrosion.

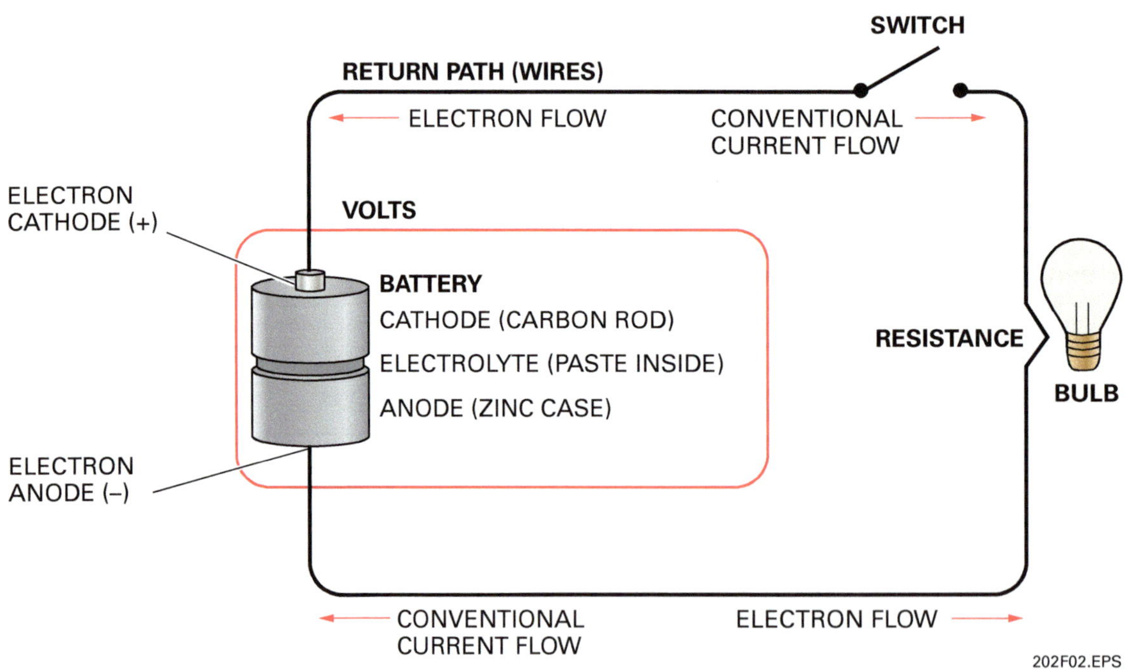

Figure 2 Battery example of corrosion.

Metals that are less susceptible to corrosion are often referred to as less active, cathodic, or **noble**. Metals that are more susceptible to corrosion are often referred to as active or more active; they are also called anodic metals. These properties can be used to help preserve a primary structure. For example, zinc, which readily corrodes, can be used to coat steel. When exposed to the elements, the zinc coating corrodes faster than the steel, thus preserving the steel. This and other methods of preventing corrosion are covered in more detail in later sections of this module.

3.0.0 Classifications of Corrosion

Materials corrode in different ways depending on their environments. One common trait of corrosion is that when a metal corrodes, the corroded material takes up more space than the original material. The following is a description of the most common classes of corrosion:

- *Uniform corrosion* – Sometimes called general corrosion, uniform corrosion occurs to large parts of a structure due to some type of harmful exposure, such as to water, acid, the atmosphere, or earth. Uniform corrosion can be prevented by removing the source of corrosion or by using an appropriate protective coating (*Figure 5*).
- *Galvanic corrosion* – This occurs when two dissimilar metals, such as copper and steel, are placed close together. *Figure 6* shows corrosion caused by steel bolts in contact with a copper pipe flange. Galvanic corrosion can be controlled by avoiding the connection of dissimilar metals, selecting noble materials, or installing sacrificial anodes or an impressed current system.
- *Localized corrosion* – Corrosion that is isolated to a small area is referred to as localized corrosion. The two types of localized corrosion are as follows:
  - Pitting corrosion is identified by a small, deep area of corrosion surrounded by an unblemished surface. It is usually caused by a tiny corrosion cell or a speck of contamination, such as a bit of salt or acid, or by a thin spot in a coating. Aluminum, stainless steel, copper alloys, and lead are especially susceptible to pitting. Pitting can be prevented by properly cleaning surfaces and by applying coatings.
  - Crevice corrosion is corrosion that occurs in a structure's crevice. Although the largest quantity of crevice corrosion occurs on carbon steel commonly used in industrial structures, the metals that are the most susceptible to crevice corrosion include aluminum, stainless steel, and titanium, as well as copper, zirconium, and their alloys. Crevice corrosion can be controlled by selecting materials that are not as susceptible to it; by designing a structure without crevices; or by using cathodic protection.
- *Stress corrosion cracking* – This type of corrosion is the result of a combination of external stress and exposure to a corrosive agent. Stress corrosion cracking of stainless and carbon steel can be the result of their exposure to chlorides, the exposure of carbon steel to caustics, or the exposure of copper alloys to ammonia.
- *Microbiologically influenced corrosion (MIC), also known as microbial or biological corrosion* – Refers to the deterioration of metals resulting from the metabolic activity of microorganisms. A number of bacteria are known to cause this type

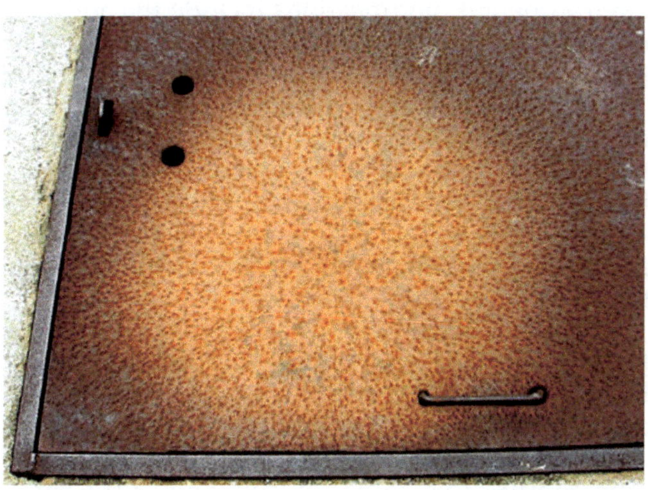

202F05.EPS

Figure 5 Uniform corrosion.

202F06.EPS

Figure 6 Galvanic corrosion.

of corrosion in carbon steels, stainless steels, aluminum alloys, and copper alloys. The bacteria live in water and soil that have a pH of 4 to 9 and temperatures of 50°F to 122°F (10°C to 50°C). They are broadly classified as aerobic (they require oxygen to become active) or anaerobic (oxygen is toxic to the bacteria). Sulphate-reducing bacteria (SRB) are anaerobic. Sulphate-reducing bacteria are responsible for most instances of accelerated corrosion damage to ships, underwater pipelines, and offshore structures. Iron- and manganese-oxidizing bacteria are aerobic. They are usually associated with aggressive pitting attacks on stainless steels at welds.

- *Macrobiological fouling* – Includes damage from zebra mussels, barnacles, and other types of marine life. This type of fouling can accelerate biological corrosion and can render underwater equipment useless. Besides causing corrosion, the attachment of barnacles to the hull of a ship can limit the ship's movement.

3.1.0 Preventing Corrosion

The task of preventing corrosion is largely centered on coatings. However, other methods are also used.

Specifiers and designers can help to prevent corrosion by selecting materials that will stand up to the environment in which they will be placed. They can also design systems that avoid potentially corrosive reactions. For example, systems should be designed to avoid the use of dissimilar metals that promote corrosion. They should also be designed to avoid friction between parts, as well as cracks and crevices that can retain dirt and moisture. Good system design also includes having all areas accessible so that they can be properly coated.

3.1.1 Protection by Coatings

Various coatings can be applied to alleviate the corrosion of ferrous and nonferrous surfaces. The necessary protection can be achieved by using one, or a combination of, the following four coating types:

- *Sacrificial coatings* – **Sacrificial coatings** are rich in zinc. The zinc acts as a sacrificial anode and corrodes to protect the surface in the event the coating is damaged. Sacrificial coatings can be applied in various ways. For example, they can be applied as coatings containing zinc powder, as a hot-sprayed zinc metal, or as a hot-dip zinc (galvanized) coating.
- *Barrier coatings* – Barrier coatings keep moisture away from a substrate surface. This moisture protection removes the electrolytic element of a corrosion cell and stops corrosion. A barrier coating is the most commonly used method for protecting underground piping and above ground concrete and steel structures.
- Inhibitive coatings – Inhibitive coatings are usually primers that assist in the control of steel corrosion through the use of pigments. Pigments that can provide an inhibitive effect on corrosion include zinc phosphate and, in some countries, red lead. The pigments react with absorbed moisture in the coating and then react with the steel to passivate it and reduce its corrosive characteristics.
- Biocide coatings – Biocide coatings are used to reduce the formation of destructive microbiological and macrobiological organisms on the hulls of ships, underwater pipelines, water intakes, and other underwater structures. Coatings of this type have included tin or copper compounds considered to be harmful to other forms of marine life.

WARNING

Lead is a known health hazard that can lead to sickness and, in extreme cases, death. If you are removing coatings containing lead, be certain to follow the rules and regulations of *OHSA Standard 29 CFR 1926.62, Occupational Health and Environmental Controls – Lead.*

3.1.2 Other Corrosion Protection Techniques

Metals that are subject to corrosion, such as iron and steel, can be protected by methods other than coatings. For the interiors of pipes, boiler tubing, and similar equipment, environmental modifications are used to reduce internal corrosion. These environmental modifications include the addition of inhibitors or the deaeration and pH control of the fluids or gases flowing through the equipment.

Also available are two electrochemical techniques that cause the metal in need of protection to act as the cathode, which essentially eliminates its corrosion. These electrochemical techniques are known as cathodic systems. They can be used for pipelines, steel structures, and concrete rebar in conjunction with barrier coatings.

One of the two cathodic systems is a passive type. It uses sacrificial anodes and galvanic action to protect the steel or iron substrate. The anodes are made of metals that are more active than the steel or iron, such as magnesium, aluminum, or zinc. Similar to a corrosion cell, and in the presence of an electrolyte for the substrate, the anodes begin to dissolve and become a source of electrons that flow through a wire connected to the substrate (or by direct contact). As a result, the entire substrate becomes the positive cathode. This prevents the activity of any smaller anodic or cathodic areas on the substrate, thereby preventing its corrosion. The anodes are available in a variety of standard and customized shapes. Because of limited electron current, sacrificial anodes are only used in certain underground or underwater applications for pipelines, ship hulls, steel structures, piers, and pilings.

The other cathodic system is the more effective of the two. It prevents corrosion of a metal through use of a direct current power supply. The power supply has a large current capability and higher voltage that boosts the electron potential difference and the resultant current flow between the anode and the cathode. This type of system is an impressed current cathodic protection system (*Figure 7*), usually referred to as an impressed current system. The negative terminal of the power supply is connected to the metal substrate to be protected. The positive terminal is connected to the anode. This configuration makes the electron potential difference very large, and a sizable electron current flow can occur. In the presence of an electrolyte, the anode becomes a very large negative source of electrons, and the metallic substrate becomes a highly positive electron receptor. The large electron flow from the anode through the power supply to the cathode cancels out any smaller anodic and cathodic activity on a large surface of the substrate, thereby preventing any corrosion. The anodes used are available in many shapes (*Figure 8*). They may be sacrificial or relatively nonsacrificial, depending on the application. These impressed current systems are used for long underground pipelines, underwater portions of offshore platforms, the underside of tanks, and other large underground steel structures. They are also used to protect structural steel and rebar in steel-in-concrete applications, such as bridge decks, parking garages, building structures, roofs, and balconies.

202F07.EPS

Figure 7 Impressed current cathodic protection system.

4.0.0 Concrete

Concrete is one of the most widely used construction materials. It is used in secondary containment, fireproofing, tank linings, flooring, and as structural members, tanks, and pipes. Concrete can even be used underwater. It is easy to use, affordable, and versatile. Since it is made up of cement (usually portland), **aggregate**, and water, it can be mixed either at the site or mixed offsite and then transported to the work area.

Although concrete is generally thought to be indestructible, it is susceptible to corrosion. Concrete has great compressive strength, which means that it can withstand a large amount of simultaneous pressure pushing on it if it is supported. However, concrete is weak under tension, which means that it cannot withstand pressure pulling it apart (*Figure 9*). To improve concrete's tensile strength, it is often reinforced with steel or fibrous materials.

Concrete is porous and can readily absorb water and other fluids if the surface is not sealed. When concrete containing water or other fluids

Flexible anode configured in concentric rings to protect a crude oil storage tank bottom before tank installation.

202F08.EPS

Figure 8 A flexible long-life anode used in under-tank installations.

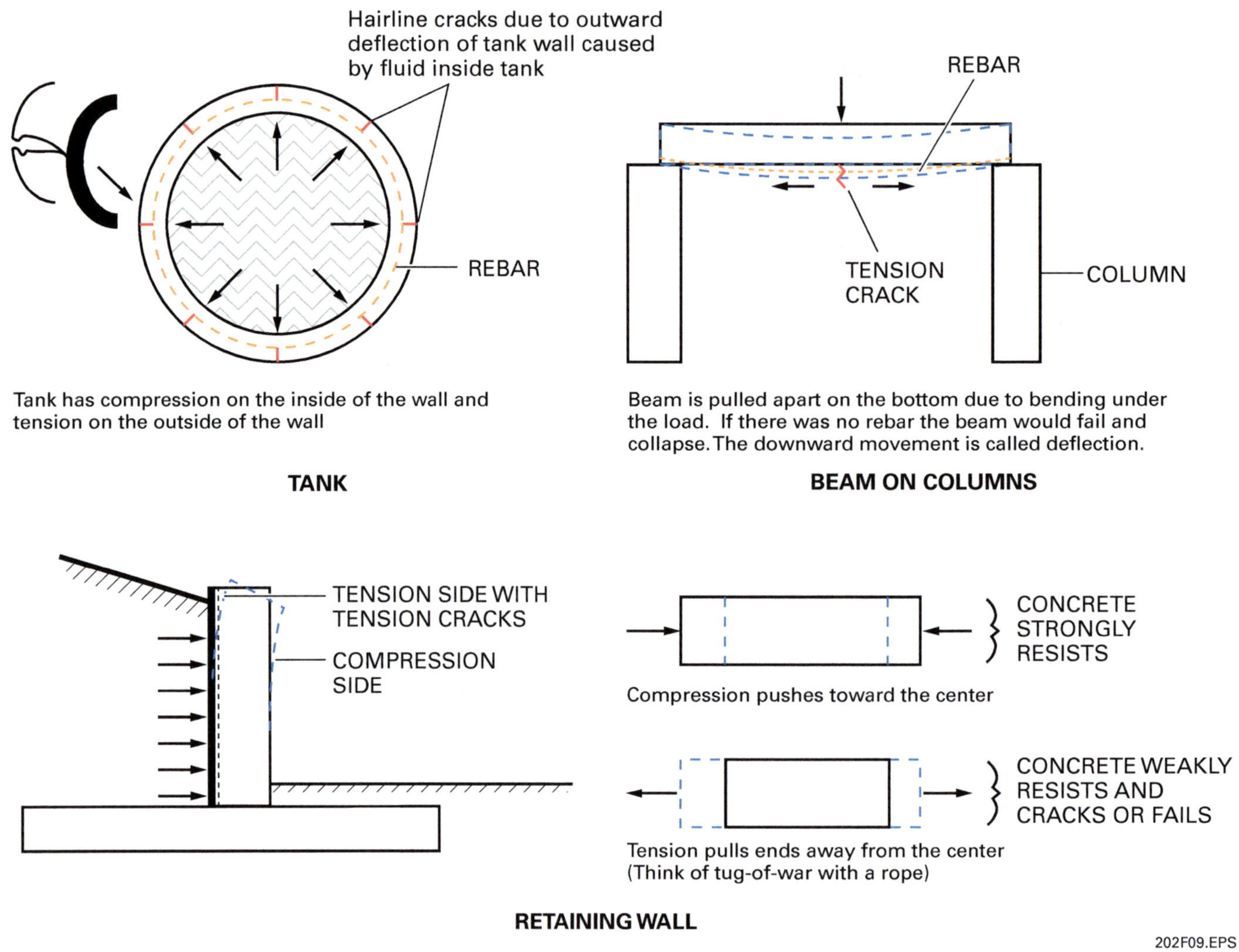

Figure 9 Compressive versus tensile strength.

is exposed to freezing temperatures, the fluid can freeze and break apart the concrete. Unsealed concrete can absorb fluids, which can cause the steel reinforcing bar (rebar) embedded in the concrete to corrode, weakening the concrete even more. It is very important to properly coat concrete and rebar that is located in an acidic environment (*Figure 10*). When this is the case, an appropriate moisture barrier should be applied to the surface of the concrete as soon as possible after it has been cured.

Even though concrete is commonly used for road surfaces and buildings, it is susceptible to damage by abrasion from foot and vehicular traffic, and from wind- and water-borne particles. To prevent abrasive corrosion, it is important to use a high-quality cement; the correct type, ratio, and size of aggregate; and potable water. Once the concrete has cured, the proper coating should be applied.

Concrete is alkaline, so it will chemically react with any acid to form salts and water. This characteristic causes the concrete to dissolve and weaken over time. Many chemicals in commercial use are acidic.

Figure 10 Concrete corrosion.

Concrete and steel rebar do not react with one another when salts, water, acids, or other contaminants are not present. However, it is virtually impossible to maintain a contaminant-free environment. Sometimes there are salts in the concrete mix. At other times, the concrete is exposed to moisture or acids. Aggregates that are high in silica react with concrete. A contaminant may be something commonly used on the concrete structure, such as the salt used to de-ice roads. Stray electrical currents could be present in the environment (this is especially true when concrete is buried in the ground). In addition, corrosion cells could be present in the rebar.

All of these situations can cause corrosion of the rebar or a reaction between the concrete and another substance. The problem with these reactions is that the corroded material takes up more room than the original component. As a result, the expansion causes the concrete to crack from within (*Figure 11*).

> **WARNING**
>
> Any extensive spalling or cracking of reinforced concrete, without any visible rust stain, can signal that a concrete structural failure may be imminent. If this condition is observed, notify a supervisor immediately and do not coat the surface.

Every precaution should be taken to avoid concrete corrosion. When mixing the concrete, use high-quality cement mix; clean, nonexpansive aggregate; and potable water. Before placing the cured concrete into service, it should be properly coated. If the concrete is contaminated, it must be thoroughly cleaned before any coating is applied.

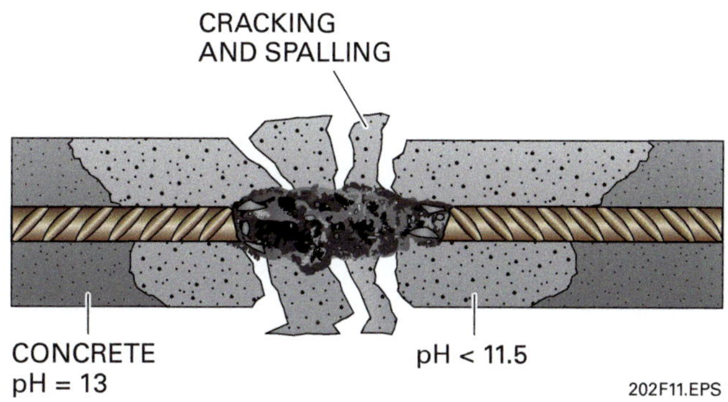

Figure 11 Rebar corrosion process.

On Site

Steel-In-Concrete Rehabilitation

Extensive cracking and spalling of the underside of NASA's assembly building roof signaled that rebar corrosion was the cause of falling debris inside the building. As part of the concrete's rehabilitation, an impressed current cathodic protection system connected to the rebar was installed to prevent any future rebar corrosion and concrete deterioration. In the case of concrete, the anodes are normally made of a noble metal, for long life.

SUMMARY

Most materials will corrode when exposed to the right environmental conditions. However, a number of steps can be taken to decrease the likelihood of rapid corrosion. In order to understand how to stop or slow down corrosion, it is first necessary to understand how and why corrosion occurs.

This module explained how corrosion cells work. It also covered the common types of corrosion and the coatings that are used to prevent or slow down corrosion. Environmental modifications and cathodic systems were also discussed.

Review Questions

1. The copper carbonate that forms on a copper surface is colored _____.
   a. dull brown
   b. blue-green
   c. black
   d. pink

2. Besides copper, a metal that commonly oxidizes is _____.
   a. gold
   b. platinum
   c. titanium
   d. aluminum

3. An electrolyte is a medium, usually water based, that _____.
   a. prevents corrosion
   b. allows current flow
   c. is a coating additive
   d. can replace coatings

4. The corrosion cell is made up of a(n) _____.
   a. anode, a cathode, a metallic pathway, and an electrolyte
   b. anode, a cathode, an electrolyte, and electrons
   c. cathode, a metallic pathway, an electrolyte, and electrons
   d. anode, a cathode, a metallic pathway, and electrons

5. The metal that is most resistant to corrosion is _____.
   a. magnesium
   b. zinc
   c. platinum
   d. copper

6. Corrosion that occurs when two dissimilar metals are placed close together is called _____.
   a. uniform corrosion
   b. galvanic corrosion
   c. localized corrosion
   d. filiform corrosion

7. Corrosion that occurs from exposure to external forces and a corrosive environment is called _____.
   a. stress corrosion cracking
   b. hydrogen-induced damage
   c. corrosion fatigue
   d. crevice corrosion

8. A type of coating that protects by corrosion is a(n) _____.
   a. inhibitive coating
   b. barrier coating
   c. sacrificial coating
   d. biocide coating

9. A highly effective cathodic system for preventing metal corrosion is the _____.
   a. biocide coating system
   b. distributed sacrificial anode system
   c. multiple barrier coating system
   d. impressed current system

10. Concrete is susceptible to freeze damage because it _____.
    a. can readily absorb water
    b. contains aggregates
    c. dissolves in water
    d. is alkaline

Trade Terms Introduced in This Module

Aggregate: Crushed rock or gravel that is screened to size for use in concrete.

Electrolyte: A nonmetal that is capable of supporting electrical current flow, such as a fluid in which salt has been dissolved.

Ions: Electrically charged particles.

Noble: A term used to describe a metal that is resistant to corrosion.

Sacrificial coating: A coating that corrodes to preserve the metal it coats.

Additional Resources

This module presents thorough resources for task training. The following resource material is suggested for further study.

Kennedy Space Center Corrosion Technology Laboratory website. http://corrosion.ksc.nasa.gov

U.S. Army Corps of Engineers Mobile District website for information about corrosion control and cathodic protection. www.sam.usace.army.mil

Chemical Bonding. Anthony Carpi, Ph.D. Visionlearning Vol. CHE-1 (7), 2003. www.visionlearning.com

Figure Credits

Photo courtesy of Hippwrap Containment, Module opener

Andy Pernick, Bureau of Reclamation, U.S. Department of the Interior, 202F03

John Troger, 202F05

Chester Ramirez. 202F06

MATCOR®, Inc., 202F07, 202F08, 202F10, 202F11

Tom Gibbons, 202F09

NASA, 202SA01

NCCER CURRICULA — USER UPDATE

NCCER makes every effort to keep its textbooks up-to-date and free of technical errors. We appreciate your help in this process. If you find an error, a typographical mistake, or an inaccuracy in NCCER's curricula, please fill out this form (or a photocopy), or complete the on-line form at **www.nccer.org/olf**. Be sure to include the exact module ID number, page number, a detailed description, and your recommended correction. Your input will be brought to the attention of the Authoring Team. Thank you for your assistance.

Instructors – If you have an idea for improving this textbook, or have found that additional materials were necessary to teach this module effectively, please let us know so that we may present your suggestions to the Authoring Team.

NCCER Product Development and Revision

13614 Progress Blvd., Alachua, FL 32615

Email: curriculum@nccer.org

Online: www.nccer.org/olf

❑ Trainee Guide ❑ Lesson Plans ❑ Exam ❑ PowerPoints Other ____________________

Craft / Level: Copyright Date:

Module ID Number / Title:

Section Number(s):

Description:

Recommended Correction:

Your Name:

Address:

Email: Phone:

Work Planning and Quality Control

69203-10

69203-10 Work Planning and Quality Control

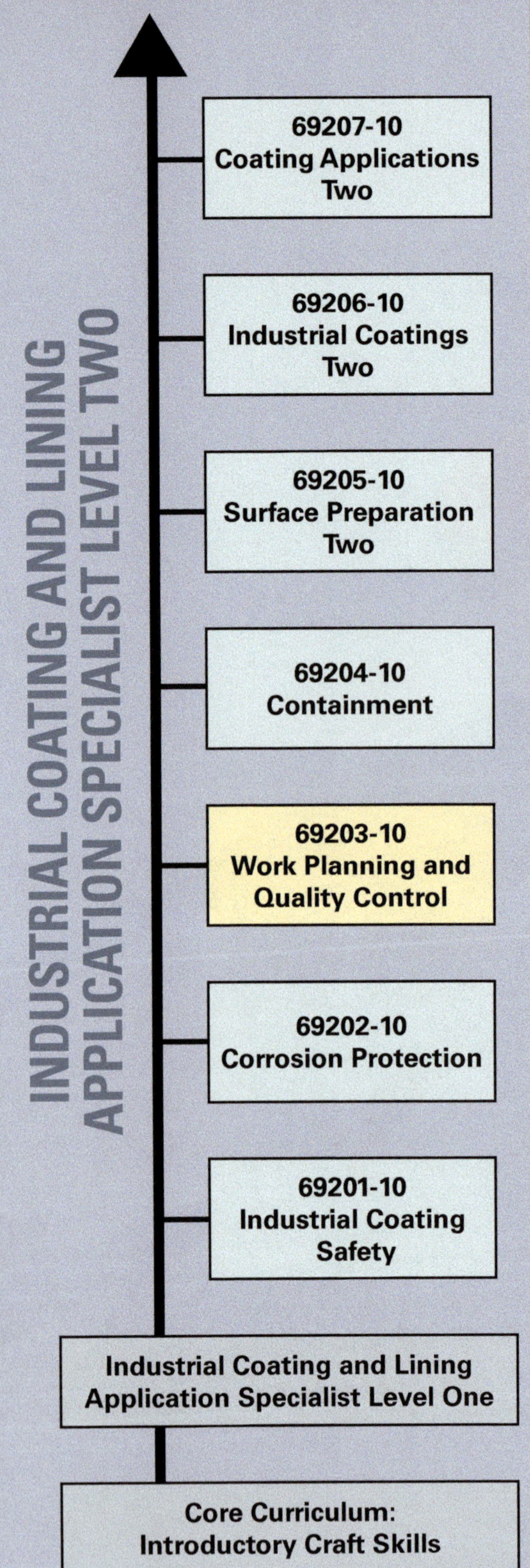

This course map shows all of the modules in *Industrial Coating and Lining Application Specialist Level Two.* The suggested training order begins at the bottom and proceeds up. Skill levels increase as you advance on the course map. The local Training Program Sponsor may adjust the training order.

Objectives

When you have completed this module, you will be able to do the following:

1. Follow and execute a work plan, including documentation of all processes.
2. Use process control on coatings projects.
3. Calculate ratios and areas for simple structures.
4. Calculate adjusted volatile organic compounds (VOCs) based on thinner additions.
5. Describe how pressure, volume, and temperature affect surface preparation and coating application operations.
6. Prepare test sections, demonstrating processes.

Trade Terms

Area
Circumference
Cylinder
Exponent
Perpendicular
Pi
Radius
Volume

Required Trainee Materials

1. Pencils and paper
2. Advanced-function calculator

Prerequisites

Before you begin this module, it is recommended that you successfully complete *Industrial Coating and Lining Application Specialist Level One* and *Industrial Coating and Lining Application Specialist Level Two,* Modules 69201-10 and 69202-10.

Contents

Topics to be presented in this module include:

Figures and Tables

1.0.0 INTRODUCTION

The first step in any industrial coatings job is the planning process. This involves planning the procedures, determining the specifications for materials, and planning quality checks.

The quality movement in manufacturing is based on the fact that rework costs money for which the manufacturer or contractor is not reimbursed. Estimates by several quality specialists suggest that from 20 to 40 percent of work is repair and rework of poor quality work. This means that the contractor must charge 20 to 40 percent more or lose a significant portion of the profit from the job. If work were performed correctly the first time, profit would be available and the contractor could charge less and get more jobs by being more competitively priced.

The object of a work plan, and of the documentation that goes with it, is to get each crew member to do the job right the first time. The checks on the process are built into the documentation. This means that daily compliance with the work plan, and with the specifications that drive the work plan, is a part of the daily tasks of the technician, who must document compliance.

2.0.0 THE WORK PLAN

A typical work plan includes manpower, materials, equipment, and scheduling requirements. A work plan for a simple job, such as coating a concrete floor, might read as follows:

- *Limits of job* – Existing uncoated warehouse floor, 50' × 100', per specification.
- *Assigned manpower* – Two men.
- *Material specification* – Low-temperature cure epoxy.
- *Scheduling* – Two weeks.

Step 1 Surface preparation (per PDS):
- Remove any oil and grease using a detergent
- Shotblast the surface
- Fill any bugholes and patch any deteriorated surface using an epoxy filler/sealer
- Check for moisture using the plastic sheet method
- Mask all walls and support posts at floor level
- Sweep and vacuum the surface

Step 2 Mix epoxy using appropriate PPE and mixing equipment.

Step 3 Check that application conditions (temperature and humidity) conform to coating requirements.

Step 4 While keeping track of pot life, apply epoxy at recommended wet film thickness (WFT) with rollers using appropriate PPE. Use a brush to cut in corners and edges as necessary.

Step 5 After minimum recoat time, inspect surface for any defects and correct DFT. Repair as required.

Step 6 Request owner inspection and approval.

Any job can be planned effectively by breaking it down into stages and requirements. A larger job may require more steps and closer examination of time constraints. Begin with the specifications, which identify the requirements for the job and the materials. The surface preparation requirements are defined, including the required inspection procedure. This ensures that the necessary profile will be available to allow coating adhesion.

The steps to carry out the job must be laid out in a manner that can be easily and quickly checked. This particular function can be quite specific for a coating job, as shown in this example:

Step 1 The stripe coat is applied and inspected by the responsible person, who may be an inspector, the site supervisor, the senior technician on site, or the owner's representative.

Step 2 Having satisfied the requirements, the rest of the area will be coated according to the specifications.

The plan will define the testing requirements for the coat, such as WFT testing. Properly written work plans establish the amount of time that the stripe coat must dry or cure before being overlapped by the area coverage, and the time the first coat must dry before any overcoating takes place.

Once the area has been coated, the work is inspected as required by the specifications. Technicians should always inspect the job before the inspector looks at it, checking all the coating characteristics against the work plan or specifications. If there is something wrong, it should be fixed. If a defect isn't fixed and the inspector finds it, the contractor might have to strip the coating and start over.

2.1.0 Process Control

The work plan should define stages at which the status of the work must be reported. This distributes control of the process. While it is common in small contracting firms for the site supervisor to give a verbal report, that doesn't serve as documentation if something goes wrong with the coating later. The question of responsibility may then go to court.

Process control includes standardizing the way work is done. This does not mean that everybody works exactly the same way, but rather that the results are the same at each stage of the process. If the specifications call for a particular type of sprayer, that sprayer must be on site and in use. Specifications drive the work plan, which drives the work process. This is especially critical when work crews are not directly controlled on a day-to-day basis by the person who writes the work plan.

Quality requirements are written into the work plan in the form of incorporated specifications. The specifications may be taken from the job specification language directly. The objective is to establish exact standards for the particular job. The work plan is frequently submitted to the owner as part of the bid and is a part of the legally enforceable agreement, given force of law by the contract.

2.2.0 Scheduling

It is important for the contractor and the owner, as well as for other contractors involved in a large-scale project, to include a realistic time schedule in the work plan. It is not always necessary to work out scheduling with other trades, but whether other trades are involved or not, the accuracy, reliability, and flexibility of the schedule are critical to the completion and profitability of the project.

The process of determining a schedule requires an understanding of the time, equipment, manpower, and weather conditions for each stage of the job. Consider, for example, that it will take two people three days to prepare the containment for a job. Since the surface preparation cannot be done while the containment is being erected, and since surface prep work is done inside the containment, the containment probably needs to be completed before the surface prep work is started. If two blasting crews are used, the surface prep should be possible in four days. Here is the resulting schedule: start containment on Monday, begin surface preparation on Wednesday, and work straight through until the following Monday.

Now the complications arise. If a windstorm delays the containment operation for two days, and the schedule has assigned the blasting crews to be on site on Wednesday, they will be standing there for two days until the containment is erected. The contractor is losing money, paying for the containment labor and the surface prep crew. These delays will also affect the dates of the coating work.

One solution to such complexities of scheduling is called the critical path method (CPM), which is shown in *Figure 1*. This is usually done with software on a computer, because it can be extremely difficult to keep track of all of the different factors and operations mentally. In CPM, each activity is assigned a duration, with allowance for problems. Antecedents and successors are assigned as well. Once the initial information is entered, the computer can keep track of changes, even if end dates for one activity change.

The same thing is done on a smaller scale in a work plan for a single job. Often the deadline for completion is set, with penalties for failure to complete the job on time. The work plan should be flexible to allow for unexpected problems while still completing the job by the established finished date.

2.3.0 Equipment Assignments

Because a contractor cannot afford to duplicate all the equipment needed for all the jobs at once, it is often necessary to schedule access to a particular piece of equipment based on availability. For example, the contractor may have only one blast pot and four teams working on jobs that will require blasting. Work plans must assign equipment based on the scheduling limits of different jobs. Using a key piece of assigned equipment without talking to the supervisor could result in failure to complete a contract and the loss of a client.

2.4.0 Safety Planning

The work plan should have a section on safety that includes the personal safety of coatings workers and others in the area. The workers must be qualified for the work they are doing and must follow all recommended safety practices. Special attention should be given to the volatile organic compound (VOC) levels in the solvents and coatings being used.

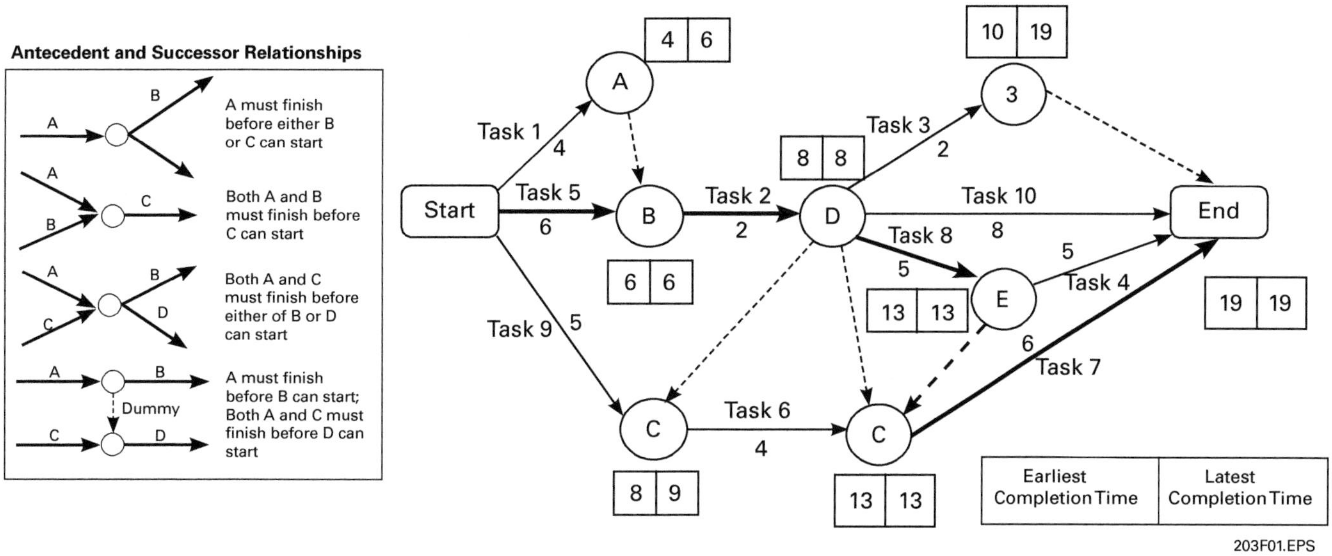

Figure 1 Critical path method.

3.0.0 QUALITY CONTROL AND QUALITY ASSURANCE

The work plan is the basis of process control and planning. It establishes the level of quality required and the process by which that level is to be reached, including applicable standards and inspection procedures. Quality can be defined as consistently meeting requirements, and it must be a companywide commitment. Everyone in the company must make quality a high priority in order to make it a reality.

Industrial coatings jobs normally have formal quality standard requirements required by the job's contract and/or specifications. It is the contractor's responsibility to perform all applicable inspections and tests needed throughout the various stages of the job in order to make sure that the quality of the work, environmental working conditions, surface preparation, coating application, and so forth, meet those specified for the job. This process is called quality control (QC). Similarly, the owner, an engineering company, or another third-party group also performs inspections and tests throughout the progress of a job to ensure that the contractor is indeed performing the job according to specifications. This process is called quality assurance (QA).

For large industrial jobs involving many people, it is common to have the quality assurance tasks performed by a third-party inspector or group of inspectors. This can reduce general disruptions and tension between owners, contractors, and material suppliers.

The reason for inspections is to ensure quality control over substrate preparations, coatings, and their application. For contract work, inspections assure that all specification requirements are met. Properly performed inspections can prevent conditions that may result in a highly expensive premature coating failure. During an inspection, conditions that might affect the final quality of the coating can be readily witnessed and easily verified.

In its simplest form, an inspection occurs when an application specialist stops after a certain portion of the work is completed and examines it for adequacy. Have any areas been missed? Are there any runs or sags? Is the blast cleaning pattern uniform and the cleanliness adequate? In the case of hand or power tool cleaning, is there any loose mill scale or rust deposits remaining? These inspections are informal but very important, and should be done before the work is formally inspected. Formal inspections require that the inspection procedures are written, and the quality of work is witnessed and documented on a regular basis. It also means that the person or group of persons performing the inspections are specially trained and are equipped with the instruments required to verify the quality of the work.

3.1.0 The Inspection Process

The scope of coating inspections should include the following:

- Inspection of materials
- Inspection of surface preparation
- Inspection of paint application
- Final inspection for acceptance

The inspection process normally begins with a pre-job conference during which the ground rules for the inspection are set. During this pre-job conference, any questions should be addressed and any potential problems identified. In many cases, test panels demonstrating surface preparation requirements and coating application are prepared and approved by all concerned prior to beginning a job. The test panels serve as inspection standards for the job as the job progresses. They can also become part of the documentation once the job is completed. As the job progresses, both the contractor's QC specialist and the designated QA inspector are responsible for witnessing, verifying, inspecting, and documenting their observations at various inspection points. As shown in *Figure 2*, the inspection process should include at least the following steps or points:

- Pre-surface preparation inspection
- Measurement of ambient conditions
- Evaluation of compressor, air cleanliness, and surface preparation equipment
- Determination of surface preparation, cleanliness, and profile
- Witness of mixing and thinning of coating materials
- Evaluation of application equipment
- Inspection of coating application(s) and cleanliness between coats
- Determination of WFT and DFT
- Pinhole and holiday testing
- Evaluation of adhesion/cure

3.1.1 Documentation

The key to quality control and an effective work plan is documentation. Daily reports on work completed, testing, and inspection allow the company to maintain the job standards. *Figure 3* is an example of a daily report form for a coatings job.

Any conflict that may arise between the specifications and the actual work being performed must be resolved. The specifications are the requirements for the job and have the force of law as part of the contract. Any deviation from the contract must be agreed to in writing by the owner. Such a change would be considered an addendum to the contract. When a problem arises and the contractor cannot comply exactly with the contract, the issue is submitted as a Request for Information (RFI) (*Figure 4*).

Another form that is used for coating work is the nonconformance report (NCR). This report is submitted by an inspector who finds that some aspect of the work is not in conformance with specifications. An NCR describes the nonconformity, and it will have to be resolved before the job can be completed.

3.2.0 Factors That May Affect Coating Quality

Surface preparation and coating application both require careful consideration of ambient conditions. Ambient conditions include air and substrate surface temperature, relative humidity (RH), dew point, and wind. Atmospheric air pres-

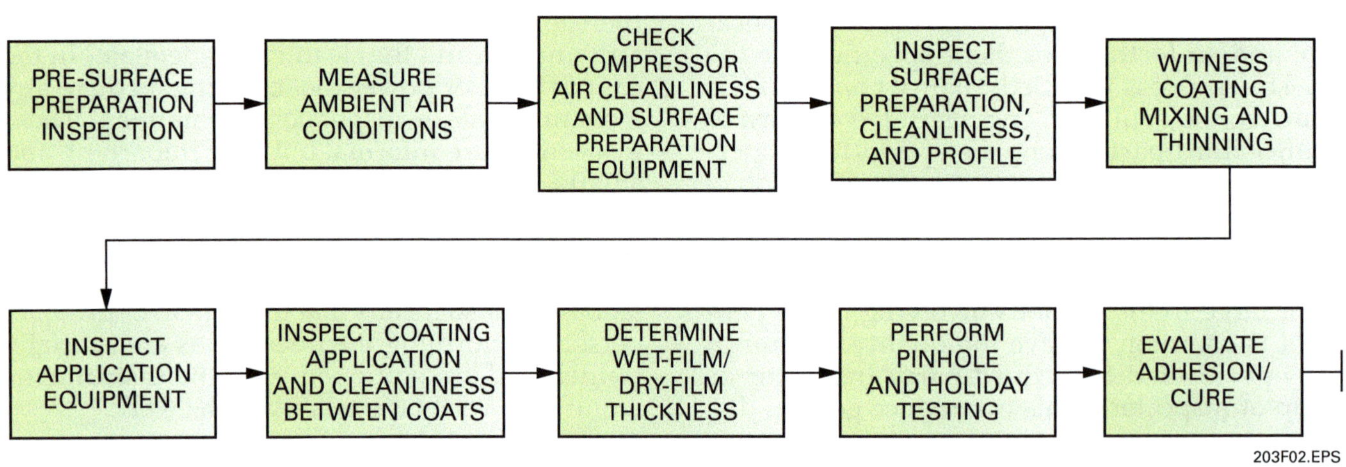

Figure 2 Sequence of inspection for a typical industrial coatings job.

Daily Report

Date ______ Report No. ______ Project No. ______ Page ______
Project Name ______ Site ______
Person Reporting ______
Description of areas and substrate ______

Work Performed ______

Conditions ______ Surface Preparation ______
Time ______ Procedure ______
Air temp. ______ Surface Condition ______
Rel. Hum. ______ Surface Profile ______
Surface temp. ______
Wind ______
Coating application ______
Coat WFT in Mils ______
Prime Coat ______ Intermediate ______ Top Coat ______
Product ______ MFR. ______
Batch No. ______ Qty ______
Thinner ______ Qty added ______
Application Method ______
Remarks ______

Signature ______

203F03.EPS

Figure 3 A sample daily report.

sure may also have some influence, depending on the elevation and/or unusual weather conditions. Sometimes a combination of otherwise acceptable factors can result in coating failures.

Normally, ambient conditions are noted at the beginning of surface preparation operations or coating operations as well as during these operations if the weather changes to any significant degree. The ambient conditions should be entered on one or more daily reports, as changing conditions warrant.

3.2.1 Air and Surface Temperatures

Air temperatures generally affect coating operations more than they affect cleaning operations. Solvent evaporation and curing rates are primarily affected by the air temperature. However, the temperatures at various locations on the surface to be coated can also affect solvent evaporation and curing rates. High air and/or surface temperatures may force rapid solvent evaporation and affect curing. Rapid evaporation during spraying, called dry spray, prevents the coating from

Request for Information

Date: RFI number:

Information Request

Project ______________ Contractor ______________ Project Type ______________

Project Name ______________ Purchase Order No. ______________

Requested By ______________

Information Requested:

______________ Representative Signature ______________ Contact Number ______________ Date

Response Needed Date ______________

RESPONSE

Action: Per conversation with Dept.

No Change

Repair

Reject

Other

______________ Engineer ______________ Contact Number ______________ Date

Resolved

______________ Representative Signature ______________ Date

203F04.EPS

Figure 4 Sample Request for Information.

wetting the surface and results in poor adhesion. High temperatures can also result in improper film formation. Conversely, low temperatures may prevent proper solvent evaporation, resulting in solvent entrapment in the coating film or improper film curing. This can eventually result in bubbling of the film.

3.2.2 Relative Humidity

Relative humidity is a measure of the amount of moisture in the air compared to a saturation level. Many coating specifications restrict coating application operations to a limited range of relative humidity. When the RH is too high, solvent evaporation is slowed, and the solvent may become trapped. In addition, the recoat time may be extended, resulting in curing problems. The extended recoat time can also cause solvents to be trapped between coats, affecting bonding of the coats. Coatings exposed to high humidity after application may show defects, including blush, poor gloss, or, in the case of epoxies, amine blush. Waterborne coatings are more likely to be affected by high humidity than other types of coatings. Some coatings, such as inorganic zinc silicate, require a minimum RH to cure properly and may require misting (spraying with water) if the humidity is especially low.

3.2.3 Dew Point

Dew point is the temperature at which water will condense, leaving moisture on a surface. The dew point is an important factor during dry abrasive blast cleaning and recoating operations. After cleaning, steel will flash rust unless the temperature of the steel surface is at least 5°F (3°C) above the dew point temperature. The same is true for recoating operations. Moisture trapped between coating layers can cause premature failure of the coating. Surface temperatures can be maintained inside some containment enclosures by infrared heating equipment. The dew point can be lowered to a level that prevents condensation by using dehumidification equipment to reduce the RH.

3.2.4 Wind

Contractors should check weather forecasts for the day and several days into the future for the possibility of high winds. Wind can affect surface preparation and coating operations in the following ways:

- Abrasives from blasting operations can be blown into coating operations.
- Wind can cause excessive drift, or overspray can be blown onto other surfaces during spray coating operations.
- Wind can cause excessive solvent evaporation after application, affecting cure; or during spray operations, resulting in dry spray.
- Blast surfaces and/or coating operations can be contaminated by blown dust, dirt, sand, salt, or sea spray.
- Wind can be a safety hazard for work performed at some height when wind speeds reach or exceed 40 mph (64 kph). In some cases, containment may have to be lowered or dismantled. For offshore platforms, waves may prevent working near sea level and/or safety boats may not be able to operate reliably.

3.2.5 Atmospheric Air Pressure

Atmospheric air pressure variations due to altitude or barometric weather changes can alter the evaporation rate of solvents. The lower the pressure, the higher the rate of evaporation and vice versa. This can affect coating drying and recoating times, as well as the pot life of some plural-component coatings.

3.2.6 Relationship Between Pressure, Temperature, and Volume

The Universal Gas Law states that pressure × volume/temperature = constant. This formula has important ramifications for blasting and painting. For the most part, temperature can be considered a constant. The important thing to remember about temperature is that increasing the air temperature increases the volume of air, and cooler temperatures decrease the volume of air. As the temperature increases during the day, a larger volume of air will be required to maintain the same pressure. Because warm air is less dense, more air is required to achieve the same pressure.

More important is the relationship between volume and pressure. During abrasive blasting operations, it is important to maintain pressure at a blast nozzle. As the blast nozzle orifice wears, the area increases according to the formula:

$$A = \pi r^2$$

Where:

π (pi) = 3.14

r = the **radius** of the blast nozzle orifice

This has a direct result on blasting efficiency. If the volume of the air supply is held constant and the orifice area increases by the square of the radius, the volume of air passing through the nozzle increases, and as a direct result, the pressure decreases. Each 1 psi (7 kPa) drop in air pressure gives about a 1.5 percent decrease in efficiency. Thus, a 10 percent drop in psi (kPa) = 15 percent loss of efficiency.

A 1/16" increase in nozzle orifice size can increase the volume of air required to maintain the same pressure by 25 percent. To maintain nozzle pressure, an air compressor used for blasting must have a cubic feet per minute (CFM) rating that is adequate to supply increasing air volume as blast nozzle orifices wear out. Additionally, to ensure that the increasing volume of air is supplied to a blast nozzle to maintain pressure, the blast hose diameter should be three to four times the diameter of the nozzle orifice size being used. Four factors can adversely affect the pressure at the nozzle:

- Length of the air hose
- Number of bends in the air hose
- Elevation of the blast nozzle above the compressor
- Inadequate air compressor CFM rating

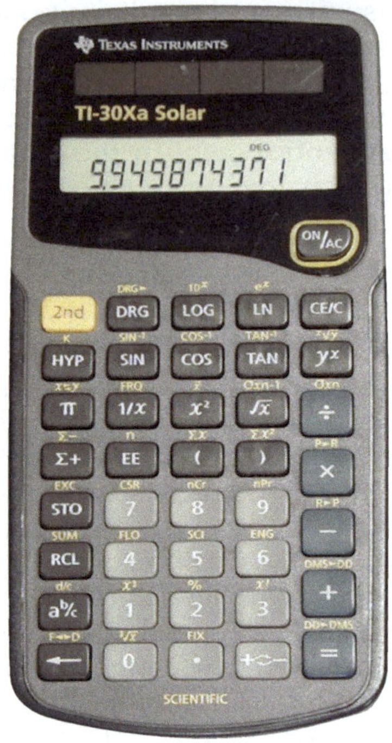

203F05.EPS

Figure 5 Typical handheld advanced-function calculator.

4.0.0 CALCULATIONS FOR THE PLANNING PROCESS

Mathematical skills are one of the absolute requirements for advancement in the trade. Planning demands the use of several mathematical procedures in order to correctly calculate materials, plan crew times, and mix materials. Ratios and proportions are methods of solving problems using comparisons. They can be used for practical calculations to determine the answers to problems involving specifications, thinner volumes, and VOCs. **Area** and **volume** calculations permit the planner to determine the sizes and material requirements for different shapes of objects to be coated. An advanced-function calculator (*Figure 5*) can be used for the calculations in the following sections.

4.1.0 Ratios

The ratio of one term A to a second term B can be written as A:B (or stated as A is to B) which is also the same as the fraction A/B. The first term becomes the numerator (top) of the fraction and the second term is the denominator (bottom):

A:B = A/B

The terms for ratios of specific quantities must be expressed in the same English or metric units of measurement, such as inches, meters, miles, kilometers, pounds, kilograms, psi, Pa, gallons, liters, and so on. Because ratios are fractions, they can be reduced to the lowest possible terms by dividing both the numerator and denominator by a common number:

20/25 ÷ 5 = 4/5

Suppose a technician has 3 gallons of a volatile solvent-based paint to be thinned using 1 gallon of solvent. This ratio can be expressed as 3 parts to 1 part, or as:

3:1 or 3/1

Where the total of the parts = 4

Because solvents are usually the source of VOCs, any time a coating is thinned significantly, the VOC level increases or decreases depending on the VOC content of each coating component. The adjusted level is calculated using a formula. The formula is based on the mix ratio parts, total parts, and VOC content of each part (component) in pounds/gallon or grams/liter (as specified in the PDSs). The adjusted VOC calculation is shown in the following example:

> **NOTE**
> Remember to express all quantities in the same units of measure (gallons, pints, liters, etc.).

Example:

Four gallons (parts) of a coating are mixed with 1 gallon (part) of activator and 1 gallon (part) of reducer, resulting in a mix ratio of 4:1:1 and a total number of parts that is equal to 6 (4 + 1 + 1).

- The VOC content of the coating (paint) is 4 lbs/gallon (= Vp).
- The VOC content of the activator is 5 lbs/gallon (= Va).
- The VOC content of the reducer is 6 lbs/gallon (= Vr).

To calculate the adjusted VOC content of the mixed coating, use the following formula:

$$\frac{(\text{paint parts} \times \text{Vp}) + (\text{activator parts} \times \text{Va}) + (\text{reducer parts} \times \text{Vr})}{\text{parts total}}$$

Substituting values yields the following:

$$\frac{(4 \text{ parts} \times 4 \text{ lbs/gal}) + (1 \text{ part} \times 5 \text{ lbs/gal}) + (1 \text{ part} \times 6 \text{ lbs/gal})}{6}$$

or

$$\frac{16 \text{ lbs/gal} + 5 \text{ lbs/gal} + 6 \text{ lbs/gal}}{6}$$

Result: An adjusted 4.5 lbs/gal of VOC for the mixed reduced coating as applied.

4.2.0 Proportions

A proportion is a name assigned to a statement where two ratios are equal. The two ratios can be stated in the form A is to B as C is to D or written as:

$A:B = C:D$ or $A/B = C/D$

When two ratios are equal, the cross-multiplication products of numerators and denominators of the ratios are equal:

$A \times D = B \times C$

If any one term of a proportion is unknown, it can be determined as shown in the following example, where the numerator Y of one ratio is unknown:

$3/5 = Y/35$

Cross-multiplication products yields:

$3 \times 35 = 5 \times Y$

or

$105 = 5Y$

Dividing both sides of the equation by 5 yields the value of the unknown term Y:

$105 \div 5 = 5Y \div 5$

or

$21 = Y$

Many problems can be solved by setting up proportions of related quantities. These quantities must be analyzed to determine where to place the terms in the ratios. Is the unknown term in the ratio larger or smaller than the known term? This ratio is stated to correspond with the ratio with two known terms. The two types of proportions used for solving different kinds of problems are direct proportions and inverse proportions.

$$\frac{\text{SMALLER}}{\text{LARGER}} = \frac{\text{SMALLER}}{\text{LARGER}}$$

OR

$$\frac{\text{LARGER}}{\text{SMALLER}} = \frac{\text{LARGER}}{\text{SMALLER}}$$

4.2.1 Direct Proportions

Two quantities are directly proportional when an increase in one causes a proportional increase in the other, or a decrease in one causes a proportional decrease in the other. When a direct proportion is set up in equation form, the numerator of the first ratio must correspond with the numerator of the second ratio. The denominator of the first ratio must correspond with the denominator of the second ratio.

For example, a 3-gallon tub of a particular epoxy coating weighs 31 pounds. The problem is to find the weight of a 55-gallon barrel of the same coating. The proportion would be stated as follows:

$3/55 = 31/X$

The two numerators correspond since the 3-gallon tub of epoxy weighs 31 pounds, and the two denominators correspond since the weight of the 55-gallon barrel of epoxy is unknown.

This example can be solved as follows:

$55 \times 31 = 1{,}705$

$1{,}705 \div 3 = 568.33$

$X = 568.33$; therefore, $3/55 = 31/568.33$

4.2.2 Inverse Proportions

Two quantities are indirectly or inversely proportional when an increase in one quantity causes a decrease in the other quantity. For example, two quantities are inversely proportional if when one quantity increases to four times its original value, the other quantity decreases four times, or becomes one-fourth of its original value.

When an inverse proportion is set up in fractional form, the second ratio is inverted (the terms are reversed). The numerator of the first ratio then corresponds with the denominator of the second ratio, and the denominator of the first ratio corresponds with the numerator of the second ratio. Once terms are inverted or reversed, cross-multiplication products are used just like a direct proportion.

For example, when adding thinner to a given amount of coating, and you want to know what the WFT will be, calculate the amount to be added by percent of solids. You can also use the relative WFTs to calculate the allowable amount of thinner. If area B has 1 quart of thinner added to a 3-gallon tub of material, and you add another quart of thinner for Area A, the result can be found using that proportion. Since the amount of thinner used is inversely proportional to the WFT of areas coated, the proportion is stated as follows:

$$\frac{\text{WFT of area B}}{\text{WFT of area A}} = \frac{\text{amount of thinner added to A}}{\text{amount of thinner added to B}}$$

$$\frac{20 \text{ mils}}{X} = \frac{2 \text{ qts}}{1 \text{ qt}}$$

$20/X = 2/1$

$2X = 20$

$X = 10$ mils

4.3.0 Expressing Rules as Formulas

A formula is a short method of writing a mathematical rule. In formulas, the values from rules are represented by letters or symbols. A letter used to replace a value is often the first letter of the word representing that value. If that letter has already been used to represent a value in the formula, it cannot be used again to represent a different value. When a rule is changed into a formula, the values and mathematical operations are replaced with letters or symbols and mathematical signs (*Figure 6*).

Since the letters or symbols in a formula represent the numbers in specific problems, these numbers can be written into the formula to replace the letters or symbols. This is called substitution. Usually, the values of all but one of the letters are known. These values can be used in the formula to find the unknown value.

Follow these steps to solve a formula by substitution:

Step 1 Determine the formula needed to solve the problem. For this example, assume that the radius of a circle is 6 inches and the diameter is unknown. The formula shown in *Figure 7* can be used to solve the problem.

$D = 2r$

Step 2 Substitute the known quantities for the letters or symbols in the formula.

$D = 2 \times 6$

Step 3 Perform the mathematical operation needed to solve the problem.

$12 = 2 \times 6$

Step 4 Check the answer.

$12 \div 2 = 6$

Step 5 Label the answer with the correct unit of measure.

D = 12 inches

NOTE

All measurements used in a formula must be the same type of unit.

If the diameter in the example was known and the radius was the unknown value, the formula would read 12 = 2r. It could be solved by dividing the diameter by 2, so that 12 divided by 2 equals D/2:

$12/2 = D/2 = 6 = r$

4.4.0 Powers

A power is the product of two or more equal factors. For example, 4 × 4 is the second power of 4 and 4 × 4 × 4 is the third power of 4. For example, 3^2 is the same as 3 × 3. The **exponent** 2 tells you that 3 is taken as a factor two times and is read three to the second power; therefore, $3^2 = 9$. Powers of numbers can be determined using the keys of the calculator shown in *Figure 7*. To raise a number to the power of 2, press the x^2 key after entering the number. To raise a number to the power

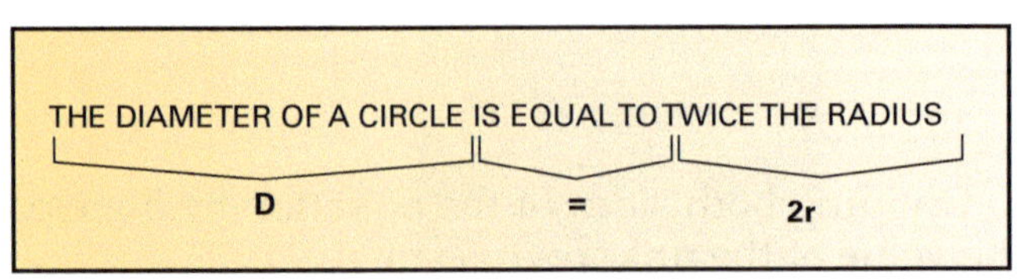

203F06.EPS

Figure 6 Changing a rule into a formula.

of 3 or more, press the y^x key after entering the number. Then, enter the number for the power desired, followed by the = key.

4.5.0 Roots

The square root of a number, if multiplied by itself, equals the number of which it is the root. The radical sign $\sqrt{}$ indicates a root of a number. The index number, which is written above and to the left of the radical sign, indicates the number of times that a root is to be taken as an equal factor to produce the given number. The index number for a square root is 2. The 2 is usually omitted in the square root sign. Therefore, the square root of 16 is written $\sqrt{16}$. It asks the question, "What number multiplied by itself equals 16?" Since 4 × 4 = 16, 4 is the square root of 16. So $\sqrt{16} = 4$.

When the square root of a fraction is to be computed, both the numerator and the denominator must be enclosed within the radical sign. For instance, $\sqrt{5/8}$ indicates that the square root of the complete fraction is to be taken. To compute the square root of a fraction, find the square root of both the numerator and the denominator. Square roots, cube roots, and other roots can be determined using the keys of the calculator shown in *Figure 7*. To obtain the square root of a number, press the $\sqrt{}$ key after entering the number. To find the cube or higher root of a number, press the 2nd (function) key and then the key after entering the number. Then, enter the number for the root desired, followed by the = key.

Example:

Compute the square root of 16/36:

$\sqrt{16} = 4$

$\sqrt{36} = 6$

$\sqrt{16/36} = 4/6 = 2/3$ reduced

Root expressions are sometimes written with two or more operations within the radical sign. For instance, the expression $\sqrt{1 + h}$ is read as "the square root of 1 plus h." It is computed by first adding 1 and h and then computing the square root of the sum. Apply the following procedure to solve problems which involve operations within the radical sign:

Example:

Compute the root of:

$\sqrt{8 + 17}$

Step 1 Perform the operations within the radical sign first.

8 + 17 = 25

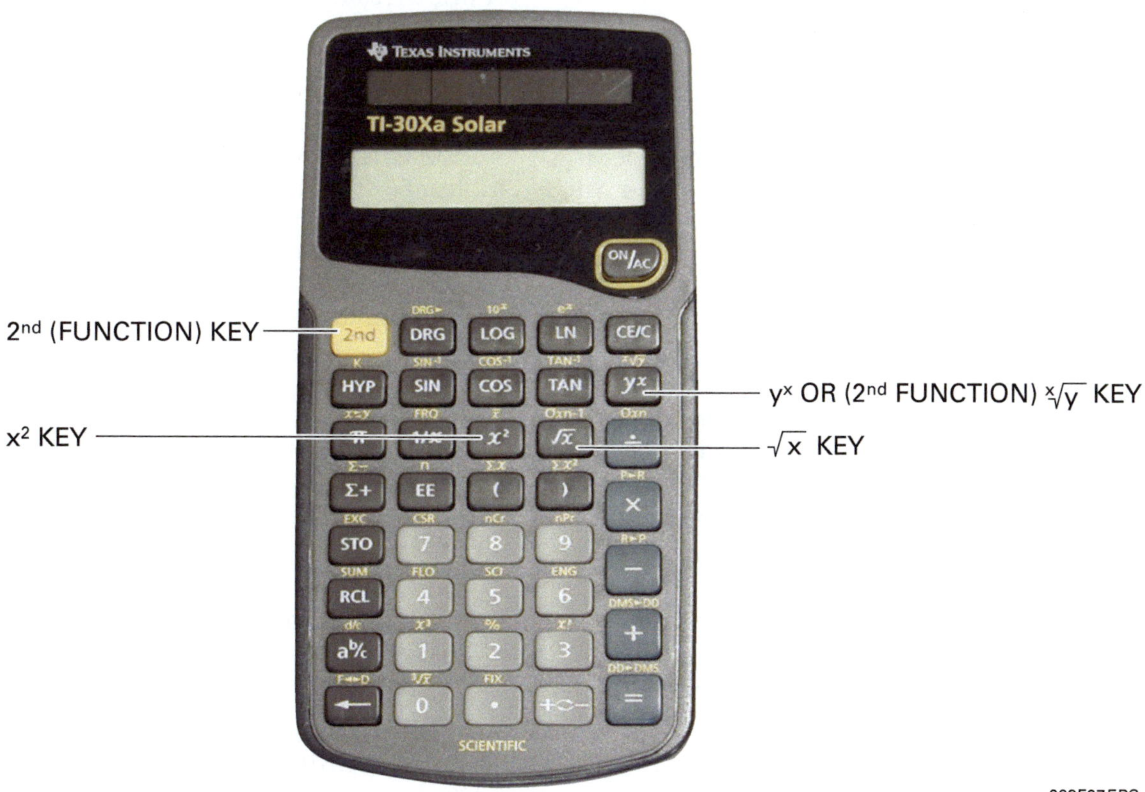

Figure 7 Calculator keys for powers and roots.

Step 2 Compute : $\sqrt{25}$: $5 \times 5 = 25$, so $\sqrt{8 + 17} = 5$

Any power of a number can be determined by multiplying the number by itself the number of times shown in the exponent. That is, the fourth power of a number, stated as n to the fourth power, is $n \times n \times n \times n$. By the same reasoning, the fourth root of a number is that number that, if multiplied by itself four times, equals the number whose root is being calculated. The cube of a number is that number to the third power ($n \times n \times n$). The cube root of a number is the number that is multiplied by itself three times to obtain that number. For example, 2^3, read as two cubed, equals 8. The cube root of 27 is 3 because $3 \times 3 \times 3$ equals 27.

4.6.0 Evaluating Formulas

Normally, formulas contain two or more arithmetic operations. It is essential that the operations are performed in the proper order to solve the formula correctly. Perform the operations in order as follows:

Step 1 Perform the operations in parentheses first. If there are parentheses within parentheses, perform the operation within the innermost parentheses first.

Step 2 Perform the operations for powers and roots as they occur.

Step 3 Perform multiplication and division operations from left to right in the order in which they occur.

Step 4 Perform addition and subtraction operations from left to right in the order in which they occur.

One way to remember this order of operations is to remember PEMDAS: parentheses; exponent (powers and roots); multiply; divide; add; subtract.

> **NOTE**
> One way of memorizing a letter sequence is to make a sentence out of it, and silly sentences tend to be easy to remember. One for the order of operations is: Please Excuse My Dear Aunt Sally.

To determine the numerical value of one letter in an expression when the numerical value of the other letter is known, use the following procedure:

Step 1 Write the expression.

Step 2 Replace each letter in the expression with its numerical value, and add a multiplication sign where multiplication is needed.

Step 3 Perform the operations in the proper order.

Example:

Find the perimeter of a rectangle when h equals 6 inches and w equals 4 inches.

Step 1 Write the formula $p = 2h + 2w$

Step 2 Replace the letters in the expression with their numerical value, and add a multiplication sign where multiplication is needed.

$p = (2 \times 6) + (2 \times 4)$

Step 3 Perform the operations in their proper order.

First multiply:

$2 \times 6 = 12$ and $2 \times 4 = 8$

then add:

$12 + 8 = 20$

$p = 20$ inches

Another use for the formula for the perimeter of a rectangle would be to determine the length of one of the sides if the perimeter and the length of another side are known. Then, the perimeter and one side could be substituted in the formula $p = 2h + 2w$. If the perimeter (p) is 20 and one of the sides is 4, then the formula could be stated as $20 = 2h + 2(4)$ or $20 = 2h + 8$. Since the same operation can be done to both sides of the equation, subtract 8 from both sides: $20 - 8 = 2h + 8 - 8$. The result is $12 = 2h$. Divide both sides by two, and the length of the unknown side is 6. Remember, you can do the same operation on both sides of an equation and the two sides will still be equal.

5.0.0 Solving Area Problems

Area is the amount of plane (flat) surface in a closed space. The area of any given surface can be calculated for practical applications, such as the following:

- Room layout
- Materials estimates
- Cost estimates
- Sizes of stock or parts

Area is measured by using standard areas of smaller units, such as the square inch and square foot. A square inch is a surface enclosed by a square that is 1 inch on each side. A square foot is enclosed by a square that is 1 foot on each side. There are 144 square inches in 1 square foot. These units can be used to solve problems in finding the area of rectangles, triangles, and circles.

5.1.0 Finding the Area of a Rectangle

A rectangle is a four-sided figure. Its opposite sides are equal, and all sides are joined at right angles. To find the area of a rectangle, it is necessary to find the number of surface units it contains. For example, if a rectangle contains 3 rows of square inches and there are 4 square inches in each row, the rectangle contains 12 square inches (*Figure 8*).

The area of any rectangle can be found by multiplying the length of one side by the width of one side, or A = lw. Area is always expressed in square units. Remember that the length or width can be obtained if the area and either the length or width are known. For example, if a rectangle is 2 feet long with an area of 10 square feet, the equation A = lw becomes 10 = 2w. Divide both sides by 2, and the result is 5 = w.

In an area formula, all measurements must be expressed as (or converted to) the same type of unit. For example, if a room is 20 feet, 6 inches long and 15 feet, 9 inches wide, the area of floor is found as follows:

A = lw

A = 20 feet, 6 inches × 15 feet, 9 inches

A = 20.5 feet × 15.75 feet

A = 322.875 square feet

To find the area of the walls, the height of the room is required, as well as the length and width. The formula for the walls of the room above is:

A = lw

A (the long wall) = 20 feet, 6 inches × 8 feet

B (the shorter wall) = 15 feet, 9 inches × 8 feet

A = 20.5 feet × 8 feet = 164 feet

B = 15.75 feet × 8 feet = 126 feet

Area of the walls in the room =
2 A + 2 B = 2(164 + 126) = 580 square feet

A square is a rectangle with four equal sides. A square that is 6 inches on each side is called a 6-inch square. The formula for finding the area of a rectangle can be used to find the area of a square. However, since each side is the same

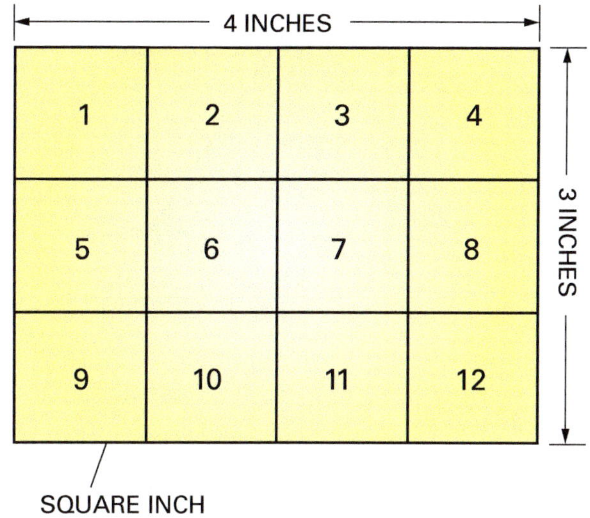

Figure 8 Rectangle.

length, the length of any side can be multiplied by itself, or $A = S^2$. This is read as "the area equals the sides squared". For example, the area of the 6-inch square is found as follows:

$A = S^2$

A = 6 × 6

A = 36 square inches

5.2.0 Finding the Area of a Triangle

A triangle is a three-sided figure with three angles. The area of any triangle can be found by multiplying the product of the base and the height by one half, or A = ½bh (*Figure 9*). The base of a triangle can be any one of its sides. The height is **perpendicular** to the base.

For example, if a triangle has a base of 15 feet and a height of 12 feet, the area is found as follows:

A = ½bh

A = 1/2 × 15 × 12

A = 90 square feet

If the base and the area are given, they can be plugged into the formula, and the result is 90 = 15h/2. First multiply both sides by 2, which gives 180 = 15h. Now divide both sides by 15, and the result is 12 = h. The height is 12 feet.

5.3.0 Finding the Area of a Circle

The area of a circle can be found by multiplying pi (pronounced *pie*) by the product of the radius multiplied by itself, or $A = \pi r^2$. Pressing the π key on the calculator shown in *Figure 10* enters the value of π (3.14159xxx). For the purposes of this

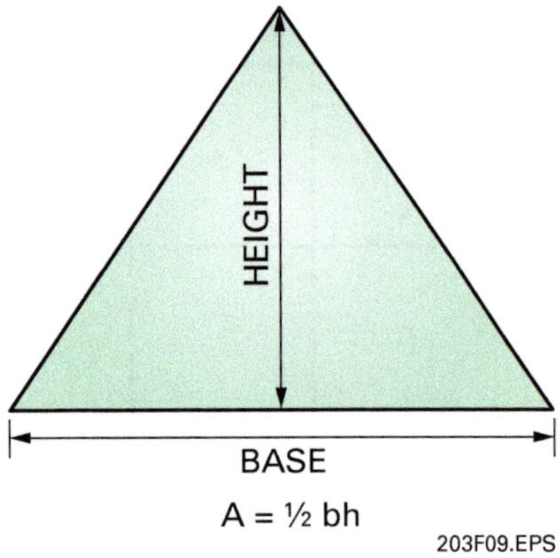

Figure 9 Triangle.

module, use the value 3.1416 for pi. For example, if a tube has an inside radius of 12 inches, the cross-sectional area of the pipe is calculated as follows:

$A = \pi r^2$

$A = 3.1416 \times 12 \times 12$

A = 452.39 or 452.4 square inches

The area of a circle can also be found by multiplying 0.7854 times the square of the diameter as shown in *Figure 10.* As an example, the tube used in the previous calculation has a diameter of 24 inches (2 × the 12-inch radius) and the area can be alternatively calculated as follows:

$A = 0.7854 \times d^2$

$A = 0.7854 \times 24^2$ (or 24 × 24)

$A = 0.7854 \times 576$

A = 452.39 or 452.4 square inches

This formula can also be used to find the pressure exerted on a surface, such as a piston. If the piston has a diameter of 6 inches and the pressure exerted on the piston head is 160 pounds per square inch, the total pressure is found as follows:

$A = \pi r^2$

$r = D \div 2 = 6 \div 2 = 3$

$A = 3.1416 \times 3 \times 3$

A = 28.2744 square inches

Pressure = 160 pounds per square inch

Pressure per square inch × area = total pressure

160 × 28.2744 = 4,523.904 pounds per square inch

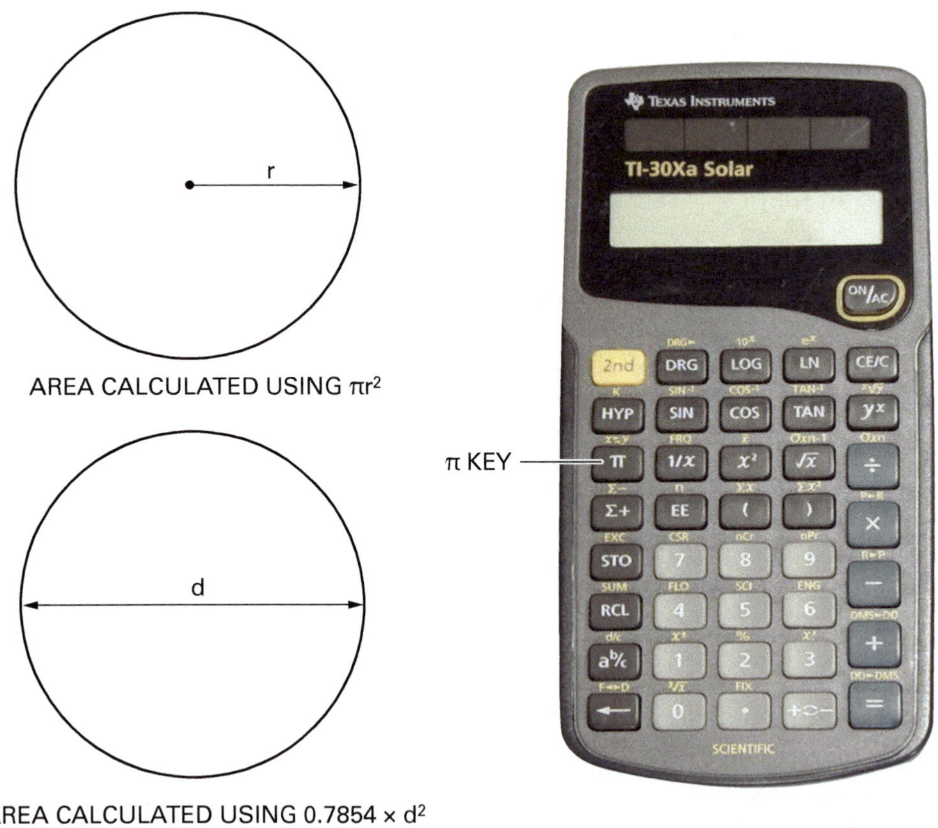

Figure 10 Formulas for circle areas and the calculator π key.

5.4.0 Finding the Area of a Sphere

The area of a sphere (*Figure 11*) can be found by multiplying pi by the square of the diameter (πd^2) or the square of two times the radius ($\pi 2r^2$). For example, if a sphere has a diameter of 20 feet, the area is found as follows:

$A = \pi d^2$

$A = 3.1416 \times 20^2$ (or 20×20)

A = 1,257 square feet

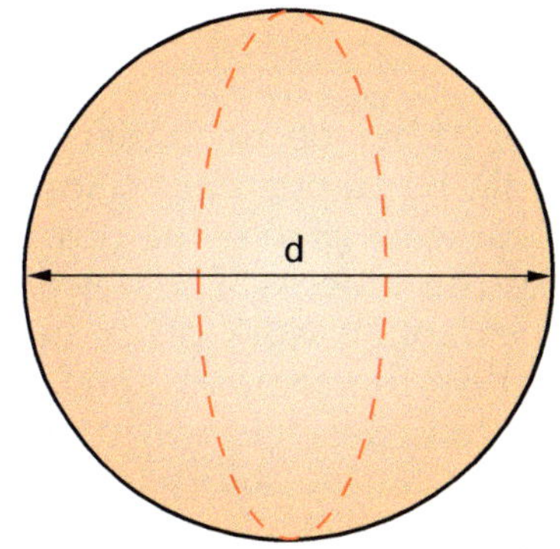

Figure 11 Sphere.

5.5.0 Finding the Area of a Cylinder

The area of a **cylinder** (like a pipe or tank) can be found by determining the **circumference** (c) and multiplying by the length or height (h) (*Figure 12*). Circumference is the distance around the cylinder and is determined by the formula πd (or $\pi 2r$). If the cylinder is capped on one or both ends, like a tank, the area of a circle using the radius (r) or diameter (d) for these is added to the area of the cylinder. For example, the area of a tank that is 20 feet in diameter and 30 feet high with a top and bottom is determined as follows:

$A = \pi d$ (or $\pi 2r$) $\times$ h for the cylinder + 2 times πr^2 (or $0.7854 \times d^2$) for the caps

$A = 3.1416 \times 20 \times 30 + 2(0.7854 \times 20 \times 20)$

A = 1,885 + 628

A = 2,513 square feet

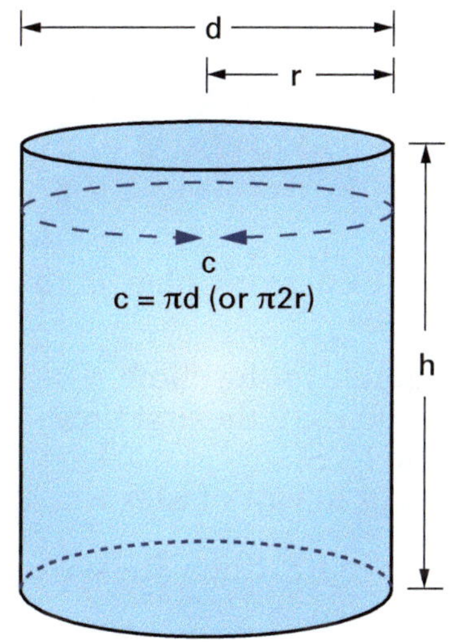

Figure 12 Cylinder.

SUMMARY

Any coating or lining job, no matter how straightforward, should be planned out before beginning field work. The work plan is the primary tool for the control of the process. The object of process control is to establish work procedures that are consistent within the company, and thus ensure that the job is done the right way each time.

Planning requires the ability to calculate ratios and proportions between quantities, thus allowing the craftworker to determine quantities without wasting material on the basis of guesswork.

Quality control and assurance are other vitally important aspects of any coating job. The inspections and tests involved are needed to make sure the level of quality specified in the work plan is met.

In both the work plan and quality control, documentation is key. Daily reports on completed work, testing, and inspection allow the company to maintain the job standards.

Review Questions

1. The object of a work plan is to make each crew member _____.
   a. accountable for lost time
   b. do the job right the first time
   c. read the PDS
   d. find ways to work faster

2. Material specifications are written into the _____.
   a. work plan
   b. PDS
   c. MSDS
   d. RFI

3. Specifications are _____.
   a. driven by the work plan
   b. the requirements for the job
   c. the blueprints package
   d. generally the same for every coatings project

4. As a blast nozzle orifice wears, _____.
   a. the volume of the air decreases
   b. the air from the nozzle heats up
   c. the pressure decreases
   d. the pressure is not affected

5. The ratio of 5 gallons of coating to 1 gallon of thinner is _____.
   a. 1:5
   b. 5:1
   c. 5:6
   d. 6:5

6. If 1 U.S. pint of thinner is added to 1 U.S. gallon of coating, the mix ratio is _____.
   a. 2:1
   b. 3:6
   c. 8:1
   d. 12:10

7. After 20 U.S. gallons of coating is thinned using 2 U.S. gallons of thinner, the mix ratio is _____.
   a. 2:20
   b. 10:1
   c. 11:10
   d. 22:2

8. The value of 6 to the second power is _____.
   a. 2
   b. 18
   c. 36
   d. 216

9. The exponent 3 means that the number is to be multiplied by itself three times.
   a. True
   b. False

10. The value of 8 cubed is _____.
    a. 16
    b. 64
    c. 512
    d. 640

11. The square root of 9 is _____.
    a. 2
    b. 3
    c. 4.5
    d. 81

12. The area of the floor of a room that is 20 feet long by 15 feet wide is _____.
    a. 250 square feet
    b. 300 square feet
    c. 600 square feet
    d. 750 square feet

13. The length of a rectangular platform that is 12 feet wide with an area of 180 square feet is _____.
    a. 15 feet
    b. 90 feet
    c. 190 feet
    d. 900 feet

14. If one side of a square is 4 inches, its area is _____.
    a. 8 square inches
    b. 16 square inches
    c. 64 square inches
    d. 512 square inches

15. A concrete pad to be sealed is 4½ feet wide and 5 feet long. Its area is _____.
    a. 22.5 square feet
    b. 25 square feet
    c. 27.5 square feet
    d. 29 square feet

16. The area of a triangle that is formed when two of the legs, joined by a right angle, are each 4 feet long is _____.
    a. 4 square feet
    b. 8 square feet
    c. 16 square feet
    d. 32 square feet

17. If the height of a triangle is 20 feet and the base is 15 feet, the area is _____.
    a. 5 square feet
    b. 150 square feet
    c. 300 square feet
    d. 400 square feet

18. The area that will be covered by sealing a triangular pad, if all three sides are 10 feet, and the distance from the middle of one side to the opposite angle is 8.8 feet, is _____. (Round to the nearest foot.)
    a. 44 square feet
    b. 50 square feet
    c. 88 square feet
    d. 100 square feet

19. The gable of your house is 50 feet across and 10 feet high. In order to cover it with acrylic coating, you need to order enough coating to cover _____.
    a. 60 square feet
    b. 100 square feet
    c. 250 square feet
    d. 500 square feet

For Questions 20-25, use the value 3.1416 for pi.

20. The area of a circle that is 6½ inches in diameter is _____.
    a. 16.33 square inches
    b. 28.27 square inches
    c. 33.18 square inches
    d. 113 square inches

21. How many square inches of plywood is required to fabricate a circular cover for an opening that is 24 inches in diameter? (Assume that the lip of the hole has to be kept uncoated, so the plywood must extend 1 inch all around.)
    a. 60 square inches
    b. 73.93 square inches
    c. 169 square inches
    d. 530.93 square inches

22. The bottom of a circular tank has to be blasted and resealed. The bottom has an area of 11,310 square feet. What is the radius? (Round to nearest foot.)
    a. 6 feet
    b. 60 feet
    c. 120 feet
    d. 3,600 feet

23. If each stroke with a sprayer covers 2 square feet of the top of a tank, how many strokes will be needed to coat a 50-foot diameter ring? (Round to the nearest stroke.)
    a. 25
    b. 100
    c. 205
    d. 982

24 The surface area of a 36-foot diameter sphere is _____. (Round to the nearest foot.)
    a. 3,346 square feet
    b. 4,072 square feet
    c. 5,063 square feet
    d. 6,494 square feet

25. The surface area of a 24-inch diameter pipe that is 30 feet long is _____.
    a. 188.5 square feet
    b. 1,230.3 square feet
    c. 340 square feet
    d. 2,262 square feet

Trade Terms Introduced in This Module

Area: The space enclosed by the outline of a two-dimensional shape.

Circumference: The line that encloses a two-dimensional shape.

Cylinder: A shape created by a circle moving in a straight line through space perpendicular to the surface of the circle.

Exponent: A number or symbol placed to the right and above another number, symbol, or expression, denoting the power to which the latter is to be raised.

Perpendicular: At a right angle to the plane of a line or surface.

Pi: The constant that defines the relationship between the diameter and circumference or the radius and area of a circle.

Radius: A straight line from the center of a circle to a point on the edge of the circle.

Volume: The space enclosed by a three-dimensional shape.

Additional Resources

This module presents thorough resources for task training. The following resource material is suggested for further study.

ANSI/ISO/ASW Q9001-2008: Quality Management Systems – Requirements. www.asq.org.

Coating Inspector Program Level 1 © NACE International, 2010.

Coating Inspector Program Level 2 © NACE International, 2010.

Corrosion Prevention by Protective Coatings. Charles Munger. Houston, TX: NACE, 1999.

IMO MSC.215(82) Performance Standard for Protective Coatings (PSPC). www.imo.org.

SSPC-PA 1, Shop, Field, and Maintenance Painting of Steel. www.sspc.org.

Figure Credits

Graco Inc., Module opener

Topaz Publications, Inc., 203F05, 203F07, 203F10 (photo)

NCCER CURRICULA — USER UPDATE

NCCER makes every effort to keep its textbooks up-to-date and free of technical errors. We appreciate your help in this process. If you find an error, a typographical mistake, or an inaccuracy in NCCER's curricula, please fill out this form (or a photocopy), or complete the online form at **www.nccer.org/olf**. Be sure to include the exact module ID number, page number, a detailed description, and your recommended correction. Your input will be brought to the attention of the Authoring Team. Thank you for your assistance.

Instructors – If you have an idea for improving this textbook, or have found that additional materials were necessary to teach this module effectively, please let us know so that we may present your suggestions to the Authoring Team.

NCCER Product Development and Revision

13614 Progress Blvd., Alachua, FL 32615

Email: curriculum@nccer.org

Online: www.nccer.org/olf

❏ Trainee Guide ❏ Lesson Plans ❏ Exam ❏ PowerPoints Other ____________

Craft / Level: Copyright Date:

Module ID Number / Title:

Section Number(s):

Description:

Recommended Correction:

Your Name:

Address:

Email: Phone:

INDUSTRIAL Coatings

Arc spray coating.

Coating a hull both inside and outside.

Corrosion is the major cause of bridge deterioration. It can be controlled with protective coatings, membranes, and sealers. Controlling corrosion lengthens the life of a bridge, promoting public safety while saving money and natural resources.

Cleaning by Abrasive Blasting

Rust Grade D

Grade D means a steel surface on which the mill scale has rusted away to the point where extensive pitting is visible.

Brush-Off Blast Cleaning — D SP-7

NACE No. 4/SSPC-SP 7 requires that after cleaning, the surface (when viewed without magnification) will be free of all visible oil, grease, dirt, dust, loose mill scale, loose rust, and loose coating. Tightly adherent mill scale, rust, and coating that pass the dull putty knife test may remain.

Commercial Blast Cleaning — D SP-6

NACE No. 3/SSPC-SP 6 requires that after cleaning, the surface (when viewed without magnification), will be free of all visible oil, grease, dust, dirt, mill scale, rust, coating, oxides, corrosion products, and other foreign matter, except for random staining present on not more than 33 percent of the surface area.

Near-White Metal Blast Cleaning — D SP-10

NACE No. 2/SSPC-SP 10 requires that after cleaning, the surface (when viewed without magnification), will be free of all visible oil, grease, dust, dirt, mill scale, rust, coating, oxides, corrosion products, and other foreign matter, except for random staining present on not more than 5 percent of the surface area.

White Metal Blast Cleaning — D SP-5

NACE No. 1/SSPC-SP 5 requires that after cleaning, the surface (when viewed without magnification), will be free of all visible oil, grease, dust, dirt, mill scale, rust, coating, oxides, corrosion products, and other foreign matter.

RUST GRADE

PREPARATION GRADE

D SP-6

SSPC photos are shown courtesy of SSPC: The Society for Protective Coatings. The digital photos provided do not sufficiently depict the photos that are available in the actual physical standards, and should not be used in the field for comparison. Actual SSPC visual standards may be purchased at http://www.sspc.org/marketplace/.

Safety

Environmental protection of a bridge structure during surface preparation and coating.

Painter on a suspended platform wearing appropriate personal protective equipment, including fall protection.

Protective clothing for abrasive blasting.

Equipment

HIGH-VOLUME
LOW-PRESSURE
SPRAY GUN

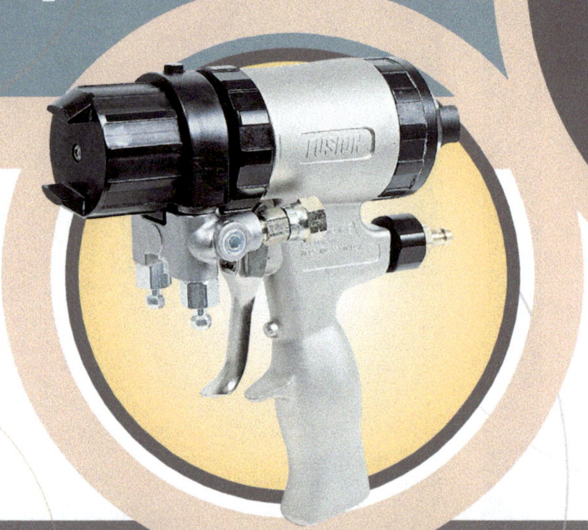

PLURAL-COMPONENT
SPRAY GUN

PLURAL-COMPONENT
SPRAY SYSTEM

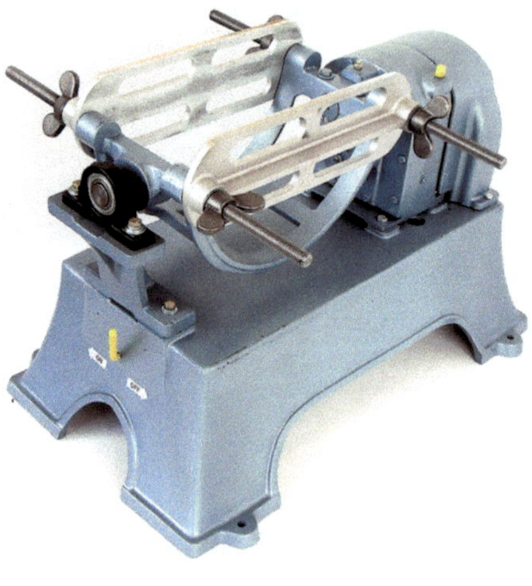

FIVE-GALLON PAINT MIXER

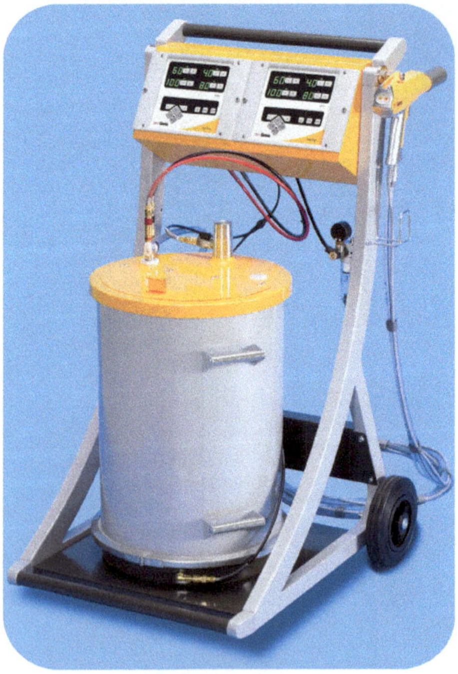

ELECTROSTATIC SPRAY SYSTEM

Page 1: Arc spray coating–Praxair Surface Technologies, Coating a hull–Rogers & Associates Consulting Services, Bridge structure enclosure: Photo courtesy Hippwrap Containment. Page 2: Abrasive blasting standard photos–Courtesy of SSPC: The Society for Protective Coatings. Page 3: Environmental protection: Photo courtesy Hippwrap Containment, Painter on a suspended platform: Graco, Inc., Protective clothing for abrasive blasting: Bullard. Page 4: Plural-component spray gun, Plural-component spray system–Graco, Inc., High-volume, low-pressure spray gun–ITW Finishing Innovations, 5-gallon paint mixer–Miracle Paint Rejuvenator Company, Inc., Electrostatic spray system–ITW Gema.

Containment

69204-10

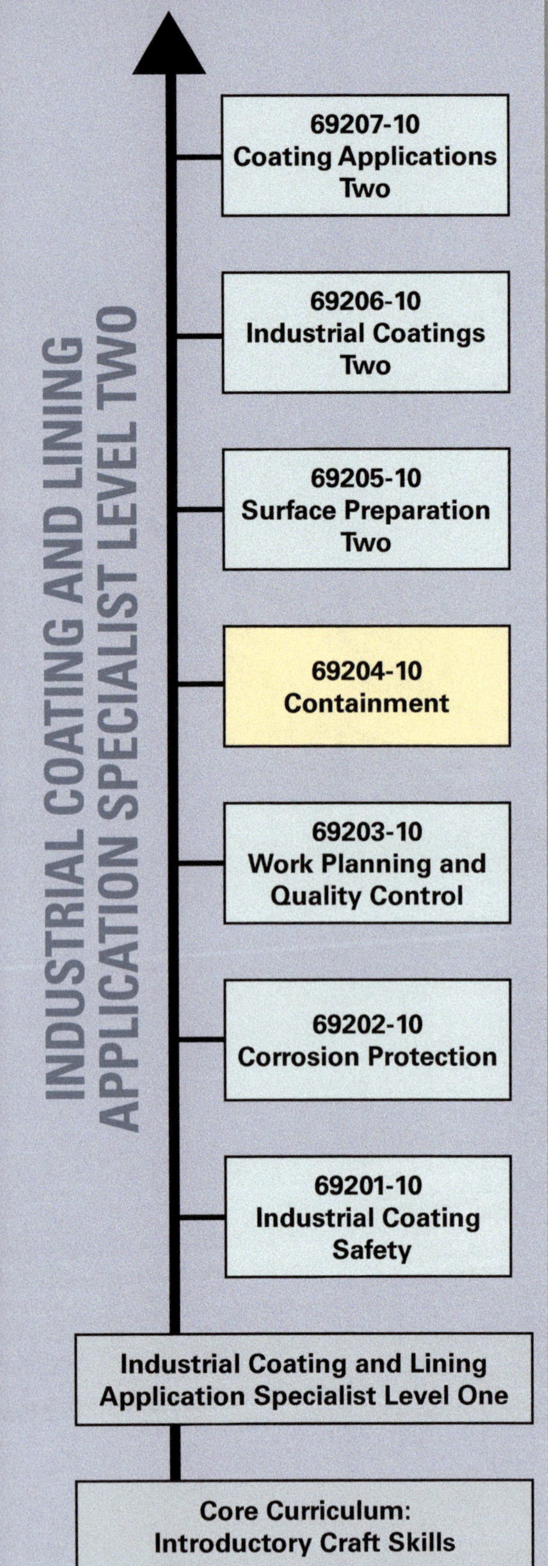

This course map shows all of the modules in *Industrial Coating and Lining Application Specialist Level Two.* The suggested training order begins at the bottom and proceeds up. Skill levels increase as you advance on the course map. The local Training Program Sponsor may adjust the training order.

Objectives

When you have completed this module, you will be able to do the following:

1. Describe which level of containment would be appropriate for a specific operation.
2. Describe the general requirements of industry standards for containment erection and dust collection.
3. List the different types of containment systems and their applications.
4. Construct a containment system for a specific application.
5. Demonstrate the installation of ventilation and humidification or dehumidification systems.
6. Demonstrate methods for verifying containment effectiveness.
7. Demonstrate methods for addressing protrusions.
8. Demonstrate procedures for maintaining containment integrity.

Trade Terms

Desiccant
Emissions
Entryway
Flame-resistant
Flame-retardant
Flash rusting
Forced airflow
Impenetrable
Impermeable
Makeup air
Negative pressure
Support structure

Prerequisites

Before you begin this module, it is recommended that you successfully complete *Industrial Coating and Lining Application Specialist Level One* and *Industrial Coating and Lining Application Specialist Level Two*, Modules 69201-10 through 69203-10.

Contents

Topics to be presented in this module include:

Figures and Tables

1.0.0 Introduction

Surface preparation and coating application methods often produce unwanted material. This material can include dust, debris, paint chips, spent abrasives, chemical vapors, and overspray. In some cases, this debris is hazardous. Even if they are nontoxic, it is important to prevent these materials from soiling areas surrounding the job site. A containment system (*Figure 1*) confines debris to a limited area and aids in its collection for disposal. A containment system includes tarps, screens, ground covers, water booms, and shrouds held in place by scaffolds and other supports. In some instances, the work area must be fully sealed to prevent debris from contaminating nearby areas.

By confining dust and debris to a limited area, containment increases the hazards for those working inside the containment area. Therefore, ventilation systems are used to control the movement of air through the work area while minimizing dust hazards and increasing visibility. Ventilation can include both mechanical and natural methods. Fans, hoods, and ducts are used for mechanical ventilation. Filters and dust collectors clean the air before it is released.

1.1.0 Developing Containment and Ventilation Systems

Industrial standards provide guidance for the selection of equipment and containment methods. Different levels of emissions control are available. The greatest level of control is required for activities that produce the greatest amount of dust and debris.

Coating application on a pipe rack and abrasive blasting of a bridge over water pose very different considerations for containment. The following factors affect the design of the system and the selection of materials and equipment:

- Contract specifications
- Federal and state environmental, health, and safety laws
- Surface preparation techniques
- Containment application techniques
- Chemicals in the coatings being removed
- Chemicals in the coatings being applied
- Work environment
- Surrounding area
- Proximity to other workers and the public

Various standards provide guidance in determining the proper materials and methods to use. Several industry organizations have identified methods for meeting containment and ventilation requirements in various situations. This module focuses on the requirements of *SSPC Guide 6, Guide for Containing Surface Preparation Debris Generated During Paint Removal Operations.* Blast cleaning produces the highest levels of dust and debris and requires the highest levels of containment and ventilation. Waterjetting produces wastewater and chemicals that must be contained (*Figure 2*). Understanding how these systems are constructed allows the installation of containment and ventilation systems when lower levels of protection are needed.

1.2.0 Health and Safety Considerations

Health and safety hazards exist on every coating job. Uncontrolled dust and debris can result in physical accidents, fire, explosion, or health problems. The solvents used in cleaning and as vehicles for many industrial coatings pose fire, explosion, and health risks. It is important to understand these hazards so that they can be

204F01.EPS

Figure 1 Containment system.

204F02.EPS

Figure 2 Waterjetting operation.

minimized through the use of containment and ventilation systems.

Physical accidents are the most common danger on a job site. Concentrated dust impairs worker visibility, which can lead to accidents. A good ventilation system removes dust from the air and reduces this hazard. Flying particles can injure eyes, skin, and lungs. Exposure to airborne hazards can be minimized by wearing the proper PPE.

Fire is another common hazard. Flammable solvents are used in coatings, resins, binders, and cleaning agents. A containment system can prevent sparks from contacting flammable liquids. Solvents evaporate easily and can become concentrated in containment areas or confined spaces. At higher concentrations, a spark or flame can cause an explosion. Explosions can be prevented with proper ventilation and monitoring.

Many solvents and other chemicals used in industrial coatings are toxic and can cause health problems. Solvents can cause skin irritations and blisters. They can also destroy the skin, leaving open sores. Solvent vapors can enter the body when inhaled. Short-term exposure to high concentrations can cause nausea, headache, unconsciousness, convulsions, and death. Long-term exposure to low concentrations can cause injury to the central nervous system, lungs, kidneys, and may eventually result in death.

Lead is also toxic, and can cause temporary or permanent health damage. Another hazard is silica dust from sandblasting. It can cause chronic lung disease (silicosis) and increase the risk of lung cancer. To minimize exposure to these hazards, always use the proper ventilation and respiratory protection.

An MSDS identifies the chemical and physical properties of hazardous chemicals (*Figure 3*). Employers must provide an MSDS for each product on a job site. Be familiar with the proper precautions for handling and using all products being employed. Also be aware of the applicable emergency and first-aid procedures. Always read the MSDS before working with a chemical to find out what type of PPE is needed.

PPE protects workers from chemical and physical hazards. Choose PPE based on the work to be done. In addition to the MSDS, OSHA and other regulatory bodies specify the types of PPE needed for each job. For example, one of the requirements for abrasive blasting is a NIOSH-approved abrasive blasting hood and cape (*Figure 4*). Some companies have PPE requirements for each task listed in their safety manual.

PPE protects workers as they perform their work. After this work has been completed, personnel must be decontaminated to keep any toxic materials confined to the containment area. A decontamination area is often set up at the entrance to the containment area. Always remove dust and debris from the hands, face, and any exposed skin, as well as from the PPE.

DID YOU KNOW?

Silicosis and Sandblasting

More than 1 million U.S. workers are at risk of developing silicosis. Of these workers, more than 100,000 are employed as sandblasters. Approximately 59,000 of the 1 million workers exposed to crystalline silica will eventually develop silicosis. OSHA estimates that most of the hazards associated with sandblasting can be prevented if the proper safety precautions are taken on job sites.

2.0.0 INDUSTRY STANDARDS

Many federal laws regulate different aspects of surface coating operations, from the control of surface preparation emissions to the disposal of waste. There are also specific environmental laws regarding toxic chemicals like lead, such as those involving pre- and post-job testing of lead levels in the blood. To help professionals comply with various regulations, trade associations have developed guidelines for performing certain tasks. Sometimes organizations work together to develop these standards.

The standards cover most situations and identify various methods for controlling debris to minimize the release of contaminants into the environment. They also describe ventilation systems that minimize worker exposure and increase visibility within the containment area. Additional precautions are required when hazardous materials are involved. Several OSHA regulations that require higher levels of containment and control also cover worker exposure to lead.

2.1.0 NACE/SSPC Standards

Trade associations are made up of skilled professionals with many years of experience. They understand the challenges and limitations that their members face in different work environments. Sometimes trade associations develop standards to show their members how to comply with new environmental or safety regulations. These standards are then adopted by everyone in the trade and are written into contract specifications.

```
                     MATERIAL SAFETY DATA SHEET
B65W511
04 00

===========================================================================
          Section 1 -- PRODUCT AND COMPANY IDENTIFICATION
---------------------------------------------------------------------------
PRODUCT NUMBER                                           HMIS CODES
                                                      Health          2
   B65W511                                            Flammability    2
                                                      Reactivity      0
PRODUCT NAME
   POLY-LON* 1900 Polyester Polyurethane (Part A), Extra White
MANUFACTURER'S NAME                            EMERGENCY TELEPHONE NO.
   THE SHERWIN-WILLIAMS COMPANY                 (216) 566-2917
   101 Prospect Avenue N.W.
   Cleveland, OH 44115
DATE OF PREPARATION                            INFORMATION TELEPHONE NO.
   01-APR-07                                    (216) 566-2902
===========================================================================
```

204F03.EPS

Figure 3 Part of an MSDS.

204F04.EPS

Figure 4 Abrasive blasting hood.

NACE International and The Society for Protective Coatings (SSPC) have developed several standards that describe containment and ventilation systems. The standards mainly provide guidelines for systems used during surface preparation activities.

Contracts often require certain tasks to be performed in accordance with an industrial standard. Containment and ventilation requirements in industrial coatings contracts commonly list compliance with *SSPC Guide 6, Guide for Containing Surface Preparation Debris Generated During Paint Removal Operations.* This is one reason why this module focuses on the requirements of *SSPC Guide 6*. In addition, blasting operations typically require the highest level of containment and ventilation. Once surface preparation has been completed, the existing containment and ventilation methods are generally sufficient for coating applications.

2.1.1 SSPC Guide 6, Guide for Containing Surface Preparation Debris Generated During Paint Removal Operations

SSPC Guide 6 provides detailed guidance for designing and implementing a containment and ventilation system. The methods and materials outlined in the guide are designed for use with steel structures. However, they can also be applied to other construction materials, such as concrete, aluminum, and wood. These methods can be used by facility owners to prepare specifications for contracts. They can also be used by contractors to design and install containment and ventilation systems. Although the ventilation systems in the guide will increase visibility and provide a level of protection for workers in the containment area, this is not a substitute for compliance with OSHA standards. Additional respiratory protection may be required, especially when lead paint or other toxic materials are involved.

SSPC Guide 6 defines terms commonly used in surface preparation. It also identifies additional industry standards and federal regulations that may be applicable to surface coating operations. Use the guide as an aid to understanding contractual and legal requirements. However, do not rely on the guide for overall compliance. Always verify that your system meets any applicable federal, state, local, or contractual requirements.

SSPC Guide 6 describes the most common methods for coating removal and debris collection, as well as procedures for assessing the effectiveness of the containment system. It also identifies the components of containment enclosures and ventilation equipment. The guide describes various procedures for the selection, construction, and assessment of containment systems, including the following:

- Selecting methods of surface preparation and debris collection
- Specifying containment systems
- Selecting methods for assessing the quantity of emissions
- Implementing containment projects

The level of emissions control needed varies for different job sites. Since there are many variables within each containment system, there are multiple combinations that could be used to design a system. To simplify this process, *SSPC Guide 6* sets four levels of containment control. The four levels, or classes, are subdivided to identify containment requirements for different surface preparation methods. This makes it easier for all parties to communicate the parameters for a particular job.

The guide also identifies several methods for verifying the effectiveness of the containment system. Some of these methods provide immediate feedback on emissions. Other test procedures provide results after laboratory analysis or project completion.

Use the guide as a basis for designing containment and ventilation systems. However, reliance on a standard is not a substitute for compliance with all applicable laws. Each job has specific considerations that must be included during planning and execution to fully protect workers, the environment, and the public from exposure. For example, monitoring emissions for lead is appropriate when lead-based coatings are being removed or applied. Identification of the appropriate methods of verifying containment effectiveness must be coordinated with local environmental heath and safety officials.

2.2.0 Worker Safety Standards

OSHA, the Nuclear Regulatory Commission (NRC), and other government agencies, both state and local, set enforceable standards to protect worker health. One of the primary concerns in surface preparation coating operations involves structures previously coated with lead. Lead-based paint inhibits the rusting and corrosion of iron and steel. It is often used on bridges, railways, ships, lighthouses, and other steel structures. Significant lead exposure can result from removing paint from surfaces that have been previously coated with lead-based paint.

WARNING

Lead can damage the cardiovascular, reproductive, hematological, and central nervous systems, as well as the kidneys. Lead can be toxic when absorbed in higher doses.

Lead overexposure is a leading cause of workplace illness. It is also one of the most common forms of overexposure. Short-term or acute overexposure is unusual, but it can quickly develop into seizures, coma, or death. Long-term or chronic overexposure to lead is more common. There is no sharp dividing line between the rapidly developing, acute effects of lead exposure and the chronic effects that can take longer to develop.

Inhalation is the most common way that lead enters the body. It is absorbed through the lungs and upper respiratory tract when workers breathe in dust containing lead. It can also be absorbed through the digestive system if lead dust enters the mouth and is swallowed. Most of the lead inhaled or ingested gets into the bloodstream and circulates through the body. It is stored in various organs and body tissues. The lead stored in the tissues can slowly cause irreversible damage—first to individual cells, and then to organs and whole body systems.

Employers have a general responsibility to provide working conditions that are free from recognized hazards. Lead hazards are addressed in three specific standards for general industry, the shipyard industry, and the construction industry. Depending on the industry, one or more of these standards may apply. The construction-industry standard specifically includes painting operations. OSHA defines construction work as "work for construction, alteration and/or repair, including painting and decorating."

Under these regulations, employers must develop and implement worker protection programs that minimize the risk of lead exposure. The programs must include containment of dust hazards, limitation of the spread of lead dust, PPE, medical monitoring, and On Site supervision. *OSHA Standard 29 CFR 1926.62, Occupational Health and Environmental Controls – Lead*, requires that employers provide the following:

- Controlled work areas and warning signs
- Dedicated work clothes
- Clean break and eating areas
- Hand-washing stations and showers
- Documented respirator use and maintenance programs

- Periodic checks of workers' blood lead levels
- Designation of a competent person to deal with hazards on the job site

OSHA defines the term *action level* as employee exposure (without respirators) to an airborne lead concentration exceeding a certain amount. The action level for lead is 30 micrograms per cubic meter of air calculated as an eight-hour time-weighted average. The action level indicates that monitoring and mitigation efforts must begin to limit maximum exposure to a certain level. In addition to lead, other OSHA standards limit employee exposure to heavy metals, such as cadmium, inorganic arsenic, mercury, and hexavalent chromium.

OSHA limits a worker's lead exposure by setting a permissible exposure limit (PEL). Ventilation and respiratory protection are required in work areas that have a concentration greater than the PEL. The current standard for lead on construction sites limits worker exposure to 50 micrograms of lead per cubic meter of air averaged over an eight-hour workday. However, respirators must be the last method of protection used. Instead, engineering and work practices should first be used to reduce employee exposure to less than the PEL. Respirators may only be used in combination with engineering controls and work practices to control employee exposure.

Engineering controls and good work practices are the most effective ways to protect employees. Engineering controls consist of equipment or modified maintenance procedures that reduce workers' exposure to hazardous dust. Engineering controls include containment and ventilation. Some ventilation methods, such as shrouds and power tool vacuum attachments, remove the dust and debris from the area in which they are generated. These methods enclose the working head of the tool in a shroud while a vacuum system removes the debris. Alternative surface preparation methods, such as wet abrasive blasting, high-pressure water blasting, and chemical stripping limit exposure by reducing dust generation. These engineering controls have tradeoffs in cost, productivity, the quality of the surface preparation, and in the durability of the new coating. The controls should be applied as is practical, without sacrificing the quality of the work or increasing costs.

Containment protects workers, bystanders, and the environment by enclosing the work area with plastic or tarps. Ventilation is required to reduce workers' exposure within the enclosure. Ventilation systems pump the contaminated air into a filtration system and replace it with fresh air.

A good work practices program ensures that contaminants are controlled as the work is performed. The three-phase approach includes good housekeeping, personal hygiene, and administrative controls. Good housekeeping practices can be as simple as removing accumulated lead dust at the end of each work shift. Lead dust on equipment, overhead ledges, and other surfaces can become airborne if disturbed by traffic or air currents. Collect lead dust regularly with a vacuum pump that has a HEPA filter, and place it in sealed, **impenetrable** bags to decrease the dust hazard.

Good personal hygiene practices eliminate the inhalation and ingestion of dust by workers on the job site. They also prevent workers from transferring the lead hazard from the job site to their personal vehicles or homes, where others can be exposed to it. Eating or smoking is not allowed in a containment area to prevent worker ingestion of lead dust. Workers must decontaminate and wash their hands before eating or smoking. Some containment systems include a decontamination zone in which workers can remove lead dust from their clothing and equipment before moving into clean areas. This confines the hazardous dust to the containment area. Administrative controls include training, supervision, and scheduling. The company must have a program that trains workers on control procedures and ensures that containment policies are followed. By balancing high- and low-exposure tasks, employers can reduce workers' overall exposure during the workday.

Nuclear power plants (NPPs) present special problems for surface preparation. The NRC requires the removal of radioactive material from surface coatings. The agency treats the radioactive material as a separate hazardous waste in addition to the hazardous waste from any lead-based paint removal. NRC regulations are based on the presence and safe handling of radioactive materials from all sources, including surface preparation. Certain NRC procedures are used to control the removal and separation of radioactive waste from other waste in order to avoid mixed waste streams. Strippable paints that adhere to the radioactive particles on the surfaces are applied to collect those radioactive particles. The strippable paints are then removed before the remainder of the surface preparation is performed.

In addition, 100 percent of any paint removal waste, including dust, must be kept out of the plant ventilation system. If the waste penetrates the ventilation system, it will cause failure of the normal activated carbon filters for the radioactive particles, rendering them useless for normal personnel protection. The vapors of the solvent methyl ethyl ketone (MEK) must also not be allowed to enter the activated carbon filters, because the vapors can cause an explosion. It is normal for NPPs to require a completely enclosed containment with zero emissions into the plant atmosphere. The containment must be ventilated into the atmosphere outside of the plant buildings, especially the containment building, the control building, and all of the areas containing the plant personnel, critical safety equipment, and fuel-storage facilities.

In the Coating Service Level 1 area of NPPs, the entire surface preparation, even when done with vacuum-shrouded tools, is performed inside a Class 1A containment system with zero emissions required, and frequently with an E1 entry condition. For power-tool cleaning, the standard is to use Class 1P containment. Chemical stripping is not permitted in most areas of NPPs to prevent mixed waste and contamination of the ventilation system's activated carbon filters. Mixed waste is the combination of hazardous chemicals and radioactive wastes. No provisions for this type of mixed waste disposal are legal in the United States; they are normally prohibited in all NPPs.

> **NOTE**
> Containment classes and entryways are described in detail in later sections of this module.

Each plant has its own rules for all of these and other operations, which are part of the licensing basis for the plant with the NRC. Anything that could possibly breach that licensing basis is prohibited, either by NRC or plant-specific regulations.

3.0.0 Debris Containment

Planning the containment and collection of debris is an important factor in the quality of the completed job. Poorly managed debris containment adds significant costs to site cleanup. Debris disposal can also be costly, especially if the debris contains hazardous materials. With advance planning, debris can be collected efficiently at the time it is created. Hazardous waste can also be separated to reduce disposal costs.

Surface preparation creates the most debris. This debris includes dust, coating chips, rust, scale, steel, concrete, and other materials. In addition, the debris from ancillary activities must also be contained. Decontamination areas should be tarped to prevent the contamination of areas outside the containment. The wastewater created from cleaning coating application equipment and other tools must be collected and disposed of properly. The dust collected by the air filtration equipment must be bagged. Lead dust must be collected in impenetrable bags.

The type and amount of debris depend on the surface being prepared and the surface preparation technique used. Select collection methods and materials based on the type and quantity of debris produced. Grinding produces a fine dust. Abrasive blasting methods produce larger particles, and the debris also includes spent abrasive media.

Most debris is dry and can be vacuumed. However, surface preparation techniques like chemical stripping and waterjetting create wet debris. Vacuuming may also be appropriate for wet debris, but the collection containers must be waterproof. The containment and collection system for spent abrasives may need to be modified if the abrasives will be reused. Collection systems for chemical stripping of debris must be made of chemically resistant materials.

> **DID YOU KNOW?**
> ***Bridge Painting***
>
> Roughly 80 percent of the more than 100,000 steel highway bridges in the United States are coated with paints that contain lead. About 1,200 bridges painted with lead-based paint, or an estimated 100 million square feet of surface area, are cleaned and repainted each year. Most often, abrasive blasting is used for cleaning.

3.1.0 Debris Collection Methods

Tailor the debris collection methods to the type and quantity of debris produced. Carefully plan debris collection. Improper collection methods cause dusts to be suspended in the air once again. Poor collection methods disperse waste into the surrounding area and negate containment efforts. The two approaches to debris collection are localized and general area collection. Local area collection uses a vacuum system and specialized equipment to collect the debris where it is produced. This reduces dust in the air, but it can limit cleaning to smaller areas. In general area collection, debris is removed from the containment area. The containment system is designed so that debris can be easily transferred into containers for disposal.

3.1.1 Localized Collection at Point of Cleaning

Localized collection at the point of cleaning uses specially equipped tools. These tools have a shroud, or cover, around the tool head that forms a small containment chamber. The shroud is typically made of rubber to conform to the surface being cleaned. On some tools, brushes form the outer edge. Debris and dust are sealed against the surface and do not escape into the air or a larger containment area. A vacuum is attached to the shroud to remove the debris from the containment chamber. The debris is collected in the vacuum system, and the exhaust air is filtered.

Various handheld vacuum tools are available, including portable shot blasters, grinders, sanders, scarifiers, and scalers. Many of these tools are approved by OSHA for use in the removal of lead paints and coatings. This limits worker exposure to airborne lead dust. OSHA defines this process as vacuum blast cleaning. Many devices come in electric and pneumatic models. Pneumatic tools are often used with a centralized vacuum system.

Portable shot blasters are self-contained abrasive blasters. Air or a blast wheel propels abrasive shot at the surface. The shot rebounds with the dust and contaminants into a separation chamber. The abrasive is recycled, and the dust and contaminants are collected for disposal. Portable shot blasters are used to remove hazardous and nonhazardous coatings from concrete and steel surfaces. Small handheld units have a 2-inch blast pattern. They can be used on walls, floors, ceilings, and overhead beams. The units are typically connected to a vacuum system.

Larger portable shot blasters have blast paths from 10 to 36 inches. They are designed to operate on concrete or steel floors, ship decks, roads, and bridges. The larger units, which are powered by gasoline or propane, have a built-in vacuum unit. The operator walks behind the smaller units and rides on the larger ones.

Grinders are operated by pressing the rotating abrasive disk against the surface of the object to be cleaned. The composition of the abrasive disk varies depending on the application. The disks are replaced when they are spent. They are used for grinding and removing hazardous and nonhazardous coatings, paint, and rust. They can be used on concrete, brick, steel, and other hard surfaces. Compact grinders are used in tight areas. Larger grinders are available in 3-, 4.5-, and 7-inch models. Some have a bullnose or floating shroud for greater visibility while working. Some larger grinders can be mounted on a cart for use on floors (*Figure 5*).

Sanders are similar to grinders, but the abrading motion may be back and forth rather than circular. On sanders, replaceable sheets of abrasive material are affixed to a moving pad or formed into a continuous belt. Some models have through-the-pad suction, which results in excellent dust control and increased visibility while working. Pneumatic sanders (*Figure 6*) are used on aluminum and fiberglass. They work well with central vacuum systems. Electric sanders are used for coarse or fine sanding of medium- and small-surface areas. Several sizes and configurations are available, including random orbit, rotary, and triangle sander models for various applications. For efficient operation, some have quick-release hook-and-loop fastenings for sanding sheets.

Needle descalers have a bundle of reinforced steel rods that operate with a hammer-type motion (*Figure 7*). They are used to remove hazardous and nonhazardous coatings, paint, stucco, and rust. Their small size allows for coating and corrosion removal in tight areas, including corners, flanges, windowsills, and door jambs.

204F05.EPS

Figure 5 Vacuum grinder with cart.

On roto-scalers or scarifiers (*Figure 8*), the hubs used for surface preparation include the rotating star-wheel cutter, hammer, abrasive wheel, and peening. These handheld tools are used for profiling and cleaning surfaces. They enable operators to remove coatings within ¼ inch of vertical surfaces. The interchangeable hubs allow them to be used in different applications. The star-wheel cutter hub removes heavy coatings such as coal tar, epoxy solids, rust scale, and cementitious coatings, from steel and concrete. A hammer-head hub removes loose coatings on steel and concrete and is usable on irregular surfaces. Abrasive wheel hubs are used to remove rust and paint from steel and to feather painted surfaces. The peening hub removes up to 20 mils (508 microns) of paint coatings, and it provides an *SSPC-SP 11* profile on steel. It also equates to providing an acid wash or broom finish to concrete. Some models can remove up to 150 square feet of paint per hour while achieving high levels of cleanliness.

A newer tool is the Bristle Blaster. It is available with pneumatic or electric drive and a vacuum pickup attachment. The bristle blasting process removes coatings and corrosion and generates an anchor profile using a specially designed rotary bristle wheel. The wheel contains hardened wire bristles with the tips bent slightly forward. As they spin at about 2,500 rpm, the bristles are bent back slightly by the round bar at the front of the tool. As the bristles pass over the bar, a high kinetic energy is imparted as they spring back and the angled tips forcefully strike the surface at a perpendicular angle. After impact, the bristles rebound, removing any corrosion or coating and leaving behind a profile that mimics grit abrasive blasting.

When these tools are used, the dust and debris are collected immediately. This prevents their release into the surrounding area and reduces airborne dust hazards. Visibility is improved, and the need for respiratory protection and containment is diminished. This is especially important in sensitive environments or when lead-based paints are present. For example, this type of procedure would be effective for removing lead-based coatings in a facility for children or establishments where food is prepared.

There are several disadvantages to localized collection at the point of cleaning. Typically, it is more time consuming than open abrasive blasting because the cleaning head is smaller. The containment chamber limits the size of the area that can be cleaned. This method is ineffective on irregular surfaces where a good seal cannot be formed. The vacuum system and shroud can obscure the blast area. Some areas may need several passes to be cleaned thoroughly. The vacuum hose limits the area that can be worked at one time. Finally, the added weight of the shroud and the vacuum system can contribute to operator fatigue.

3.1.2 General Area Collection

In general area collection, the debris is contained in a structure that encloses the tools, the workers, and the work area, rather than just the tool head. Bulk collection and channeling are the two methods for general area collection. In bulk collection, debris is allowed to fall into a sealed containment chamber, from which it is periodically removed. In a channeling arrangement, the containment system is designed to funnel debris into collection areas.

204F06.EPS

Figure 6 Vacuum sander.

204F07.EPS

Figure 7 Needle descaler.

204F08.EPS

Figure 8 Roto-scaler or scarifier with a vacuum shroud.

> **CAUTION**
>
> If the debris contains hazardous materials such as lead, OSHA may limit the use of certain collection techniques. Once dust has settled out of the air, avoid collection techniques that cause the dust to become airborne. Collection methods that create airborne dust, such as sweeping, are prohibited.

For bulk collection, debris is allowed to fall into a sealed containment area. The debris must be cleared from the containment area regularly. Collection should occur at least daily, and more often if debris accumulates.

Sweeping and vacuuming are common methods used to gather debris for bulk collection. Brooms, buckets, shovels, and brushes are used to manually remove debris from the walls and floor of the containment structure. Vacuums, vacuum trucks, and conveyors can also be used. The collected debris is then placed in appropriate containers for disposal. In the United States, containers approved by the Department of Transportation (DOT) must be used if the debris contains hazardous materials.

Channeling systems are generally used on bridges or other elevated structures. This method uses gravity to channel the debris into the collection system. The flow of the debris is directed to a central removal location through a series of traps, panels, augers, or other suitable equipment. Spent abrasives and debris are channeled to a recycling unit if the blast media will be reused. Typically, the channeling system is part of the main containment structure. However, there are applications for independent systems.

Once the debris has been channeled to the removal area, it can be containerized immediately, stored, or moved to another location. Manual or mechanical equipment can be used to place the material in containers for disposal or recycling.

There are several advantages to a channeling system. Debris is removed from the work area continuously as it is produced. This reduces trip, slip, and fall hazards in the work area. The channeling system also eliminates dust production resulting from sweeping and various other collection methods. Collecting debris in a central area makes cleanup and disposal easier.

Although channeling has a number of advantages, it is a very expensive method of debris containment. It is also subject to clogging. Channeling usually depends on gravity, so it will not be effective on all job sites. In addition, open areas beneath work areas create a hazard. Objects that are inadvertently dropped into the channeling system, such as tools, may be difficult to retrieve. Always take precautions to prevent unwanted materials from falling into the channeling system.

4.0.0 Selecting Containment Systems

The type of containment system needed depends on many different factors, including the physical area surrounding the system and the type of work being performed. The facility owner or representative usually specifies the level of emissions control required. Emissions control includes control of plumes of airborne dust, debris, spills, and leaks. Facilities may also identify the type of environmental monitoring required to verify containment effectiveness. Many contracts use *SSPC Guide 6* classifications to identify required levels of containment.

Greater emissions control is required for sensitive areas. Repainting a bridge over water typically calls for greater emissions control than repainting a pipeline in an industrial complex. Increased control is generally required when the work area is near population centers, pumps and motors, agricultural or food production areas, bodies of water, parks, or wildlife habitats. In addition, some industrial sectors, including pharmaceuticals and computer technologies, require high levels of emissions control if their products are sensitive to dust.

The work being performed is also an important factor when selecting a containment system. Abrasive blasting creates a large amount of debris spread over a wide area. Cleaning with vacuum power tools produces less debris, and the debris is contained as it is produced. Coating application does not create solid debris, but chemical fumes and vapors may need to be controlled. Methods that do not use abrasives generally produce the least amount of dust and debris.

The combustibility and/or toxicity of the waste are other factors to consider when selecting a containment system. Note that the toxicity of the waste does not depend on the quantity of waste. A small volume of lead-contaminated debris requires greater control levels than a larger amount of nontoxic debris.

4.1.0 Containment Classes

SSPC Guide 6 classifies containment systems depending on the degree of emissions control required. The guide defines four containment classes, ranked in order of decreasing emissions control. The four classes are defined as follows:

- *Class 1* – Provides the highest level of control.
- *Class 2* – Provides a high level of control.

- *Class 3* – Provides a moderate level of control.
- *Class 4* – Provides a minimal level of control.

SSPC Guide 6 lists the general requirements for each containment class. It describes the requirements for containment materials, penetrability, **support structures**, joints, and **entryways**. The containment systems for Classes 1 and 2 require fully sealed joints and various types of entryways depending on the surface preparation and/or the coating method to be used. Classes 3 and 4 allow for partially sealed joints and an overlapping or open-seam entryway.

The guide also identifies the ventilation system requirements for each containment class. The air makeup, input airflow, air pressure, air movement, and exhaust dust filtration are identified for each class.

These containment and ventilation requirements were described in Level One. To fully determine the containment requirements, the surface preparation techniques used must also be considered.

4.2.0 Surface Preparation Technique Subclasses

Surface preparation techniques vary widely in the type and amount of debris generated. Abrasive blasting produces a significant amount of dust and larger particles of rust, old coatings, mill scale, and spent abrasive material. Waterjetting creates a large wastewater stream contaminated with coating debris and other chemicals. Chemical stripping produces sludge from old coatings and solvents. Hand and power tool cleaning produces dust and air-borne paint chips.

Containment and ventilation must be tailored to the type and amount of waste created. Dust is a concern during abrasive blasting and power- tool cleaning. Dust control is not a major consideration for waterjetting and chemical stripping, but water and chemical permeability are. Specifying the level of emissions control for a project is insufficient in and of itself. This is because the types of debris produced differs for each surface preparation method. Both the surface preparation method and the level of emissions control must be considered when developing containment systems.

> **WARNING**
>
> Always follow the local Department of Health (DOH) and United States Environmental Protection Agency (EPA) requirements for wastewater treatment and disposal.

The containment classification in *SSPC Guide 6* includes both a number for the level of emissions control and a letter for the surface preparation method. Surface preparation methods are grouped into four types. Each type is identified by a letter as follows:

- *Type A* – Abrasive blast cleaning
- *Type W* – Water blasting or waterjetting (with and without abrasives or vacuum shrouds)
- *Type C* – Chemical stripping
- *Type P* – Power-tool cleaning (with and without shrouds and hand-tool cleaning)

SSPC Guide 6 includes four tables that list containment and ventilation requirements. Each table lists the requirements for one type of surface preparation. For example, Table A in *SSPC Guide 6* covers the containment and ventilation requirements for abrasive blast cleaning. Within each table, the rows reflect the containment class. *Tables 1* and *2* show the containment and ventilation requirements for abrasive blast cleaning. This covers blast cleaning that uses compressed air to propel abrasives through nozzles. The use of expendable and reusable abrasives is included. The abrasives can be natural minerals, mineral slag, or manufactured material.

To read the tables, first select the correct table for the surface preparation method being used. Then move down the table to select the level of emissions control required. Read across the table to identify the requirements for a specific containment classification. The tables specify the types of materials needed for containment and ventilation systems. Containment system considerations include containment materials, penetrability, support structures, joints, and entryways. Ventilation considerations include air makeup, input airflow, air pressure, air movement, and exhaust dust filtration.

Tables 3 and *4* show the containment and ventilation requirements for water blast cleaning. This includes all types of waterjetting systems. These systems propel water and/or other materials in a cleaning action. The water stream may include cleaning chemicals, rust inhibitors, or abrasives. The abrasives may be mixed in separate tanks or added to the water jet at the nozzle.

Table 1 Containment Systems for Abrasive Blast Cleaning

Containment Classification	Containment System				
	5.3.1 Containment Materials	5.3.2 Penetrability	5.3.3 Support Structure	5.3.4 Joints	5.3.5 Entryway
Class 1A	A1 – Rigid A2 – Flexible	B1 – Air impenetrable	C1 – Rigid C2 – Flexible	D1 – Full seal	E1 – Airlock E2 – Resealable
Class 2A	A1 – Rigid A2 – Flexible	B1 – Air impenetrable	C1 – Rigid C2 – Flexible	D1 – Full seal	E2 – Resealable E3 – Overlap
Class 3A	A1 – Rigid A2 – Flexible	B1 – Air impenetrable B2a – Air penetrable (tightly woven)	C1 – Rigid C2 – Flexible	D1 – Full seal D2 – Partial seal	E3 – Overlap E4 – Open seam
Class 4A	A1 – Rigid A2 – Flexible	B2a and B2b – Air penetrable	C3 – Minimal	D2 – Partial seal	E4 – Open seam

204T01.EPS

Table 2 Ventilation Systems for Abrasive Blast Cleaning

Containment Classification	Ventilation (5.4 of SSPC – Guide 6)				
	5.4.1 Air Make-up	5.4.2 Input Air Flow	5.4.3 Air Pressure	5.4.4 Air Movement	5.4.5 Exhaust Dust Filtration
Class 1A	F1 – Controlled	G1 – Forced G2 – Natural	H1- Instrument verification H2- Visual verification	I1 – Minimum specified	J1 – Air filtration
Class 2A	F1 – Controlled F2 – Open	G1 – Forced G2 – Natural	H2- Visual verification	I1 – Minimum specified	J1 – Air filtration
Class 3A	F1 – Controlled F2 – Open	G1 – Forced G2 – Natural	H3 – Not required	I2 – Not specified	J1 – Air filtration
Class 4A	F2 – Open	G2 – Natural	H3 – Not required	I2 – Not specified	J2 – No controls on exhaust

204T02.EPS

Table 3 Containment Systems for Water Blasting

Containment Classification	Containment System				
	5.3.1 Containment Materials	5.3.2 Penetrability	5.3.3 Support Structure	5.3.4 Joints	5.3.5 Entryway
Class 1W	A1 – Rigid A2 – Flexible	B1 – Air impenetrable AND B3a – Water impermeable	C1 – Rigid C2 – Flexible	D1 – Full seal	E2 – Resealable
Class 2W	A1 – Rigid A2 – Flexible	B3a – Water impermeable	C1 – Rigid C2 – Flexible	D1 – Full seal	E3 – Overlap
Class 3W	A1 – Rigid A2 – Flexible	B3a – Water impermeable	C3 – Minimal	D2 – Partial seal	E4 – Open seam
Class 4W	A1 – Rigid A2 – Flexible	B3a – Water permeable	C2 – Flexible	D2 – Partial seal	E4 – Open seam

204T03.EPS

Table 4 Ventilation Systems for Water Blasting

Containment Classification	Ventilation (5.4 of SSPC – Guide 6)				
	5.4.1 Air Make-up	5.4.2 Input Air Flow	5.4.3 Air Pressure	5.4.4 Air Movement	5.4.5 Exhaust Dust Filtration
Class 1W	F2 – Open	G1 – Forced G2 – Natural	H2 – Visual verification	I1 – Minimum specified	J1 – Air filtration
Class 2W	F2 – Open	G2 – Natural	H3 – Not required	I2 – Not specified	J2 – Not required
Class 3W	F2 – Open	G2 – Natural	H3 – Not required	I2 – Not specified	J2 – Not required
Class 4W	F2 – Open	G2 – Natural	H3 – Not required	I2 – Not specified	J2 – Not required

204T04.EPS

Tables 5 and *6* show the containment and ventilation requirements for power tool cleaning. This includes various power-operated tools, such as wire brushes, sanders, scrapers, grinders, descalers, needle guns, and rotary peening equipment. The power tools may have a localized exhaust or a vacuum system that contains the exhaust. The level of emissions control may be reduced if vacuum systems are used.

Tables 7 and *8* show the containment and ventilation requirements for chemical stripping. This covers the use of any chemical stripping solutions or slurries applied to the paint surface. The effectiveness of these chemicals may be enhanced by

Table 5 Containment Systems for Power Tool Cleaning

Containment Classification	Containment System				
	5.3.1 Containment Materials	5.3.2 Penetrability	5.3.3 Support Structure	5.3.4 Joints	5.3.5 Entryway
Class 1P	A1 – Rigid A2 – Flexible	B1 – Air impenetrable	C1 – Rigid C2 – Flexible	D1 – Full seal	E2 – Resealable
Class 2P	A1 – Rigid A2 – Flexible	B1– Air impenetrable B2a or B2b – Air penetrable	C1 – Rigid C2 – Flexible	D1 – Full seal	E3 – Overlap E4 – Open seam
Class 3P	A1 – Rigid A2 – Flexible	B2a or B2b – Air penetrable	C3 – Minimal	D2 – Partial seal	E4 – Open seam

204T05.EPS

Table 6 Ventilation Systems for Power Tool Cleaning

Containment Classification	Ventilation (5.4 of SSPC – Guide 6)				
	5.4.1 Air Make-up	5.4.2 Input Air Flow	5.4.3 Air Pressure	5.4.4 Air Movement	5.4.5 Exhaust Dust Filtration
Class 1P	F1 – Controlled	G1 – Forced G2 – Natural	H2 – Visual verification	I1 – Minimum specified	J1 – Air filtration
Class 2P	F2 – Open	G2 – Natural	H3 – Not required	I2 – Not specified	J2 – Not required
Class 3P	F2 – Open	G2 – Natural	H3 – Not required	I2 – Not specified	J2 – Not required

204T06.EPS

Table 7 Containment Systems for Chemical Stripping

Containment Classification	Containment System				
	5.3.1 Containment Materials	5.3.2 Penetrability	5.3.3 Support Structure	5.3.4 Joints	5.3.5 Entryway
Class 1C	A1 – Rigid A2 – Flexible	B1 – Air impenetrable AND B3a – Water impermeable AND B4 – Chemical resistant	C1 – Rigid C2 – Flexible	D1 – Full seal	E3 – Overlap
Class 2C	A1 – Rigid A2 – Flexible	B3a – Water impermeable AND B4 – Chemical resistant	C1 – Rigid C2 – Flexible	D2 – Full seal	E3 – Overlap
Class 3C	A1 – Rigid A2 – Flexible	B3a – Water impermeable AND B4-Chemical resistant	C3 – Minimal	D2 – Partial seal	E4 – Open seam

204T07.EPS

Table 8 Ventilation Systems for Chemical Stripping

Containment Classification	Ventilation (5.4 of SSPC – Guide 6)				
	5.4.1 Air Make-up	5.4.2 Input Air Flow	5.4.3 Air Pressure	5.4.4 Air Movement	5.4.5 Exhaust Dust Filtration
Class 1C	F2 – Open	G1 – Forced G2 – Natural	H3 – Not required	I2 – Not specified	J1 – Air filtration
Class 2C	F2 – Open	G2 – Natural	H3 – Not required	I2 – Not specified	J2 – Not required
Class 3C	F2 – Open	G2 – Natural	H3 – Not required	I2 – Not specified	J2 – Not required

204T08.EPS

applying a cover of paper or cloth after their application. The softened paint is removed by peeling or scraping, along with additional washing or flushing of the surface as required.

4.3.0 Selecting *SSPC Guide 6* Containment

The level of emissions control required and the corresponding containment class are generally identified in the contract or bid documents. Facility managers must consider the environmental concerns in their area. If the containment class has not been determined, work with the facility manager to identify concerns.

In addition to the level of emissions control and the surface preparation method, other factors must be considered when designing a containment system. If the debris contains hazardous chemicals, additional precautions must be taken.

DID YOU KNOW?

Lead Stabilizers

Several products are available that can stabilize lead-based coatings. These products bond with the lead in the coatings, making it nonsoluble. These processes make the resulting debris nonhazardous under the U.S. EPA definition, which can significantly reduce disposal costs. **However, OSHA requirements for removal of lead paint may still apply.**

Some stabilization coatings are applied over the existing coating before removal. Other products are chemicals that are added to the abrasive blast mixture before wet-abrasive blasting. **Although the debris is rendered nonhazardous, containment is still needed in order to collect debris and contain emissions.**

The complexity, size, and elevation of the structure itself must also be considered. Containment for the joists and trusses on a bridge over water is more complex than containment for a steel floor. The location of the structure and its proximity to other buildings may affect containment considerations. The construction of the structure and its loadbearing capacity may also limit some choices.

Finally, the local climate will make some containment systems more effective than others. Wind, rain, and sun can destroy the containment structure or compromise penetrability. This is especially important if the project will take a significant amount of time. On longer projects, containment systems must be replaced when they become worn.

Use the following procedure to identify the proper level of containment for a particular operation:

Step 1 Identify the level of emissions control required and select the corresponding containment class.

Step 2 Identify the surface preparation method used, and select the corresponding letter designation.

Step 3 Look up the containment requirements for that class on the tables to determine the containment and ventilation system components.

Step 4 Identify additional considerations that affect containment selection.

Step 5 Select containment materials that meet the requirements stated in the guidelines and as dictated by local conditions.

Step 6 Select a ventilation system that is compatible with containment and with applicable regulatory requirements.

5.0.0 CONTAINMENT ENCLOSURE COMPONENTS

The containment enclosure surrounds the work area. It is constructed by joining panels over a support structure. The panels are secured to the support structure and to each other at the seams. The enclosure must have an entryway so that workers and equipment can enter and exit the structure.

Wind load requirements vary by structure and can be found in the job specifications. Always be aware of the potential for excessive wind loads. To ensure containment integrity, containment systems must be disassembled prior to reaching specified wind speeds.

The containment structure must be specifically designed to meet the needs of a particular project. The panels and framework of the structure can be custom made or reusable. Reusable components must be cleaned before reuse. All components are subject to wear and should be inspected before reuse. On larger projects, the containment system may need to be replaced due to wind loads and wear. Sometimes containment is erected for a section of the project. After preparation and coating operations have been completed, the containment area is dismantled and moved to the next section. After relocation, the framing and sides are reinstalled before cleaning and coating operations begin again.

A wide selection of materials is available for containment construction—from plastic wrap to plastic sheeting. Scaffolding can be used for the framework, or additional wooden framing can be built. It is easier to design a containment structure by identifying the properties needed in each of the components of the structure.

The five components identified in *SSPC Guide 6* form the basis for all containment enclosures. These components are identified as Types and are assigned unique letters. The subdivisions within each Type are assigned a number. This allows easy cross-reference between the tables and the body of the guide. The tables list alphanumeric codes, which are further detailed within the body of the guide. The guide uses the following types:

- Containment Material (Type A)
- Containment Material Penetrability (Type B)
- Support Structure (Type C)
- Joints (Type D)
- Entryway (Type E)

The codes reference other sections in the guide in which different materials are identified. For example, Type B1 refers to air impenetrable materials, and Type B4 signifies chemically resistant materials. Acceptable materials are described in the guide. Once the requirements for the desired containment class have been located on the tables, use the codes to find descriptions of acceptable materials.

5.1.0 Containment Material (Type A)

The use of containment for protective coatings has greatly increased over the past ten to fifteen years. To serve that need, several companies now specialize in producing containment structures and materials. A wide variety of materials are used to form the walls, ceiling, and floor of a containment enclosure. *SSPC Guide 6* identifies two general types of containment material, rigid and flexible, designated as Types A1 and A2.

Type A1, rigid containment materials, consists of single panels, interlocking panels, or modular fabrications. These panels may be constructed of plywood, aluminum, rigid metal, plastic, or similar materials. Several panels can be joined together to form the size and shape of the containment area needed. The panels interlock, forming a tight seal. However, the size of the containment structure is limited to the size of the panels or multiples of that size.

Rigid materials are more durable and can typically be reused. The panels must be cleaned between projects, especially if a hazardous substance has been removed. These systems can be expensive. For these reasons, the use of rigid panels is limited. All containment classes allow the use of either rigid or flexible materials. However, the contract may specify Type A1 materials.

Rigid panels are used to form a floor under structures in order to access them. For example, rigid panels can be used to access bridges (*Figure 9*), offshore oil platforms, or ships. One manufacturer produces 8.5-foot by 20-foot sections of panel that are joined together with stainless steel pins. These sections are especially useful for constructing Class 1 containment for bridge recoating. The modular platform forms a solid working surface. Traversing systems can be moved along the bridge structure as coating operations proceed. Auger conveyor systems can also be added to remove spent abrasives. This reduces the weight load on the structure.

204F09.EPS

Figure 9 Bridge painting.

Flexible materials are allowed for all containment classes and are commonly used. Type A2 materials include screens, tarps, drapes, plastic sheeting, or similar materials. Many companies make tarps using woven polypropylene, polyethylene, nylon, vinyl, and other fabrics. *Figure 10* is an example of custom-made funnels that capture spent media and debris for recycling or disposal. Most are custom made to contractor specifications and can be imprinted with the contractor's logo. Factors to consider when selecting a flexible material include strength, color or opacity, and flame and ultraviolet resistance. Clear or lightly colored materials allow sunlight into the enclosure, which can provide additional warmth in cold weather.

A material's weight and thickness are one indication of strength. Thicknesses range from 2 mils (50 micrometers) to 100 mils (2.5 millimeters). Tarps less than 6 mils are suitable for indoor use. A 10-mil tarp can be used outdoors in moderate to heavy wind loads.

The heavier vinyl tarps are made in several weights, from 10 ounces to 19 ounces per yard (*Figure 11*). Vinyl tarps are made of polyvinylchloride reinforced with nylon or polyester threads. This type of material is very strong and resists tearing. Typically, the tarps are treated so that they are UV- and flame-resistant. They are waterproof and chemically **impermeable** when heat-sealed. Applications include use in abrasive blasting curtains, **negative pressure** enclosures, debris chutes and funnels for chemical strippers, and ground covers for waterjetting.

Woven polyethylene tarps are coated so that they are flame- and UV-resistant (*Figure 12*). They are generally lighter than vinyl, ranging from 5 to 10 ounces per yard. They are air impenetrable and are suited for lead-abatement structures and negative-pressure enclosures. Woven polyethylene with a burst strength of more than 600 psi should

204F10.EPS

Figure 10 Spent media and debris collection by funneling.

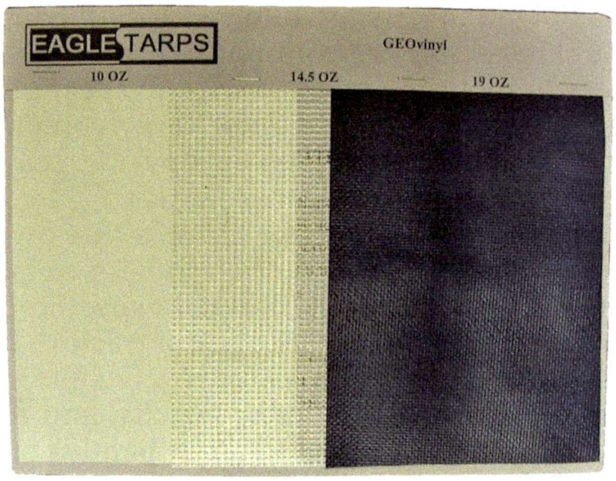

204F11.EPS

Figure 11 Vinyl tarps.

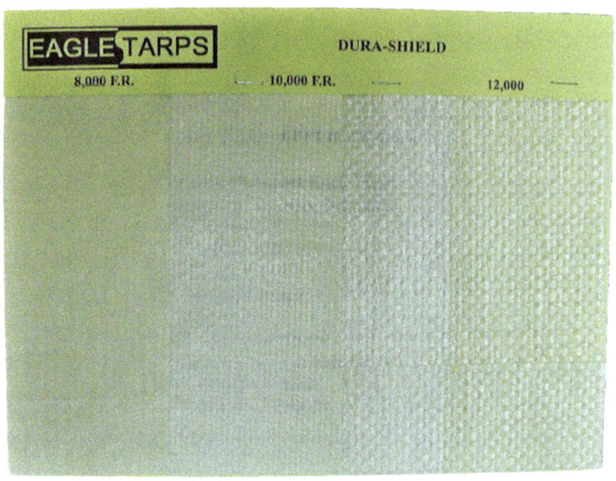

204F12.EPS

Figure 12 Polyethylene tarps.

be used in areas subject to high winds. It can also be used as debris chutes and winter enclosures. The lighter color and clear fabric allow more light to pass through, increasing visibility within the enclosure.

Reinforced laminate (*Figure 13*) is a **flame-retardant** material. Flame-retardant tarps stop burning once the source of fire has been removed, whereas **flame-resistant** tarps may resist burning for a while but will continue to burn once ignited. Flame-retardant tarps should be used when ignition sources are present. This type of tarp is useful in winter conditions when propane heaters are used.

5.2.0 Containment Material Penetrability (Type B)

Air penetrability and water permeability are separate factors in selecting containment materials. Some containment materials must be impenetrable to air or impermeable to water and chemicals. These factors are divided into six Types in *SSPC Guide 6* as follows:

- *Type B1* – Air Impenetrable
- *Type B2a* – Air Penetrable – Tightly Woven
- *Type B2b* – Air Penetrable – Open Mesh
- *Type B3a* – Water Impermeable
- *Type B3b* – Water Permeable
- *Type B4* – Chemical Resistant (and Water Impermeable)

When dust emissions or vapors must be avoided or negative pressure maintained, an air impenetrable, or Type B1, material should be specified. Acceptable materials include coated woven materials that are free of holes or openings, plastic sheeting, and rigid panels. Any holes or openings reduce containment effectiveness. To create an airtight containment, 9- and 12-mil plastic sheeting is often used. The sheeting is heat shrinkable, and the seams are sealed with a heat gun.

Material that is tightly woven but is not coated is classified as Type B2a. Some types are flame resistant (*Figure 14*). These materials are capable of retaining airborne particles, but because they are not coated, they are not air impenetrable. Screens (*Figure 14*) are composed of breathable curtains, and they are not suitable for negative-pressure enclosures. Applications include sandblasting

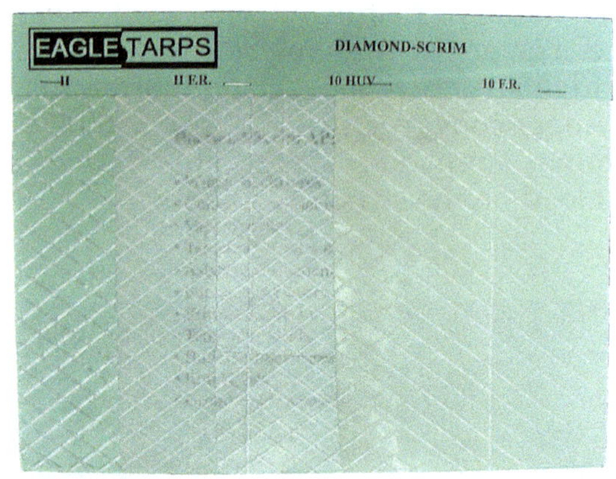

204F13.EPS

Figure 13 Reinforced laminate tarps.

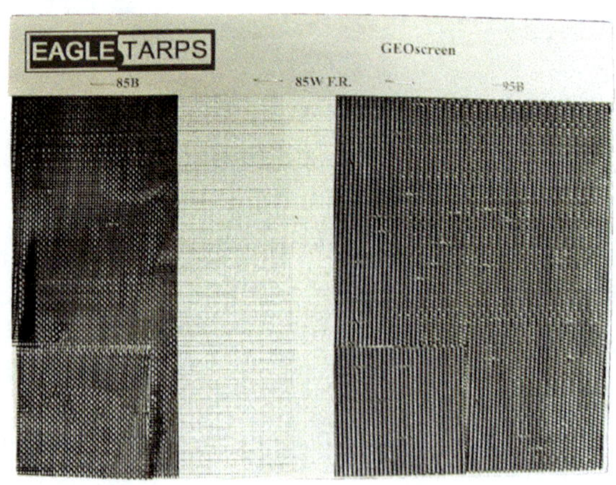

204F14.EPS

Figure 14 Tightly woven materials.

enclosures in which hazardous materials are not present, paint overspray curtains, and scaffolding enclosures.

Air-penetrable, open-mesh material (*Figure 15*) is formed or woven to allow airflow, but it can retain some airborne particles. These materials are listed as Type B2b in *SSPC Guide 6*. Particles larger than the size of the openings in the mesh are collected, and finer particles and dust are released. These materials are also known as debris netting because they are mainly used to collect larger debris within the containment enclosure.

When wet surface preparation and coating methods are used, the material's permeability to water is a factor. Materials that are impermeable to water are classified as Type B3a. Water-permeable materials are classified as Type B3b, which includes open mesh material that is formed or woven with openings of less than 25 mils. These materials collect debris while allowing water to pass through. Water-permeable materials should only be used when the debris particles are larger than 25 mils or when the wastewater does not contain chemicals.

The final category is Type B4, or chemical-resistant materials. These materials will not break down when exposed to chemical stripping solutions and are impermeable to the water used for rinsing chemical stripping agents. Heavier vinyl tarps are waterproof and chemically impermeable when heat sealed. These tarps should be used in chemical stripping, waterjetting, or wet abrasive blasting operations.

5.3.0 Support Structure (Type C)

Containment materials must be supported in some manner. Support structures have three classifications in *SSPC Guide 6*. Rigid structures are Type C1, flexible structures are Type C2, and minimal support structures are Type C3.

204F15.EPS

Figure 15 Open mesh materials.

Rigid structures (Type C1) do not allow the containment materials to move. Scaffolding and wooden framing are often used to form rigid support structures. An example of wooden framing is shown in *Figure 16*.

This category also includes moveable rigid enclosures. They are fixed-frame units that cover a small area. Once surface preparation and coating operations have been completed for the small area, the containment is moved to the next section.

Flexible support structures (Type C2) allow minor movement. They consist of cables, chains, or similar systems to which the containment materials are affixed. The framing can be attached to the structure or supported by cables. These systems are often used on water towers (*Figure 17*). Metal framing is attached to the top of the tower, and concrete anchor blocks are placed in the ground. Cables are run from the top to a bottom anchor ring. The ring is fastened to the anchor blocks with turnbuckles, as shown in *Figure 18*.

204F16.EPS

Figure 16 Wood frame support structure.

204F17.EPS

Figure 17 Water tower containment.

Minimal support structures (Type C3) involve little or no supporting structure. The containment materials are supported by cables or chains attached to the structure itself or to the ground.

5.4.0 Joints (Type D)

Most containment structures are formed from separate panels that must be joined together and attached to the support structure. The method used for sealing the joints is classified as Type D. Fully sealed joints are Type D1 and partially sealed joints are Type D2.

Type D1 requires that all mating joints between the containment materials and the structure floor, ceiling, or ground are sealed. To seal the joints, flexible materials must be overlapped and sealed by stitching, taping, caulking, or heat sealing. If the structure is waterproof or chemically resistant, the materials used to seal the joints must also be waterproof or chemically resistant. A containment structure is only as strong as its weakest link. The joints can be the weakest link in the enclosure, so they must be carefully sealed.

For Type D2, the joints do not need to be completely sealed. Tarps are typically made with grommets, and ties are used to mate two panels together. The use of overlapping seams is recommended but not required. Some panels have strapping built into the fabric at five-foot intervals (*Figure 19*). They have holes in the strapping every four inches to allow the panels to be stitched together with cable.

204F18.EPS

Figure 18 Bottom anchor ring turnbuckles.

In some systems, the edges of the tarps slide into tracks on the framing. One company provides an aluminum framing system with channels. Tarps are anchored to the framing by rope welt hems that slide into channels on the frame. The sides are raised and lowered to permit movement into and out of the containment area.

5.5.0 Entryway (Type E)

The other weak link in the enclosure is the entryway. The entryway must be designed to keep debris contained, but it must allow the movement of materials, equipment, and personnel into and out of the enclosure. Entryways must be accessible from both directions. Four types of entryways are listed in *SSPC Guide 6*.

The highest level of containment is provided by Type E1, an entryway through airlocks with resealable doors. This system has access doors that can be resealed repeatedly. An airlock is a chamber between the containment area and the outside environment. One door leads from the containment area into the airlock. Another door leads from the airlock to the outside. Only one door can be opened at a time. This area can be used for decontamination purposes, or a separate decontamination area can be designated on the inside of the containment enclosure.

Type E2 entryways have resealable doors. They must be capable of being repeatedly resealed and should be accessible in both directions. The door provides a good way to contain debris. However, dust and airborne particles may escape when the door is opened. Take care not to leave the door open or open it when there is a large amount of dust in the air.

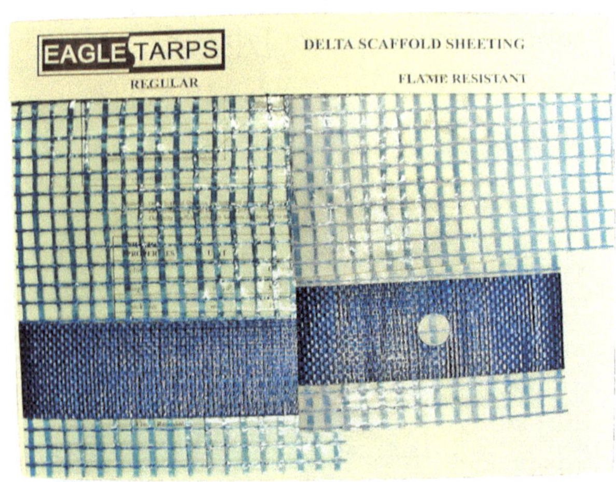

204F19.EPS

Figure 19 Tarp with strapping.

Lower levels of containment allow the use of Type E3, an entryway through overlapping door tarps. Multiple overlapping door tarps are used to prevent the escape of dust through the entryway. A tarp is hung on each side of the opening. One tarp at a time is lifted so that at least one tarp remains between the containment area and the open environment.

Type E4 provides the lowest level of containment. Entry to the work area is made through an unsealed seam in the containment materials. The door should be positioned so that it will not be opened by wind or containment operations.

6.0.0 Containment Ventilation System Components

A containment area seals in debris and other emissions. However, debris and dust remain trapped within the work area, increasing worker exposure. As the level of containment increases, airflow is restricted. Ventilation systems provide improved air movement through the work area, which protects worker health and aids visibility. Ventilation systems can also include heating, dehumidification, and dust collection. They are also a key part of the overall containment structure. A poorly designed system vents debris outside the containment area. A properly designed system provides dust collection before the air is discharged.

There are several factors to consider when designing a ventilation system, including the input, or **makeup air**; the input airflow; the air pressure in the enclosure; the movement of air inside the enclosure; and the exhaust system. *SSPC Guide 6* specifies parameters for each of these components. Each component is identified as a type, as it is with containment components. The types refer to other sections in the guide in which parameters for that component are explained. The guide identifies ventilation components by the following types:

- *Type F* – Air Supply
- *Type G* – Input Airflow
- *Type H* – Air Pressure Inside Containment
- *Type I* – Air Movement Inside Containment
- *Type J* – Powered Exhaust and Dust Filtration

Once the containment class has been identified, reference the tables in *SSPC Guide 6* to identify recommendations for each ventilation component. The tables list codes that refer to the body of the guide for each containment class. For example, reading down *Table 2* for Class 1A, Abrasive Blasting, it lists Type J1 under the column for Exhaust and Dust Filtration. The description of what is required to meet Class 1A ventilation standards is listed in Section 5.4.5 of the table under Type J1.

6.1.0 Air Supply (Type F)

Air that enters a containment enclosure is called intake or makeup air. This air replaces or makes up for the air that was exhausted from the enclosure. Ambient air outside of the containment area is typically used as a supply source.

Under certain conditions, debris can escape from the containment enclosure at the intake points. For example, abrasive blasting produces substantial dust, which can escape the enclosure through the air intake. Where this is a concern, *SSPC Guide 6* specifies the use of Type F1, which requires the use of baffles, louvers, flap seals, filters, and ducts on supply air points. These measures will prevent the release of debris. A fan may or may not be used in combination with these devices.

Type F2 is recommended when the release of debris from air intake points is not a concern. Type F2 allows open-air entry points without the use of ducts, valves, or baffles. All classes of water blasting and chemical stripping do not require controlled air entry points.

6.2.0 Input Airflow (Type G)

Air can be forced into the containment enclosure using fans or blowers (*Figure 20*). Systems that require **forced airflow** are coded as Type G1. When forced input is used, the input air must be properly balanced with the exhaust capacity. If the air input exceeds the volume of air exhausted, it increases the air pressure in the containment area. This is known as positive pressure, which should be avoided in containment systems. With positive pressure, contaminated air will escape the system at any available point, including improperly sealed joints and entryways. Slight negative pressure, which occurs when exhaust air exceeds input air, must be maintained.

Containment systems that do not require forced-air input are coded as Type G2. Air flows into the system naturally without the use of mechanical devices. Natural air inflow includes methods that use exhaust systems. In these systems, air is withdrawn from the enclosure using mechanical devices. Clean air flows into the containment structure as the air within the enclosure is exhausted.

204F20.EPS

Figure 20 Portable blower.

6.3.0 Air Pressure Inside Containment (Type H)

The air pressure inside the containment area depends on the relationship between the volume of air being exhausted and the volume of air being pumped in, or makeup air. Optimally, slightly more air is exhausted than enters the system. This is known as negative air pressure, which helps prevent emissions. Negative air pressure must be maintained within some containment systems. This can be measured with instruments or through visual verification.

Containment systems that require instrument verification of negative pressure are coded as Type H1. *SSPC Guide 6* requires that an average negative pressure of 0.03" water column (7.5 Pa) relative to ambient conditions should be maintained throughout the enclosure during surface preparation and cleanup operations. A manometer (*Figure 21*) or magnehelic gauge can be used to measure the air pressure. Instrument verification is limited to some Class 1 abrasive blasting systems.

Think About It

Negative Pressure

Exhaust air systems can be adjusted to exceed input airflow in order to avoid positive pressure in a containment system. What will happen if the negative pressure in a containment enclosure is too high?

Visual verification of negative pressure (Type H2) is usually sufficient. For visual verification of negative pressure, observe the walls, ceiling, and floor of the containment enclosure. If they are concave or curved inwards, there is negative pressure in the structure. If the structure is constructed of rigid materials, smoke or other visible means can be used to observe airflow patterns. Systems that do not require a specified degree of negative pressure are coded as Type H3.

6.4.0 Air Movement Inside Containment (Type I)

Air movement inside the containment area is especially important when hazardous dusts are generated. OSHA requires engineering controls to limit worker exposure to lead dust. The engineering controls also improve visibility, remove dust-laden air from the work area, and help prevent emissions. When minimum air movement is specified, it is coded as Type I1 in *SSPC Guide 6*. When a minimum air movement is not specified, it is coded as Type I2.

Mechanical ventilation is used to create air movement within the enclosure. Air movement is specified by establishing a minimum velocity for the movement of air. The velocity can be measured in several ways, including the following:

- In cross draft or downdraft
- Air velocity past the worker
- The volume needed to ventilate a given volume of containment

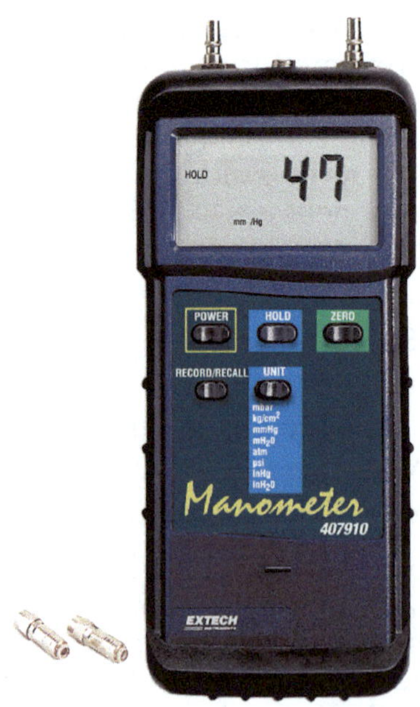

204F21.EPS

Figure 21 Manometer.

Ventilation systems are specified by the volume of air moved over time, typically expressed in cubic feet per minute (CFM) or cubic meters per minute (CMM). The rate of input airflow depends on the size of the area and the amount of debris generated. Additional considerations include the number of operators, the type of work being performed, the type of materials being removed or applied, and the amount of rust and scale on the surface being cleaned. The general rule of thumb is to provide a complete air change every three minutes. This level provides sufficient visibility during blasting operations. This level is normally sufficient to lower the concentration of toxic chemicals to safe levels. However, toxic concentrations should be monitored and adjustments made as needed.

Even if no minimum air movement is specified, it may still be necessary to use engineering controls, such as mechanical ventilation, to reduce airborne lead exposure or prevent emissions of lead particles.

WARNING

OSHA requires the use of feasible engineering controls for removal of lead paint. Ventilation should be used as the first step to reduce lead exposure to safe levels. **SSPC Guide 6 guidelines may not provide adequate reduction in worker exposure to hazardous materials in all situations.** Make sure that adequate precautions are taken to minimize worker exposure to airborne hazards.

6.5.0 Powered Exhaust and Dust Filtration (Type J)

Powered exhaust and dust filtration are used to control dust emissions. Exhaust fans remove the air from the containment area. Portable dust collectors filter the air before it is released (*Figure 22*). Several types of dust filters, baghouses, and dust collectors are available. When powered exhaust and dust filtration is specified, it is coded as

On Site

Containment Airflow Patterns and Velocity Tests

Handheld anemometers, such as the one shown here, can be used to measure ventilation air velocity. They are used inside containment enclosures to measure air velocity in the working area; at enclosure air inlets; and at the low-pressure side of exhaust outlets. They are also available with a time- and date-stamped data-logging capability if a record of periodic measurements is required.

Smoke tubes are used to test airflow patterns and air velocity inside containment areas. The smoke tube kit shown here can provide up to a one-day supply of smoke for multiple tests from a single tube. The plastic tube version works when two ampules of chemicals, which are contained in the plastic tube attached to a rubber aspirator bulb, are crushed. The vapors are mixed and produce smoke only when air is pushed through the tube by squeezing the rubber aspirator bulb. Unless other tests are to be conducted, the tube can be removed from the aspirator and both ends plugged by rubber stoppers until other tests are required during the day. Glass tube versions are also available.

PLASTIC TUBE KIT

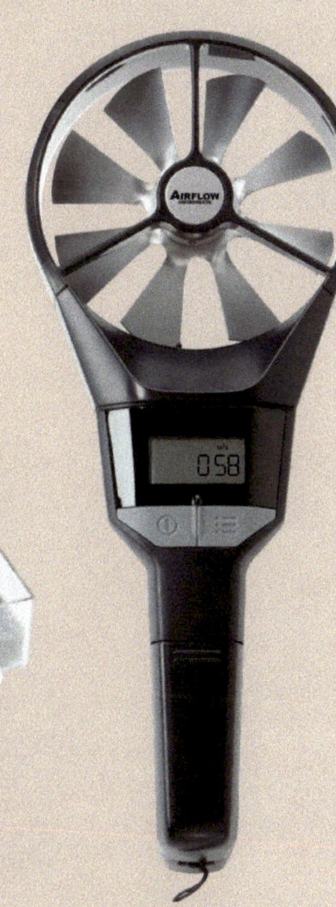

ANEMOMETER

204SA01.EPS

Type J1 in *SSPC Guide 6*. When powered exhaust and dust filtration is not specified, it is coded as Type J2. A powered exhaust system, in combination with a fully sealed containment area, must be used to create negative pressure as required by Types H1 and H2.

A powered exhaust system uses fans and blowers to move the air. Flexible hoses carry the air from the containment area to dust collectors, baghouses, or filters. The filters remove dust and other particles, which are then collected for disposal. The filtered air is then vented outside the containment area. The dust collectors must be appropriately sized for the type and size of the particulate matter and the volume and velocity of the air moved through the system. Machines that recycle abrasives often have a built-in dust collection system.

7.0.0 Containment Effectiveness

Containment systems are designed to keep dust and debris within the enclosure. It is important to measure how well the system works. The system's effectiveness is measured by monitoring the amount of dust and debris that are released from the enclosure. If the system is not working well, measurements will help identify weaknesses so that the necessary adjustments can be made.

There are several methods for measuring emissions of dust and debris. Some methods provide immediate information on the emissions being released. Other methods report results the following day. Some methods monitor during operations, while others are performed after operations have been completed. The project's overall effect on the environment can be measured by testing the water and soil before and after the project and then comparing the two results.

204F22.EPS

Figure 22 Portable dust collector.

SSPC Guide 6 identifies six methods for testing the effectiveness of containment structures. The testing procedures detailed in the standard include the following:

- *Method A* – Visible Emissions
- *Method B* – Ambient Air Monitoring for PM-10
- *Method C* – Occupational Monitoring of Area Emissions for Lead
- *Method D* – EPA Ambient Air Monitoring for Toxic Metals
- *Method E* – Soil Analysis for Toxic Metals
- *Method F* – Water and Sediment Analysis for Toxic Metals

One or more of these methods may be required on a given project. Some methods should not be used on a particular project. For example, when hazardous materials are being removed, ambient air monitoring of particulate matter or monitoring of visible emissions provides data of limited value. Monitoring for the specific hazardous material provides more meaningful data. The contractor or the facility owner should consult local and state authorities to determine the most appropriate testing methods and acceptance criteria.

7.1.0 Method A – Visible Emissions

The simplest way to check the effectiveness of the containment area is to check for visible emissions. This method provides immediate feedback and shows where the system is not working well. Trained observers periodically check for emissions from the containment enclosure. The observations are rated either as a percentage or on an opacity scale. Observations should accurately represent operations.

The two methods for rating visible emissions are general surveillance and opacity. The general surveillance method averages random emissions over the observation period. Visible emissions are observed and recorded at a given frequency and duration. The recorded emissions are averaged over time. Acceptable limits are set either as a percentage of the total time or as a number of minutes in an eight-hour work shift. The frequency of emissions is rated by level as shown in *Table 9.*

The observations must occur within the established boundary or property line. They must occur during normal operations, such as surface preparation and cleanup. The frequency and duration of the observations should be established under the guidance of local authorities.

Table 9 Frequency of Emissions

Level	Percentage	Minutes in 8-Hour Workday
0	0	0
1	1	5
2	5	24
3	10	48
5	Unlimited	Unlimited

204T09.EPS

The opacity measurements are made by a trained and certified observer. The observer compares the opacity of the emissions to a calibrated scale. The scale ranges from 0 to 100 percent in 5 percent increments. Measurements are typically made for a given interval or during a period of time. For example, an observation every 15 seconds for 30 minutes is commonly specified. The acceptance criteria are set in the contract. Local regulations may provide guidance on the required acceptance levels.

7.2.0 Method B – Ambient Air Monitoring for PM-10

The emission of particles 10 micrometers (0.39 mils) or less in size (PM-10) is regulated by the EPA under the *National Ambient Air Quality Standards* and varies from location to location. The particulate size monitored depends on the filter used in the monitor.

Typically, high-volume air samplers equipped with PM-10 monitors (*Figure 23*) are used to measure emissions. The number and location of the monitors depends on the wind direction and the proximity to homes, playgrounds, bodies of water, and other sensitive areas. These monitors operate continuously throughout the duration of the project. Monitoring should start a few days before the work begins to establish background levels. If the background levels vary significantly, continuous background monitoring may be necessary. Total solid particulate (TSP) monitors are also available and may be required in some areas.

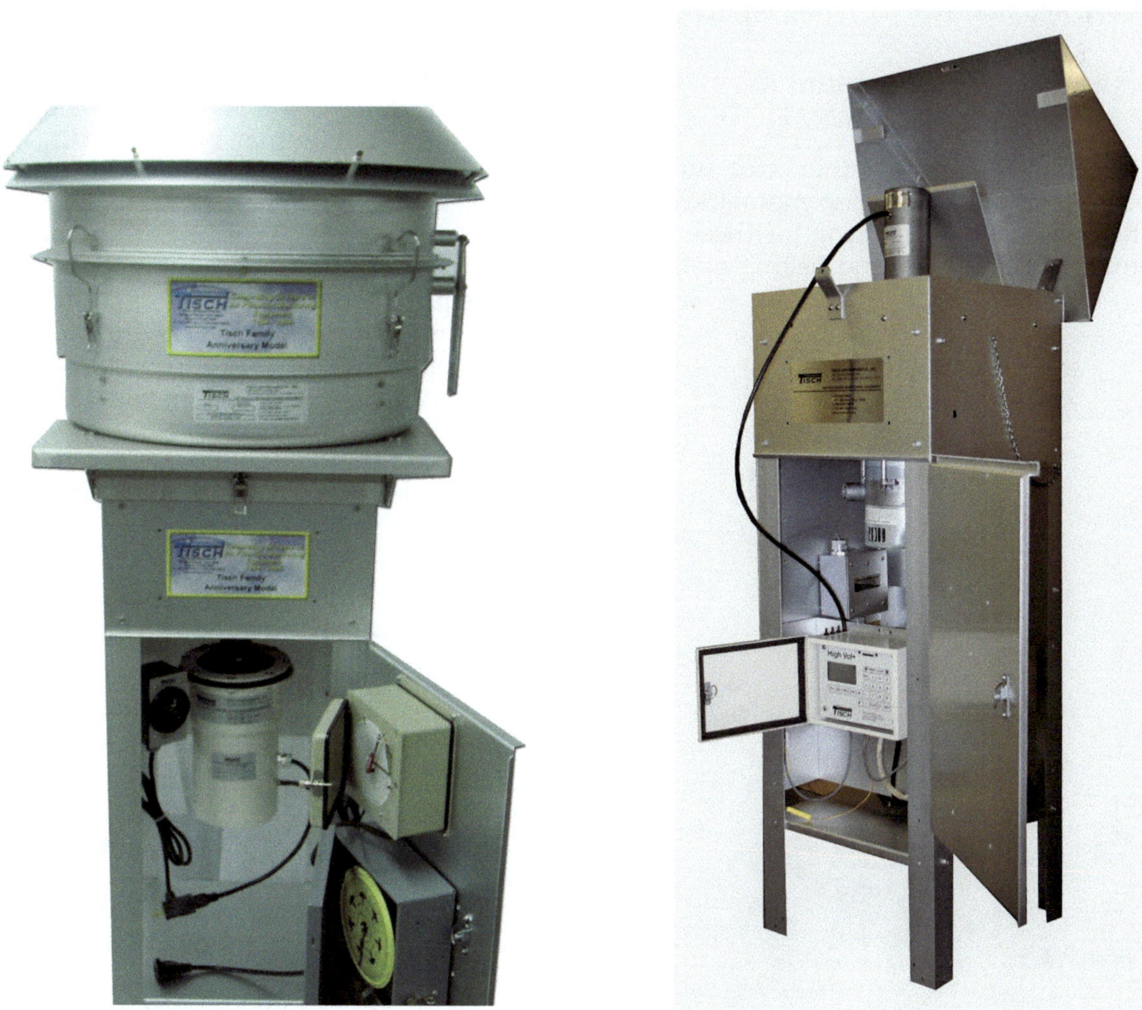

204F23.EPS

Figure 23 PM-10 monitoring instruments.

7.3.0 Method C – Occupational Monitoring of Area Emissions for Lead

Lead dust is generated when lead coatings are removed, unless it was stabilized prior to removal. The inhalation of lead dust is harmful to workers. OSHA requires a monitoring program for airborne lead emissions. Lead monitoring must be performed in accordance with regulatory requirements.

There are several acceptable methods for lead monitoring. High-volume air sampling equipment can be used to continuously monitor lead emissions. Personal monitors located outside of the containment area can also be used (*Figure 24*).

The action levels for lead are 30 micrograms per cubic meter of air, according to OSHA standards. Additional OSHA requirements must be followed, including isolating the area with signs and barricades when lead levels have been exceeded. The additional areas that must be monitored and isolated include the containment area, dust collector, and abrasive recycling equipment.

7.4.0 Method D – EPA Ambient Air Monitoring for Toxic Metals

The EPA regulates lead and other toxic metals. When coatings containing toxic metals are removed or applied, air monitoring for these metals is required. The location and testing methods should be based on wind direction and proximity to sensitive areas, such as homes, playgrounds, and bodies of water. Background monitoring for the duration of the project may also be required.

High-volume air sampling equipment is equipped for the collection of total suspended particulates. These are analyzed for lead and other toxic metals. The *National Ambient Air Quality Standard* for lead is 0.15 micrograms per cubic meter as a 90-day average. Because paint removal and coating operations are not typically conducted over a 90-day period, daily criteria must be established. *SSPC Guide 6* contains formulas for calculating the daily average. Testing methods and acceptance criteria should be set under the direction of authorities having jurisdiction.

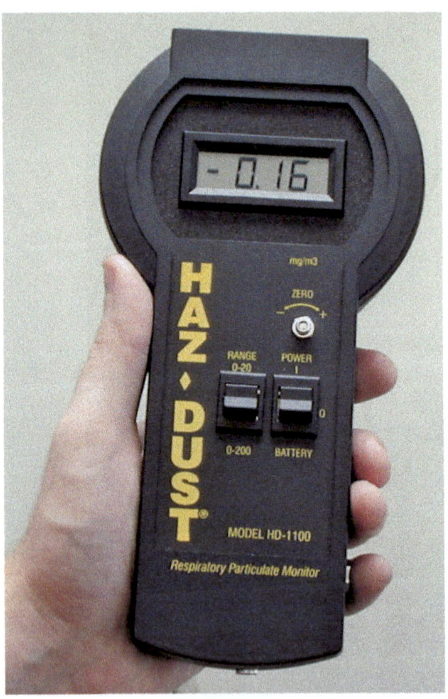

204F24.EPS

Figure 24 Personal lead monitoring equipment.

7.5.0 Method E – Soil Analysis for Toxic Metals

When coatings containing toxic metals are removed or applied, the soil under the project area must be tested. The soil should be tested before the project begins to establish background levels. The soil should also be tested after project completion to determine the effectiveness of the containment. If the soil is contaminated with toxic metals, remediation may be required.

Before the project begins, select several test sites beneath the structure and several from 10 to 100 feet away from the structure in each of the four directions. Long structures, such as bridges, may require additional test sites. Record the specific location of each test site. Sample the soils and have them tested at an approved testing laboratory. After the project has been completed, resample the same areas and have the new samples tested. Any allowable increase should be determined under the direction of authorities having jurisdiction.

7.6.0 Method F – Water and Sediment Analysis for Toxic Metals

When coatings containing toxic metals are removed or applied, bodies of water near the project area may need to be tested. This testing method is also a comparison of the level of toxic metals before and after the job. Water and sediment samples in adjacent water bodies should be taken before the job begins. Record the sampling locations carefully. Resample these areas after completion of the project. The samples must be sent to an approved testing laboratory for analysis. Any allowable increase should be determined under the direction of authorities having jurisdiction.

Water sampling may only be beneficial for sedentary water bodies or if drinking water intake is nearby. Lead and other toxic metals would be swept away in fast moving water, and the results would be inconclusive.

7.7.0 Method G – Visual Assessment of Site Cleanliness

Visual assessment of site cleanliness is a straightforward method that should be performed daily. Before the project begins, inspect the area for any existing debris, and have it removed before the project begins. Starting with a clean site decreases possible accidents and injuries from trip hazards. It also makes it easier to identify any new emissions. Any questions about the responsibility for cleanup must be resolved in advance. If possible, take photographs of the area to document the conditions before the job begins.

The site should be inspected daily, usually after operations have been completed. Check areas around the containment enclosure for any emissions. Pay particular attention to joints, seams, entryways, debris collection areas, and ventilation exhaust areas. Any emissions should be noted and cleaned up. Modify the containment system to reduce or eliminate any emissions. It is important to perform the inspections daily so that any leaks or damaged containment can be fixed as soon as possible. Inspect the area at the end of the project, after the containment structure has been removed. The area should be in better condition than it was before the project started. Attention to site cleanliness is important for the firm's reputation.

7.8.0 Containment Integrity and Protrusions

Testing containment effectiveness is unproductive unless the results are reported and measures are taken to correct problems. Check with a supervisor about the procedures for recording and reporting test results. Prompt corrective action reduces costly, time-consuming cleanup when coating operations are completed. A simple check for negative- or positive-pressure containment is to look for a concave or convex appearance of the walls, floors, or ceilings. This check can be done at any time, and it provides a visible indication that ventilation is occurring.

Materials that are accidentally released from containment enclosures are called emissions. Parts of the structure that extend beyond the

On Site

Containment Integrity Monitoring

For hazardous emissions, job specifications and/or authorities having jurisdiction may require periodic monitoring and data logging of negative containment air pressure to ensure that adequate pressures are maintained for ventilation and emissions capture. This is especially true for rigid containment that cannot be visually checked for negative pressure. The preferred method uses a data-logging differential-pressure monitor so that a permanent record of the containment integrity is created. The unit shown here can measure differential pressure between containment interior and exterior air pressures to as low as 0.001" of water column (wc) (0.25 Pa) and display the result in wc or in Pascals. Typically, pressure differentials of 0.02" to 0.04" wc (5 to 10 Pa) are required for negative-pressure containment. The time- and date-stamped data-logging function can range from once every second to once every 12 hours as required. For nonhazardous emissions, positive-pressure containment may be used in some rare instances. Pressure monitors can also be used for periodic positive-pressure monitoring. For negative or positive containment monitoring, a decreasing pressure differential signals that additional leaks are beginning to occur or that exhaust filters are becoming clogged.

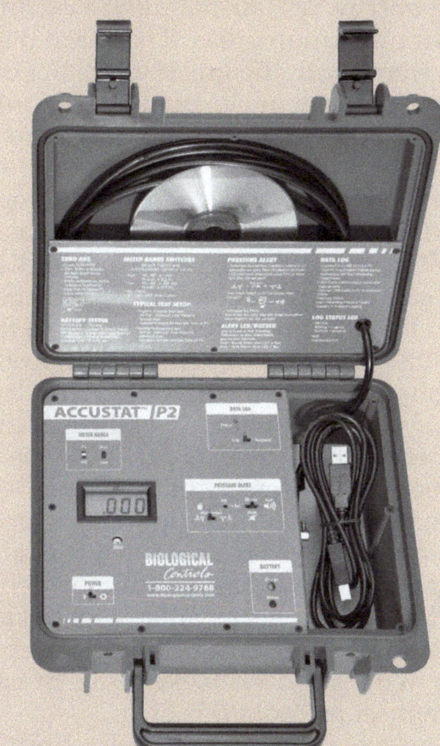

204FSA02.EPS

containment, such as pipes, are called protrusions. Protrusions must be properly sealed when the containment is erected to prevent emissions. Breaches in containment integrity can release hazardous (fugitive) emissions. Fugitive emissions must be cleaned up and may result in contract penalties. In addition, there are significant fines and penalties for exceeding EPA limits on emissions of hazardous materials, which include metals such as lead, cadmium, and chromium. The release of any amount of hazardous materials is considered a spill and must be reported to federal authorities. Cleanup of contaminated water or soil is costly. Monitor emissions closely, and adjust the containment system to minimize emissions.

Weather, winds, and surface preparation operations can damage containment. Sharp edges and pointed objects, such as bolts, can require extra covering to prevent damage to containment materials. Maintain containment integrity by inspecting each component regularly. Make sure that each one is functioning properly. Replace damaged or worn containment materials promptly. Reseal joints and protrusions if they begin to leak. Some contractors use expanding foam or silicone sealants to seal joints. On longer projects, the entire containment structure may need to be replaced. Report problems to the supervisor, who will decide whether to repair or replace the component or system.

Fugitive emissions occur most frequently at joints, entryways, air handling systems, and debris accumulation areas. These are the points in the systems with the greatest interaction between the containment enclosure and the outside environment. Make sure that these areas are inspected daily.

Controlling excessive emissions from the dust filtration system may be as simple as changing the filters more frequently. Air handling equipment can become fouled and may need cleaning or service. Check the volume of air in the exhaust and intake systems. An increase in negative pressure should control emissions from the air handling system. The capacity of the filtration system may need to be increased to handle the treated air volume and the density of the dust.

The debris accumulation area can easily become overfilled. Make sure that this area is large enough to handle the volume of debris generated. Have containers available for packing debris for disposal. Regular or daily cleanup of debris, along with good housekeeping practices, will limit emissions from this area.

Good work practices can also limit emissions from entryways. Make sure that dust and debris are removed from clothing before exiting the containment structure. If necessary, designate a decontamination area near the doorway. Tools and equipment should also be cleaned before they are removed from the containment area.

The containment structure may need to be modified to reduce excessive emissions or raise the containment level. For example, if the entryway is an open seam, add overlaying door traps. Flaps or baffles over air intake points may be needed to reduce the release of dust.

At the end of the project, review the test results. Assess the overall effectiveness of the containment structure and materials. In areas where containment was less effective, consider alternative methods and materials for future projects. Containment systems are complex structures with many components. Experience is required in order to build and maintain successful containment systems.

8.0.0 Environmental Humidity/Temperature Control

Environmental conditions can have a significant impact on cleaning and coating operations. These factors include air temperature, relative humidity, dew point, wind velocity, and airborne contaminates. Corrosion protection, coating application, appearance, and rate of curing are all affected by relative humidity and temperature and by the surface temperature of the structure. If the temperature is too low, for example, the coating will not dry or cure properly. If the temperature is too high, the coating will not wet the surface, which will result in poor film formation. Moisture also affects surface preparation and coating application. Moisture in the air condenses if surface temperatures fall below the dew point. This can cause rust-back of dry abrasive blasted steel. Moisture between coats can cause premature coating failure.

The maximum service life of coatings is based on ideal conditions for surface preparation and coating application. Coating manufacturers recommend upper and lower temperatures and relative humidity limits for application on the coating technical data sheet. However, ideal conditions are difficult to achieve. Outdoors, projects are subject to natural weather cycles and possible dew formation at night. For indoor projects, plants must often remain functioning during coating operations. Different areas within the plant can have very different conditions. Temperatures may be extremely high near the ceiling. Around certain machines, such as paper machines, the relative humidity can reach nearly 100 percent.

The surface conditions should not be allowed to deteriorate between preparation and coating application. Optimally, ferrous surfaces should be coated when the surface condition is indicated in the job specification to prevent rust-back. In some areas, the standard practice is to prepare an area and coat it in the same day. A limited area is prepared and coating applied. The disadvantage of this method is that the crew must change from surface preparation to coating application operations each day. Setup and cleanup of both operations can be time consuming.

The time between surface preparation and coating application can be extended if the area is kept warm and dry. Temperature and humidity control equipment is incorporated as part of the containment system to maintain optimal surface conditions for coating application. This equipment is also needed in severe environments that have excessively high or low temperatures or high humidity. For wet abrasive blasting or hydroblasting, **flash rusting** can occur within one-half to two hours or longer after cleaning with water, unless a coating-compatible inhibitor is used. However, some research suggests that light flash rusting does not cause coating failures.

8.1.0 Dehumidifiers

There are several advantages to using dehumidification equipment. Crews can work in the most favorable conditions, optimizing surface preparation and coating application. This extends the time for applying the coating. The work season may begin earlier in the spring and continue later in the fall. The project is not at the mercy of weather conditions, and project completion time can be more accurately predicted.

> **NOTE**
> Dehumidification can mask the presence of chlorides on steel. The substrate should be tested for the presence of chlorides after surface blasting. Nonvisible contaminants can cause premature coating failure.

There are two types of dehumidifiers, refrigeration dehumidifiers and absorption-type **desiccant** dehumidifiers. Refrigeration dehumidifiers cool the air and collect the resulting condensation. Absorption-type, or desiccant, dehumidifiers use a material with a high affinity for water to remove moisture from the air.

Refrigerant dehumidifiers work on the principle that colder air holds less moisture. A typical cycle of refrigeration and dehumidification is shown in *Figure 25*. A refrigerant gas is pumped through a closed pressurized system. The compressor pressurizes the gas, which heats it. The hot gas is cooled into a liquid in the condenser.

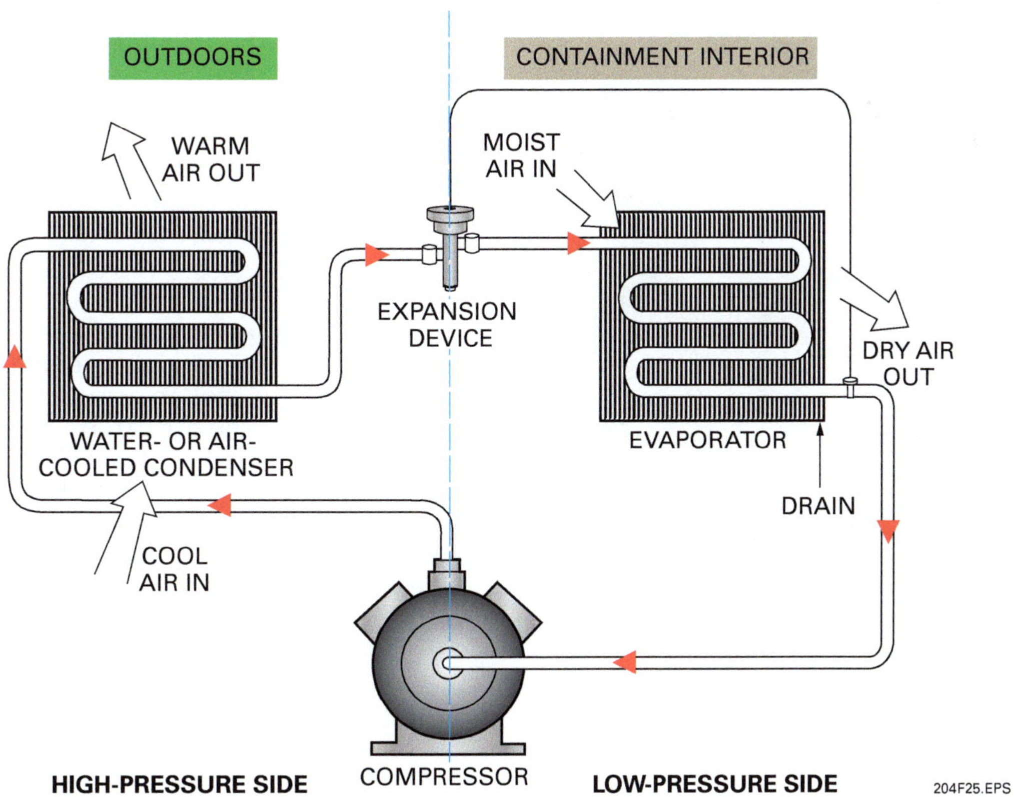

Figure 25 Basic refrigeration/dehumidification cycle.

As the liquid passes through a valve, it expands and cools to just above freezing. The humid air is drawn over the evaporator and is cooled. Moisture in the colder air condenses on the evaporator fins and is drained away from the system.

In a desiccant dehumidifier (*Figure 26*), the humid air is drawn through the desiccant material. The desiccant attracts and retains some of the water vapor. The drier air is then released. The desiccant is circulated to another area where it is exposed to heated air. The heated air draws off the moisture, recharging the desiccant. In general, a desiccant dehumidifier has four major components: a desiccant holder, a process air fan, a regeneration air fan, and a heater.

One of the limitations of a refrigerant dehumidifier is that it does not work well in colder temperatures. At air temperatures of 45°F (7°C) or lower, moisture may freeze on the evaporator fins, causing them to work less efficiently, if at all.

204F26.EPS

Figure 26 Desiccant dehumidifier.

Also, refrigerant dehumidifiers lower the temperature of the exhaust air. Desiccant dehumidifiers remove moisture without freezing and produce dry air that is warmer than the ambient air. Desiccant dehumidifiers are used more often in the spring, fall, and winter. Refrigeration dehumidifiers are used more often in the summer.

8.2.0 Heating Units

In some climates, heating units are needed to maintain the temperature in the enclosure at optimum levels for coating application. Heaters can be part of the ventilation or dehumidification system. Portable vented heaters are often used. Typically, heaters use electricity, steam, or bottled propane. They are generally operated in banks to cover an entire surface.

CAUTION

Large frame structures, such as bridges and towers, are too big to heat effectively. They should not be coated in cold temperatures.

Always use caution with heaters that burn fuel oil or propane to avoid the buildup of carbon monoxide. The exhaust from these units can be a health hazard and can interfere with the proper curing of some coatings. Heaters can introduce moisture as a byproduct of combustion. The exhaust can introduce contamination, and the heating will not be even.

WARNING

Carbon monoxide is a gas produced by combustion. Breathing carbon monoxide can cause serious illness or death.

Dust collectors circulate a large volume of air in and out of the containment area. In the winter, this can rapidly cool an area. If the materials are not hazardous, the air can be recirculated. This allows the warmed air from the enclosure to be reused after filtration. An additional heater may be necessary to heat the space.

SUMMARY

Surface preparation and coating application projects can produce a significant amount of dust and debris that must be controlled. Containment enclosures limit the spread of dust and debris to the surrounding environment. Containment systems can be complex structures that include the containment enclosure, debris collection and ventilation equipment, dehumidifiers, dust filtration, and monitoring systems.

The type of operations involved and the materials removed or applied affect the selection of the type of system developed and the materials used. Greater levels of containment and monitoring are required when toxic materials, such as lead, are being removed or applied. Construct containment systems using the methods described in *SSPC Guide 6*. Certain types of containment and materials are inappropriate for some operations. Be sure to select materials and equipment that will adequately contain the types of debris generated. The contained debris must be collected and packaged for disposal or recycling. Air handling and air conditioning equipment may be required to raise or lower temperatures, decrease humidity, or remove dust.

Finally, the effectiveness of the containment system must be measured. Any emissions should be cleaned up and the system modified to reduce or eliminate emissions. By following these procedures, safety problems and environmental violations can be minimized. In addition, leaving the job site in better condition than it was before the project began will enhance the firm's reputation and your own status as a professional applicator.

Review Questions

1. Which surface preparation technique produces the most dust and debris?
   a. Chemical stripping
   b. Power tool cleaning
   c. Abrasive blasting
   d. Waterjetting

2. Long-term exposure to solvents can cause _____.
   a. lung disease
   b. kidney damage
   c. unconsciousness
   d. convulsions

3. The guidelines for performing certain tasks detailed in industry standards are _____.
   a. required by law
   b. enforced by OSHA
   c. optional practices
   d. written into contract specifications

4. In the United States, *OSHA Standard 29 CFR 1926.62* sets specific requirements for worker exposure to _____.
   a. airborne dust
   b. toxic metals
   c. solvent vapors
   d. lead

5. Using a tool with a vacuum shroud to control lead dust is an example of recommended _____.
   a. housekeeping practices
   b. engineering controls
   c. ventilation systems
   d. good work practices

6. Localized collection at the point of generation reduces dust generation but also reduces the _____.
   a. amount of debris collected
   b. quality of surface preparation
   c. size of the cleaning area
   d. amount of time required

7. An example of general area debris collection is _____.
   a. a channeling system
   b. wet abrasive blasting
   c. a dust filtration system
   d. a vacuum-shrouded needle descaler

Match the containment classification with the corresponding *SSPC Guide 6* code.

8. _____ A moderate level of emissions control for waterjetting
9. _____ The highest level of emissions control for abrasive blasting
10. _____ A high level of emissions control for power tool cleaning
11. _____ A minimal level of emissions control for water blasting
12. _____ The highest level of emissions control for chemical stripping
13. _____ A moderate level of emissions control for power tool cleaning

    a. Class 1A
    b. Class 1C
    c. Class 2P
    d. Class 3P
    e. Class 3W
    f. Class 4A
    g. Class 4W

14. *SSPC Guide 6* includes four tables that list _____.
    a. containment and ventilation requirements for each class
    b. types of containment materials and their applications
    c. opacity levels and allowable emissions
    d. dust filtration requirements

Match the containment enclosure components with the corresponding *SSPC Guide 6* Type.

15. _____ Chemical resistant containment materials

16. _____ Flexible support structures

17. _____ Partially sealed joints

    a. Type A1
    b. Type B4
    c. Type C2
    d. Type D2

18. In a negative pressure system, _____.
    a. more air is pumped from the system than is supplied
    b. more air enters the system than is pumped out
    c. air is pumped in through the exhaust system
    d. the air supply is well balanced

19. Air pumped into a containment enclosure is called _____.
    a. fresh air
    b. exhaust air
    c. resupply air
    d. makeup air

20. The most appropriate testing methods for containment are determined by _____.
    a. daily test results
    b. visual observation of emissions
    c. testing soil and water samples
    d. methods set by local and state authorities

Trade Terms Introduced in This Module

Desiccant: Material that attracts and retains water vapor.

Emissions: Materials that escape the containment area, including airborne plumes of material, abrasives, paint chips, debris, spills, or leaks of water.

Entryway: The area of a containment structure by which workers may enter and leave a containment and ventilation system.

Flame-resistant: Materials that resist burning for a while but that continue to burn once ignited.

Flame-retardant: Materials that stop burning once the source of fire has been removed.

Flash rusting: The rusting that occurs from the time the waterjetting process starts to the time the water used in the process dries.

Forced airflow: Air that is supplied to a containment area through the use of fans and blowers.

Impenetrable: Materials that dust and wind will not flow through or penetrate.

Impermeable: Materials that water will not flow through or penetrate.

Makeup air: Air that replaces the air that has been exhausted from the containment enclosure.

Negative pressure: A condition in an enclosure in which exhaust air exceeds input air.

Support structure: The framework upon which the containment material is mounted.

Additional Resources

This module presents thorough resources for task training. The following resource material is suggested for further study.

Coating Inspector Program Level 1. © NACE International, 2010.

EPA Fact Sheet EPA-747-F-98-006, Generators' and Transporters' Responsibilities for Management and Disposal of Lead-Based Paint Debris. United States Environmental Protection Agency (7404), December 1998.

Industrial Lead Paint Removal Handbook. Vol. 1. 2nd Edition. Kenneth A. Trimber. Pittsburgh, PA: The Society for Protective Coatings, 1993 [*SSPC 93-02*].

OSHA Standard 1910.1025, Toxic and Hazardous Substances, Appendix A, Substance Data Sheet for Occupational Exposure to Lead. Occupational Safety & Health Administration.

OSHA Technical Manual. Controlling Lead Exposures in the Construction Industry: Engineering and Work Practice Controls. Occupational Safety & Health Administration, January 1999.

PaintSquare: Portal to the *Journal of Protective Coatings and Linings* and the *Journal of Architectural Coatings* (JPCL and JAC). www.paintsquare.org.

Preventing Lead Poisoning in Construction Workers: NIOSH Alert. DHHS (NIOSH) Publication No. 91-116a, April 1992.

Paints & Coatings Resource Center (PCRC). www.paintcenter.org

SSPC Course C-3, Lead Paint Removal. www.sspc.org.

SSPC Course C-5, Lead Paint Removal Refresher. www.sspc.org.

SSPC-Guide 6, Guide for Containing Surface Preparation Debris Generated During Paint Removal Operations. www.sspc.org.

SSPC-Guide 7, Guide for the Disposal of Lead-Contaminated Surface Preparation Debris. www.sspc.org.

SSPC-Guide 18, Specifier's Guide for Determining Containment Class and Environmental Monitoring Strategies for Lead-Paint Removal Projects. www.sspc.org.

SSPC-Guides on Environmental Protection. www.sspc.org.

Figure Credits

Photo courtesy of Hippwrap Containment, Module opener

Eagle Industries, 204F01, 204F10, 204F22

Topaz Publications, Inc., 204F02, 204F11–204F15, 204F17–204F19

Sherwin-Williams Co., 204F03

Photo courtesy of Bullard, 204F04

CS Unitec, Inc., 204F05

Nilfisk-Advance America, Inc., 204F06–204F08

SSPC: The Society for Protective Coatings, 204T01-204T08

Picture courtesy of Detroit Tarp Inc., 204F09

Courtesy of Reef Industries, Inc., 204F16

W.W. Grainger, Inc., 204F20, 204F21, 204SA01 (plastic tube kit)

TSI Incorporated, 204SA01 (anemometer)

Tisch Environmental Inc., 204F23

SKC Inc., 204F24

ACCUSTAT P2, portable room differential pressure monitor manufactured by Biological Controls, 204SA02

ARID-Dry by CDIMS, 204F26

NCCER CURRICULA — USER UPDATE

NCCER makes every effort to keep its textbooks up-to-date and free of technical errors. We appreciate your help in this process. If you find an error, a typographical mistake, or an inaccuracy in NCCER's curricula, please fill out this form (or a photocopy), or complete the online form at **www.nccer.org/olf**. Be sure to include the exact module ID number, page number, a detailed description, and your recommended correction. Your input will be brought to the attention of the Authoring Team. Thank you for your assistance.

Instructors – If you have an idea for improving this textbook, or have found that additional materials were necessary to teach this module effectively, please let us know so that we may present your suggestions to the Authoring Team.

NCCER Product Development and Revision

13614 Progress Blvd., Alachua, FL 32615

Email: curriculum@nccer.org
Online: www.nccer.org/olf

❑ Trainee Guide ❑ Lesson Plans ❑ Exam ❑ PowerPoints Other ____________________

Craft / Level: Copyright Date:

Module ID Number / Title:

Section Number(s):

Description:

Recommended Correction:

Your Name:

Address:

Email: Phone:

Surface Preparation Two

69205-10

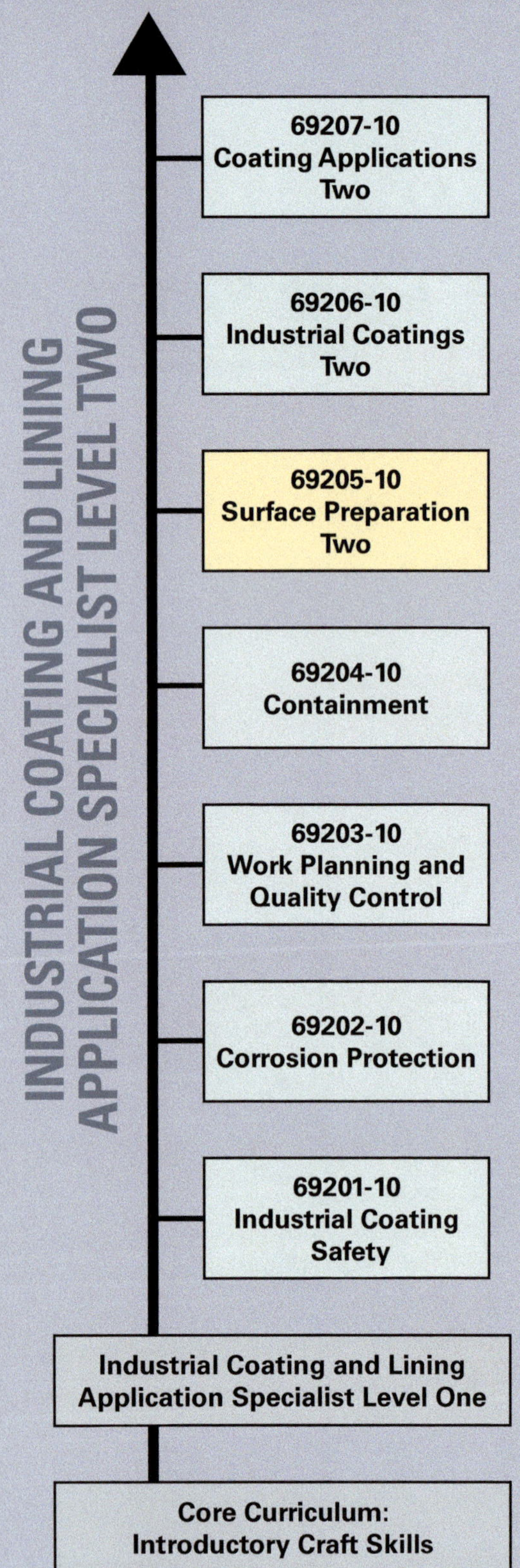

This course map shows all of the modules in *Industrial Coating and Lining Application Specialist Level Two.* The suggested training order begins at the bottom and proceeds up. Skill levels increase as you advance on the course map. The local Training Program Sponsor may adjust the training order.

Objectives

When you have completed this module, you will be able to do the following:

1. Identify and document surface conditions of steel substrates as depicted in the SSPC/NACE/ISO visual guides.
2. Examine and document preparation requirements for steel, nonferrous metals, nonmetallic surfaces, synthetic coatings, and existing coatings.
3. Demonstrate the use of a dull putty knife to confirm compliance with *SSPC-SP 2, SSPC-SP 3, NACE No. 4/SSPC-SP 7*, and *NACE No. 8/SSPC-SP 14*.
4. Demonstrate proper procedures for selecting, verifying the cleanliness of, and storing abrasives, including abrasive additives.
5. Explain how to maintain and troubleshoot abrasive blasting field equipment, and demonstrate field procedures for verifying cleanliness of abrasives and related documentation procedures.
6. Demonstrate procedures for measuring and documenting nozzle aperture, nozzle air pressure, and compressed air cleanliness.
7. Demonstrate the procedures for determining surface profile and surface cleanliness and associated documentation using a visual comparator, visual guide, dial surface profile gauge, replica tape, transparent tape test according to *ISO 8502-3*, ultraviolet light for detection of hydrocarbon contamination, and commercially available equipment to test for soluble salts.
8. Use the written standard and the accompanying visual guides (*NACE VIS 9/SSPC-VIS 5/* and *NACE VIS 7/SSPC-VIS 4*) to classify various examples of surface profiles.
9. Demonstrate the use and documentation of inspection procedures for concrete coating work using moisture meters, RH probes, anhydrous CaCl *ASTM F1869*, plastic sheet method *ASTM D4263*, ICRI surface roughness comparators, and concrete profiler putty.
10. Describe and demonstrate the proper use and documentation of environmental monitoring procedures using psychrometers, thermometers, and psychrometric tables.

Trade Terms

Feathering
Hydrocarbon
Hygrometer
Microsiemens
Patch test
Sieve test
Tie coat

Prerequisites

Before you begin this module, it is recommended that you successfully complete *Industrial Coating and Lining Application Specialist Level One* and *Industrial Coating and Lining Application Specialist Level Two*, Modules 69201-10 through 69204-10.

Contents

Topics to be presented in this module include:

Figures and Tables

1.0.0 INTRODUCTION

The goal of surface preparation is to provide the coating with a surface to which it can adhere. Although surface-coating application is the most visible part of the job, surface preparation is the most vital part of the industrial coating process. Eighty percent of all premature coating failures are related to improper surface preparation. An understanding of how coatings bond to surfaces is necessary for determining the proper surface-preparation method.

Coatings adhere best to surfaces that are clean of all contaminants, such as dirt, dust, grease, oil, and moisture. In addition, coatings bond best to surfaces that have some degree of roughness to them. This roughness is called surface profile, or tooth. Often the project specification will state the desired depth for the surface profile. This roughness helps to hold the coating to the surface in much the same way that a jigsaw puzzle is held together. On steel surfaces, surface profile is often called the anchor profile (*Figure 1*).

The most critical aspect of good coating bonding is surface preparation. In *Industrial Coating and Lining Application Specialist Level One*, a number of surface-preparation methods, such as using hand and power tools, abrasive blasting, waterjetting, and chemical stripping were covered. This module covers the appropriate selection of these surface-preparation methods.

Before beginning any surface-preparation work, inspect the surface to be prepared. The goal is to prepare the surface to the level specified in one of the industry specifications. Throughout this module, a number of visual industry guides will be used that are based on the associated written standards to prepare steel surfaces for coatings. These written standards take precedence over the visual guides:

- *SSPC-VIS 1, Guide and Reference Photographs for Steel Surfaces Prepared by Dry Abrasive Blast Cleaning*
- *SSPC-VIS 2, Standard Method of Evaluating Degree of Rusting on Painted Steel Surfaces*
- *SSPC-VIS 3, Guide and Reference Photographs for Steel Surfaces Prepared by Hand and Power Tool Cleaning*

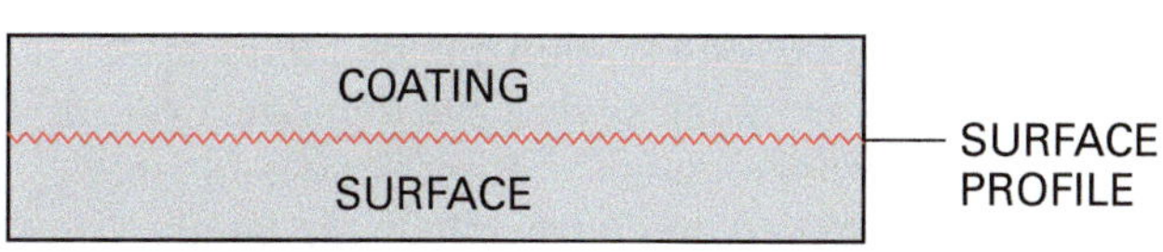

Figure 1 Coating bonding.

- *NACE VIS 7/SSPC-VIS 4, Guide and Reference Photographs for Steel Surfaces Prepared by Waterjetting*
- *NACE VIS 9/SSPC-VIS 5, Guide and Reference Photographs for Steel Surfaces Prepared by Wet Abrasive Blast Cleaning*
- *ISO 8501-1 through ISO 8501-4, Preparation of Steel Substrates Before Application of Paints and Related Products – Visual Assessment of Surface Cleanliness*

These guides, as described in *ASTM D2200*, are visual representations of the preparation requirements of the standards and will be available at training centers and on the job. Trainees will need to consult these specifications often. Ask instructors to point out the fine details in the pictures and to compare the pictures to an actual surface, noting similarities and differences.

2.0.0 SURFACE PREPARATION TESTS

The goal of surface preparation is to provide a clean, slightly roughened surface to which coatings will adhere. It is often difficult to determine visually whether a surface is adequately prepared for a coating, so a number of tests may have to be performed. Always record what test was performed, when it was performed, where it was performed, who performed it, and the results in the project documentation.

2.1.0 Dull Putty Knife Test

Sometimes the project specification will require that all old coatings be removed from a surface. Other times it will specify that old coatings that are tightly adhering to the surface may remain after testing for adherence with a dull putty knife. Vigorously scrape the coating with the blade of a dull putty knife, being careful not to gouge the surface or bend the blade. If the coating cannot be removed, it is considered to be tightly adhered to the surface.

> **NOTE**
>
> All industry standards call for the use of a dull putty knife, not a sharpened putty knife, to determine coating adherence. There is no standard for defining a dull putty knife; however, the SSPC Surface Preparation commentary states that a dull putty knife is a knife in new condition without sharpening.

2.2.0 Cleanliness

For coatings to bond tightly, surfaces must be free of dirt, dust, grease, oil, and other contaminants. Gross contamination is obvious visually, but it is often difficult to detect minute contamination. Testing for contaminants such as dust and **hydrocarbons** may be required. Hydrocarbons refer to any petroleum products.

To test for dust on metals or synthetics, use *ISO 8502-3* transparent test tape (*Figure 2*). Apply the prescribed length of tape to the surface and rub the tape (without moving it on the surface) to ensure firm adhesion. Remove the tape and mount it on a viewing card (such as an unlined white index card). Use the magnifier to view the number of dust particles on the tape, comparing your tape to the supplied sample (*Figure 2*). Repeat this test in various areas of the surface as specified in the project specification. Excessive dust can be cleaned with a tack cloth or by washing with water.

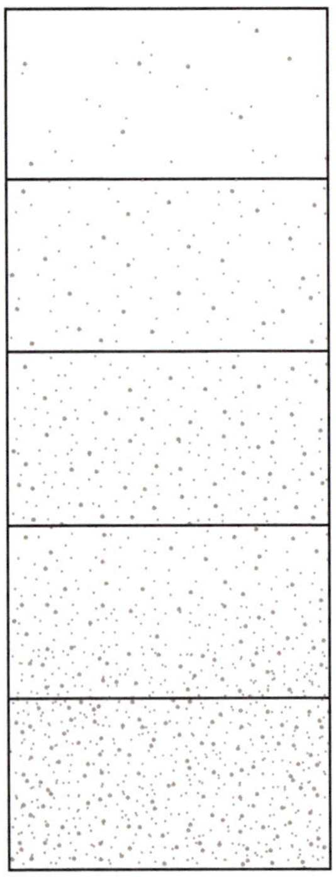
SUPPLIED SAMPLES

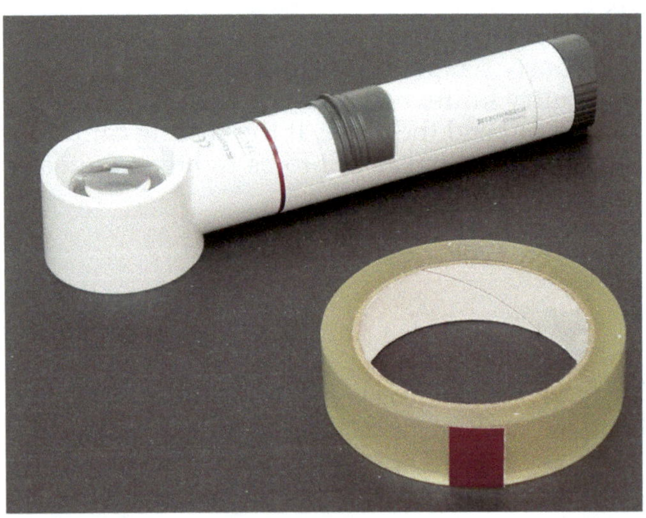
TRANSPARENT TAPE

205F02.EPS

Figure 2 *ISO 8502* transparent test tape.

To test whether metal or synthetic substrates are free of grease or oil, place a drop of water in various spots. If the water beads, as it would on a freshly waxed car, the surface is likely contaminated with grease or oil. To remove grease and oil, wash the surface with detergents, degreasers, or a permitted solvent.

To test for grease or oil on concrete, apply a few drops of hydrochloric acid solution, also known as muriatic acid. If the solution bubbles, the surface is clean. Hydrochloric acid is acidic and concrete is an alkaline or base. The bubbles are a chemical reaction between the acid and base. Light oil or grease contamination on concrete can be cleaned with detergents, solvents, or degreasers followed by a clean water rinse. Hydrochloric acid is not a degreaser. Heavy contamination can be removed with a scraper or by flame cleaning. Flame cleaning may melt the grease or oil and allow it to be deeply absorbed into the concrete. It may be necessary to replace heavily contaminated concrete. Concrete can be tested for oil or grease beneath the surface by shining a heat lamp onto the surface of the concrete. Locate the lamp 2 feet (0.61 meters) above the surface for 8 hours. As the concrete heats, the oil or grease will rise to the surface.

Oil and grease contamination can also be detected using an ultraviolet (UV) light. A UV light must be used in a darkened area. Shine the light onto the surface being tested. Oil or grease contamination areas will become fluorescent under the light and will appear to shimmer. Be careful not to shine the UV light into anyone's eyes. A momentary flash from the ultraviolet light can cause temporary blindness. Looking directly into the ultraviolet light can cause serious eye damage.

WARNING

Looking directly into unshielded ultraviolet lamps can cause severe eye damage. Consult the safety engineer or other knowledgeable person about any ultraviolet lamp being considered for use.

2.3.0 Soluble Salts Tests

Soluble salts are a major problem in the coating process. Unacceptable levels of soluble salts on any surface may cause premature coating failure. Soluble salts corrode metal, including rebar buried in concrete. There are a variety of commercial kits available for testing for soluble salts. Always follow the directions for the kit being used, but generally the steps to manually test for salts include the following:

Step 1 Determine the areas to be tested. Usually each test area will be approximately a 4-inch (10 centimeter) square.

Step 2 Dispense the prescribed amount of deionized or distilled water into a clean test tube or beaker. Be certain to use only deionized or distilled water during the test, because tap water may contain small amounts of salts.

Step 3 Moisten the patch or cotton ball from the kit with deionized or distilled water.

Step 4 Holding the patch or cotton ball with clean tweezers, vigorously scrub the test area. Don't use bare fingers to hold the patch or cotton ball because salts from skin can be transferred to the material.

Step 5 Swish the patch or cotton ball in the test tube or beaker of deionized or distilled water to rinse soluble salts out of the material.

Step 6 Dip the test strip into the rinse water and use the supplied scale to determine whether soluble salts are present.

Some test kits will substitute electrodes and a meter in place of test strips. In this case, the electrodes are placed in the rinse water and the test device is turned on. The device measures the amount of current flow through the solution. The higher the salt content of the solution, the more current flow there will be. The operation of metered kits varies, so always follow the manufacturer's instructions.

DID YOU KNOW?

pH

The pH scale measures a solution's alkalinity or acidity. An alkali is sometimes called a base. A strong acid will have a pH well below 7. Deionized or distilled water has a pH of 7.0 and is considered neutral, meaning it is neither acid nor base. A strong base has a pH well above 7. Any solution with a pH below 7 is an acid; any solution with a pH above 7 is a base.

On concrete, test for the presence of salts with universal indicator paper. Universal indicator paper measures the pH of a solution between the values of 0 to 14 based on a range of color changes. The pH indicates whether the surface is alkaline or acidic. If the pH level is below 10, it could indicate salt contamination. This test is not as accurate as the above tests because a pH level below 10 could also mean that the concrete surface has been exposed to an acidic solution during a cleaning process. However, this test is still valuable for concrete surfaces because some coatings bond better based on the pH of the concrete, which should be tested as described in *ASTM D4262*. The results of this test will determine if the surface needs to be treated to achieve the desired pH.

When soluble salts are detected, they must be removed per the specifications. Solvents will not dissolve salts. A vigorous detergent wash and thorough rinse with clean water may remove them, but it is more likely that the surface will need to be washed with a commercial product that is specially designed to dissolve salts.

2.4.0 Moisture Tests and Controls

Moisture on any surface can cause coatings to fail. Both metal and concrete are affected by moisture, so steps must be taken to keep moisture at the desired level. One common way to control moisture on metal and in concrete is by using dehumidification (DH) units. These units must be used in a

DID YOU KNOW?

Water

While water is a very poor conductor of electricity, don't stand in a puddle of water when using a power tool. Pure water is a poor conductor, but most water is not pure. Even clean tap water is full of harmless impurities that allow current flow. The more impurities in the water, the more current flows. Remember that the next time you are tempted to stand in a muddy puddle to grind just one more spot.

confined area. They control the humidity in the air and can also dry out a saturated concrete surface.

Moisture on bare metal can cause flash rust or bloom to develop in as little as 30 minutes. This is usually a problem when waterjetting or wet abrasive blasting bare metal. For this reason, rust inhibitors are often used with water washing. Some rust inhibitor products also remove salts and other contaminants. Irregular development of flash rust on a surface indicates the surface is still contaminated. Rust inhibitors will become part of the coating system and may not be compatible with some coatings that will be applied. Always check the coating PDS or manufacturer to determine if a particular inhibitor is acceptable.

Moisture in concrete is problematic. Too little moisture in concrete can prevent the coating from bonding tightly to the surface. Too much moisture can lead to coating failure and promote corrosion of metal rebar in the concrete. The type of coating to be applied determines the desired moisture content. There are a number of ways to measure the moisture in concrete. The first is to use a moisture meter (*Figure 3*) or a relative humidity (RH) meter (*Figure 4*). A moisture meter measures the amount of moisture in the concrete surface, while an RH meter measures the moisture relative to the air temperature. Some of these devices are tiny (*Figure 4*) and can be left in the slab permanently or semi-permanently (*ASTM F2170*).

Another way to measure moisture in a concrete surface that has been used in the past, but is not currently recommended, is the plastic sheet method (*ASTM D4263*). In this method, a plastic sheet with a minimum thickness of 4 mils (0.1 mm) is taped to the concrete surface. The sheet is allowed to sit for at least 16 hours. When the sheet is removed, the underside of the plastic and the concrete surface are inspected for moisture.

Another method of assessing moisture in concrete is the calcium chloride test (*ASTM F1869*). Calcium chloride easily absorbs moisture, so a weighed amount of calcium chloride is placed on the surface of the concrete and then covered with a plastic dome. This separates the calcium chloride from moisture in the air. It is allowed to sit for the time stated in the project specification. After this time, the calcium chloride is removed and weighed again. The difference between the dry and wet weights is moisture. This is applied to a scale to find out the moisture content of the concrete.

2.5.0 Nonporous Surface Discontinuity Tests

A job specification may require that the surfaces of nonporous materials such as ferrous metals, nonferrous metals, glazed ceramic, and certain plastic materials be inspected for discontinuities. The specification may require that this nondestructive inspection be conducted in accordance with *ASTM E165*. Discontinuities are surface-breaking defects resulting from cracks, seams, laps, cold shuts, shrinkage, lamination, through-leaks, or lack of fusion. Liquid penetrant inspection (LPI), also called dye penetrant inspection (DPI), uses two types of penetrant based on sensitivity. One is a visible type, usually red in color, which is used for low-sensitivity inspections us-

205F03.EPS

Figure 3 Moisture meter.

205F04.EPS

Figure 4 RH meter.

ing a visible light source. The other is a fluorescent type that, when excited by a UV light source, causes the dye to produce a brilliant yellow-green light against a dark background. Fluorescent dyes are more sensitive to small defects and usually require darkened ambient lighting conditions unless a very high-power UV light source is used.

Application and observation procedures for visible or fluorescent dyes are essentially the same. First, the surface must be cleaned to remove any surface contaminants and loose scale or coatings that could cause irrelevant or false indications. Any cleaning methods or materials used must be compatible with the penetrant. Then the penetrant is applied by brush or spray and allowed to soak into any flaws (5 to 30 minutes). The time allowed depends on the penetrant, the material being tested, and the size of the flaws sought. After the soak-in time is reached, the penetrant is removed from the surface by a compatible cleaning material (water or solvent based). After removal of the penetrant, a compatible dry or wet developer is applied to the surface. The developer draws the dye from the discontinuities to the surface of the material for inspection purposes. Using the appropriate light source, inspection of the surface should take place after a 10-minute development time. If too much time is allowed to elapse, the indications may bleed out and hinder the inspection. If a coating will be applied after inspection, the surface should again be cleaned and prepared by methods compatible with the coating.

WARNING

Do not look directly into an unshielded ultraviolet lamp or shine it at anyone else. Doing so can cause severe eye damage. Consult the safety engineer or other knowledgeable person about the safety considerations for any ultraviolet lamp that is to be used.

2.6.0 Surface Profile Tests

Surface profile and surface cleanliness are the two most important aspects of surface preparation. Surface profile is achieved on metals and synthetics by mechanical means or abrasive blasting. Surface profile on concrete is achieved by abrasive blasting or etching with a chemical such as hydrochloric acid. The standard test methods for field measurement of surface profile of abrasive blast cleaned steel are described in *ASTM D4417* and *NACE SP 0287.* The tests must be conducted in ordered areas across the entire blast area and the results averaged.

Surface profile can be measured by one of the following methods:

- Surface profile comparators (*Figure 5*) and lighted magnifier
- Replica tape (*Figure 6*) and spring-loaded thickness gauge
- Dial and digital surface profile gauges (*Figure 7*)
- Profile putty (*Figure 8*)

Each comparator has 5 profile grades ranging from 0.5 to 5.5 mils (13 to 140 microns) for the applicable abrasive profile. Using surface profile comparators requires skill. Comparators provide a variety of sample surface profiles. After selecting a comparator that matches the blast method used for the surface to be inspected, examine the surface through the center hole with a lighted magnifier (usually a five to seven times magni-

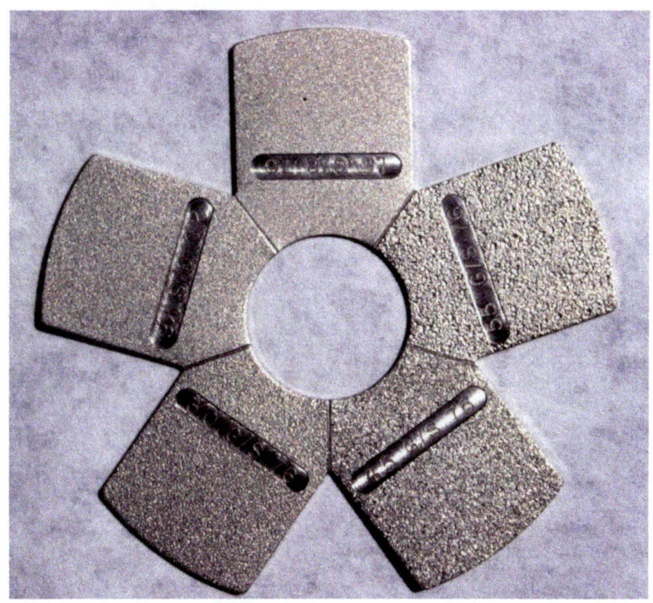

205F05.EPS

Figure 5 Surface profile comparators.

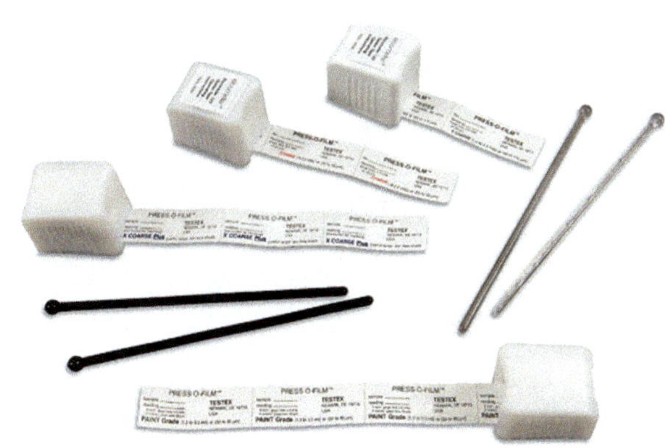

205F06.EPS

Figure 6 Replica tape.

fier). By comparing the surface to the comparator, determine the grade of the surface profile by selecting the comparator profile that is nearest to the profile of the blasted surface.

Surface profile can also be measured with replica tape. One side of the replica tape is a thick Mylar film, while the other side is an easily moldable foam surface that is pressed against the substrate surface to form an impression of the surface profile. There are two types of tape that are commonly used. One is coarse, for 0.8 to 2.0 mils (20 to 50 microns) surface profile. The other is extra coarse, for 1.5 to 4.5 mils (40 to 115 microns) surface profile.

A piece of tape is applied to the surface, foam side down. A hard, rounded object, such as a clean burnishing stick or swizzle stick, is used to crush the foam to the cleaned surface, causing the foam to form an impression of the actual surface profile.

The tape is removed from the surface and a spring-loaded thickness gauge of the proper diameter is used to measure the thickness of the foam and the plastic. The thickness of the Mylar film (2.0 mils or 50 microns) is subtracted from the micrometer reading, and the result is the depth of the surface profile.

A digital or dial surface profile gauge can also be used to measure surface profile. The instrument base rests on the tops of the surface profile peaks, while the spring-loaded tip projects into the valleys. For accurate measurements, surface profile gauges must be checked prior to use by pressing the gauge on a flat surface, such as glass, and checking for a reading of zero. When finished with a set of readings, the gauge should again be checked for a zero reading. Do not slide the gauge on a profiled surface when taking readings because this will dull the fine point of the plunger and result in erroneous readings. An advantage of the digital gauge is that it can store a large number of readings and perform automatic averaging. It also has wireless communication so that the data can be downloaded to computers for storage.

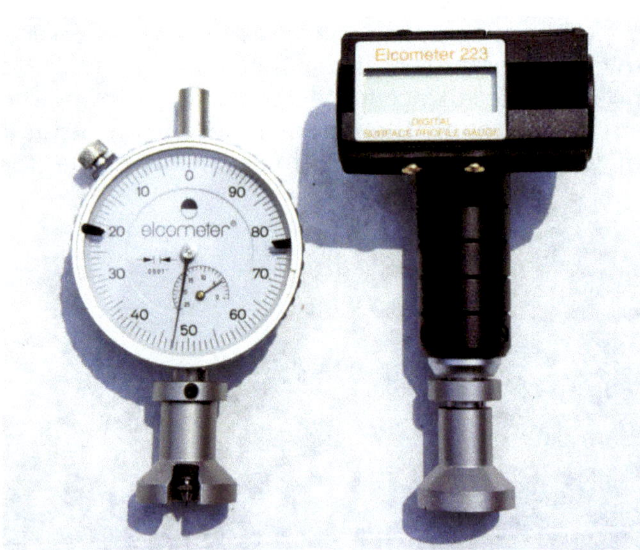

205F07.EPS

Figure 7 Dial and digital surface profile gauges.

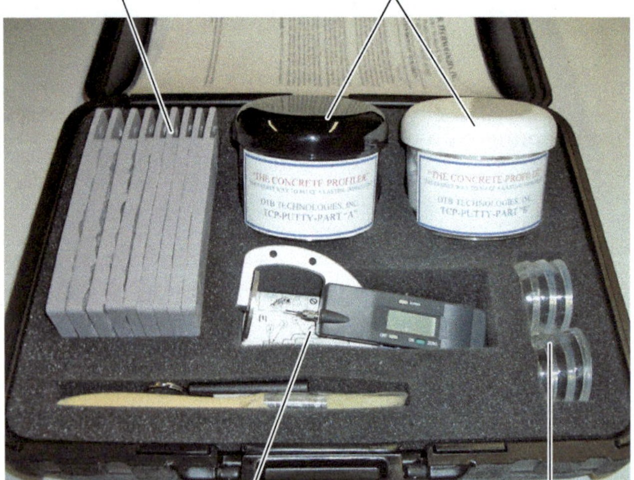

205F08.EPS

Figure 8 Concrete profiler kit.

As shown in *Figure 8*, one manufacturer supplies a kit for determining concrete surface profiles. The concrete profiler (TCP) kit consists of nine International Concrete Repair Institute (ICRI) surface profile chips, a special spring-loaded digital concrete profile gauge, and a two-part putty. The chips cover a range of about 27 mils (0.7 mm) to 241 mils (6.1 mm) and are listed as follows:

- *Number 1*
  For surface profiles of about 27 mils (0.7 mm)
- *Number 2*
  For surface profiles of about 32 mils (0.8 mm)
- *Number 3*
  For surface profiles of about 38 mils (1 mm)
- *Number 4*
  For surface profiles of about 50 mils (1.3 mm)
- *Number 5*
  For surface profiles of 66 mils (1.7 mm)
- *Number 6*
  For surface profiles of 124 mils (3.1 mm)
- *Number 7*
  For surface profiles of 175 mils (4.4 mm)

- *Number 8*
  For surface profiles of 210 mils (5.3 mm)
- *Number 9*
  For surface profiles of 241 mils (6.1 mm)

The putty is mixed, applied to the surface, and allowed to set (usually around ten minutes). When this replica is removed, it provides an impression of the surface profile. Unlike replica tape that can be crushed, TCP is fully cured in 24 hours, providing a permanent record of the surface profile. The replica can be compared visually to the ICRI chips and/or the replica profile can be measured at ten different locations with the gauge to determine the lowest and highest reading with the difference representing the profile range.

If the surface profile is measuring as less than was specified, a deeper profile can be attained by blasting with a harder or larger abrasive or at a greater air pressure. If the surface profile is greater than specified, repair work may not be possible. In this case, more coating will be needed to have adequate coverage.

2.7.0 Environmental Conditions

During surface-preparation operations, the environment of the work area may need to be controlled. Containment and ventilation are used to protect the environment from hazardous or unsightly debris and to protect workers from the effects of dust and debris. While these are valid reasons to control environmental conditions, there are two environmental conditions that directly affect the results of surface preparation. These conditions are temperature and humidity.

Moisture on a surface will prevent most coatings from bonding with the surface. To prevent moisture from developing on a surface, it's necessary to control both the temperature and relative humidity of the area in which the surface is located. Relative humidity is the percentage of moisture that the air at the current temperature is holding compared to the maximum amount it can hold. Relative humidity is related to temperature, with warm air having the ability to hold more moisture than cold air.

Humidity in the air can condense on the surface of any substrate given the correct conditions. When a cold soft drink is poured into a glass on a warm day, moisture condenses on the outside of the glass. That's because the temperature of the glass is colder than the air temperature, so as the warm air hits the cold glass, moisture is released from the humid air. When hot water is poured into a glass on a warm day, moisture doesn't condense on the outside of the glass. The temperature of the glass is higher than the air temperature.

These same principles are applied to controlling moisture during surface preparation and coating operations. The surface to be coated can be held at a temperature higher than the atmosphere using heating units, or the humidity can be removed from the atmosphere with dehumidification units (*Figure 9*).

Air temperature can be measured with either an electronic or analog thermometer. To measure relative humidity, a mechanical or electronic **hygrometer** can be used. Some electronic devices will measure both temperature and relative humidity. *Figure 10* shows an electronic hygrometer and thermometer. Relative humidity can also be measured using a sling psychrometer (*Figure 11*). An analog sling psychrometer consists of two thermometers, the bulb of one covered with a wick that is moistened with water, and the other left open to the air. The thermometers are mounted on a light frame that is connected to a handle. The psychrometer is twirled by its handle for about 20 seconds, and then the wet and dry bulb temperatures are read. The test is repeated until the same results are achieved two or more times. Using these readings and a psychrometric

205F09.EPS

Figure 9 Dehumidification unit.

table (*Figure 12*), or psychrometric calculator, the relative humidity of the air can be determined. Psychrometric tables and charts are available using Celsius or Fahrenheit temperatures. With either type, the relative humidity is read at the intersection of the dry bulb temperature vertical line with the wet bulb temperature diagonal line. In the chart, for example, the intersection of the vertical line from a dry bulb temperature of 78°F with the diagonal line from a wet bulb temperature of 65°F falls exactly on the 50 percent relative humidity curve. Other temperature combinations may fall in between each of the relative humidity curves and the humidity values have to be approximated visually. The digital psychrometer (*Figure 13*) does the entire process electronically and does not require any slinging. Humidity, wet/dry bulb temperature, and dew point are all displayed by the unit. However, other electronic psychrometers may require slinging like the analog version, but the relative humidity and other information is electronically determined and displayed.

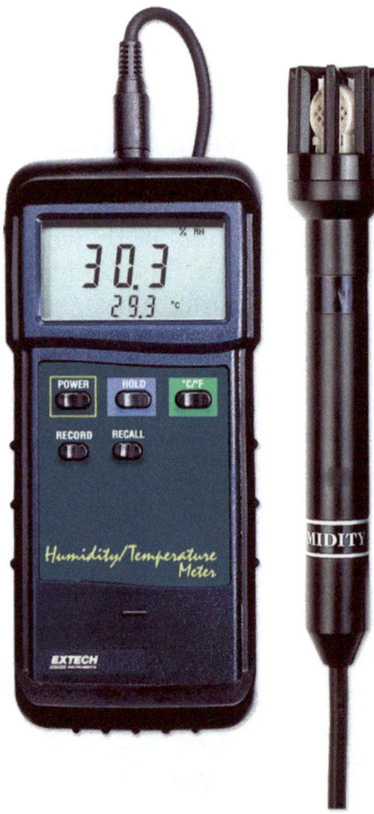

ELECTRONIC HYGROMETER

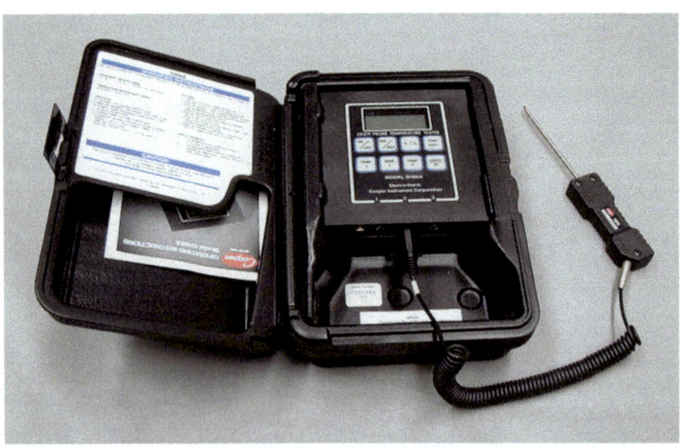

ELECTRONIC THERMOMETER

205F10.EPS

Figure 10 Electronic hygrometer and thermometer.

3.0.0 Evaluating Surfaces

The first step to any surface-preparation operation is to study the project specification to determine the degree of cleanliness and the surface profile that is needed before a coating may be applied. Next, carefully inspect the existing surface. Then, determine what surface preparation methods must be applied to achieve the desired results, and what types of repairs are necessary before the surface can be coated.

During this phase, it is important to document the findings, as well as any evidence that supports the findings. Examples of evidence include the following:

- High-quality photographs that are clearly marked with the location in which they were taken and that show the condition on the surface
- Video clips of any tests as they are performed and the results of the tests
- Coating chips
- Samples of contamination

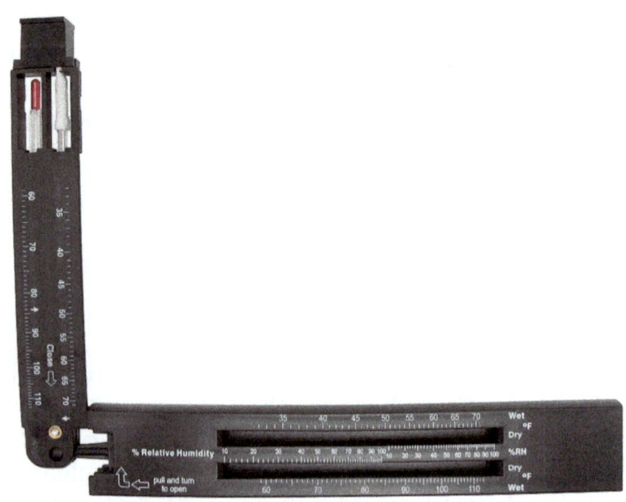

205F11.EPS

Figure 11 Analog psychrometer.

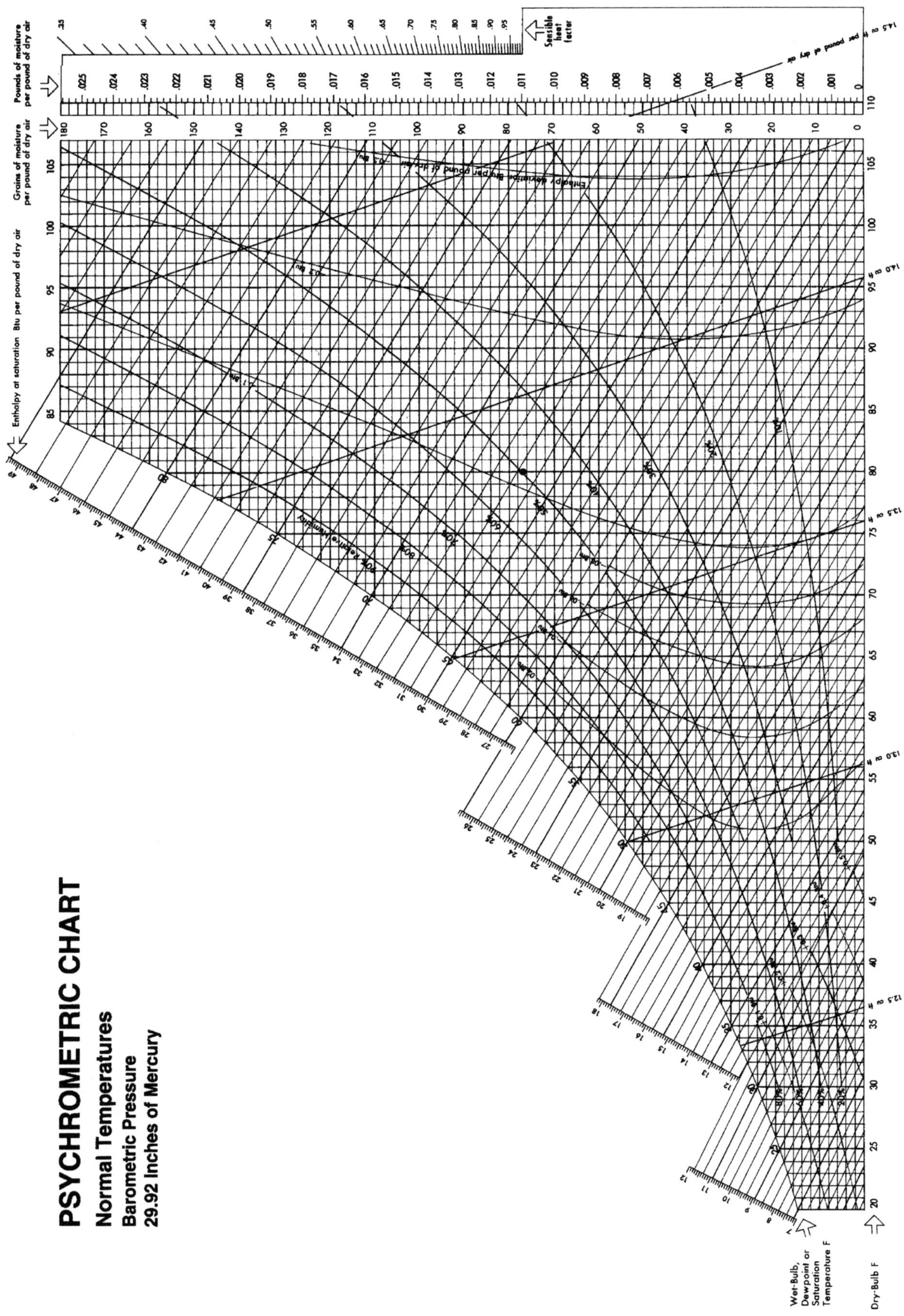

205F12.EPS

Figure 12 Psychrometric chart.

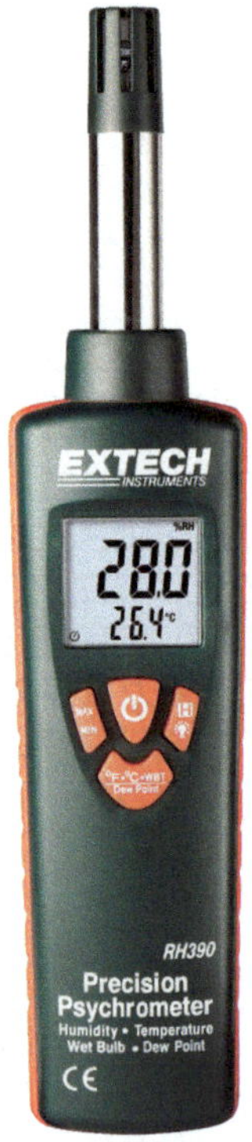

205F13.EPS

Figure 13 Digital psychrometer.

It is very important that all findings are clearly and completely documented. In some cases, these documents may be introduced into a court of law should problems develop later. Be sure to specify the exact location of any tests, and how they were performed. When using still photographs, be certain to document the location of the photo. When using video clips, pan across the area and then zoom in on the surface so the viewer can determine the location of the test. While documenting surface conditions use the following guidelines:

- If notes are handwritten, be sure they are legible.
- When using pictures or videos, be sure the notes discuss each image.
- Avoid misspellings. If you don't know how to spell a word, look it up.
- Use clear and definite language. For example, "Visible oil and grease contamination over 80 percent of the surface" is clearer than "Most of the surface is dirty."
- Always document who performed a test, how, and the results. For example, "Hydrocarbon contamination was detected on 50 percent of the surface of Bay A with an ultraviolet light. Test was performed by Violet Ultra (Coatings We Do, Inc.) and Harry Jones (owner's representative)." Document four times a day.
- Sometimes it's necessary to document why an operation was performed, such as, "Could not evaluate surface due to a uniform layer of mud on the surface. Will evaluate after power washing." Remain factual.

The goal of surface preparation is to remove all visible oil, grease, dust, dirt, and other foreign matter from the surface to meet specifications. This module presents surface-preparation methods for the following surfaces:

- Steel
- Galvanized steel
- Weathering steel
- Stainless steel
- Copper and aluminum
- Masonry and concrete
- Synthetic substrates
- Existing coatings

Although each one of these surfaces has unique characteristics and uses different methods for surface preparation, use the following list of do's and don'ts during the pre-preparation inspection of any surface:

- Do understand the project specification requirements.
- Do document all findings, including measurements as necessary.
- Do take clear, high-quality photographs if allowed.
- Don't confine testing to only the best or worst areas of the surface. Determine rust grades using *SSPC-VIS 1* or *ISO 8501-1* for unpainted steel, *SSPC-VIS 2* or *ISO 8501-2* for previously painted steel, *SSPC-VIS 4* or *ISO 8501-4* for waterjetting, and *SSPC-VIS 5/NACE VIS 9* for wet abrasive blast cleaning.

3.1.0 Steel

Steel and its alloys are the most commonly used metal, because steel is low-cost and readily available. Steel is made by melting iron and other elements together, and then pouring it into ingot molds to cool. The ingots can be hot-rolled in a mill to many shapes and sizes. As the hot-rolled steel cools, it reacts with oxygen in the air, so mill scale forms on the steel surface. Cold-rolled steel has little, if any, mill scale formation on the surface.

Defects on the surface of steel can affect the quality of the coating system. Some surface defects are a result of the manufacturing process of the steel, and some are caused by the assembly process. Manufacturing defects include scabs, rollovers, hackles, and laminations. Fabrication defects include weld spatter, skip welds (not necessarily a defect, except with relation to surface preparation and coatings), rough welds, sharp corners, bends, angles, and edges (again, not necessarily a defect, but a problem for coating).

Steel fabrication defects on existing structures may have been remedied before the initial coating was applied as required by specifications, but carefully inspect the surface in case repair or modification work occurred after the initial coating operation.

Gouges need to be documented and reported. Large or unusual holes or cracks in existing steel structures may be a sign of metal fatigue, which coating specialists are not qualified to diagnose or repair. Whenever unusual defects are noticed, ask the project engineer or supervisor for help before any repairs are made.

Liquid-applied coatings pull away from sharp edges and angles as they dry, so the coating, unless DFT certified, is thinner in these areas.

The project specification may call for all welds, bolts, edges, angles, rivets, and repairs to be brush coated with a primer to ensure that the coating is applied to the specified thickness. Even when the specification does not call for the application of a stripe coat, it may be wise to consider it around edges and angles.

When the existing structure is already coated, look for surface defects by carefully examining the coating. Stains that leach through the coating, coatings that blister and peel, and bumps under the coating are all signs of possible corrosion. On some coated surfaces, the surface defects may not be identifiable. In this case, re-inspecting the steel surface after the old coating has been removed will be necessary. Pipe, in particular, near a marine environment, may be contaminated by chlorides and show black rust. If chlorides are not removed, it may cause osmotic blistering. These are bubbles formed beneath the paint film. Methods of removal include ultra high-pressure waterjetting or special removal agents.

3.1.1 Galvanized Steel

Galvanized steel has been coated with zinc. The zinc coating for new steel is applied during the manufacturing process as molten zinc (hot dipped) or by electroplating. Zinc is a sacrificial coating that corrodes to protect the steel substrate, so the zinc coating will need periodic maintenance.

When any zinc coating needs to be blasted, it must first be tested for chromate, also known as hexavalent chromium. Chromate is derived from the metal chromium, which is commonly used in the galvanizing process, as well as zinc paints and primers. Chromate is a toxic cancer-causing salt, so workers must take adequate steps to prevent exposure to the dust and release of any dust or debris into the environment. High chromate levels may be found in newly galvanized products, so any new parts added to an existing structure—even small parts like nuts and bolts—must be tested for chromium. To test a surface for chromate, perform the following procedure:

Step 1 Apply lead acetate solution to the surface.

Step 2 Wait 5 seconds.

Step 3 Examine the surface. If a dark spot appears, the surface is chromate-free.

WARNING

Hexavalent chromium causes cancer. Hexavalent means the chromium can form six chemical bonds. Trivalent means the chromium can form three chemical bonds. The coating industry has been working to eliminate the chromium from coatings with a limited amount of success. Coatings without hexavalent chromium do not last as long as coatings that have it.

3.1.2 Weathering Steel

Weathering steel, also known as Cor-Ten® steel or A-588, is an inspection challenge because specifications often refer to visual guides, of which there are no pictures. On uncoated weathering steel, the surface corrosion becomes a protective barrier that prevents more corrosion. Adherent corrosion and non-adherent corrosion appear exactly the same in color and texture.

During surface preparations on uncoated weathering steel, avoid removing adherent corrosion. During the pre-inspection, test the adherence of the corrosion by scrubbing with a wire brush or tapping with a hammer. Any corrosion that is easily removed in this manner is non-adherent and needs to be removed.

Coated weathering steel is inspected in a similar manner to coated steel. Look for stains leaching through the coating, coatings that are blistering and peeling, and bumps under the coating.

3.1.3 Stainless Steel

The surface of stainless steel oxidizes when exposed to air to form a thin, invisible, chromium-containing oxide film that is the protective barrier. If the surface is scratched, more oxide quickly forms. Stainless is inspected and repaired as steel is inspected and repaired.

NOTE

In marine environments, stainless steel must be coated because chloride salts attack and destroy the oxide film faster than it can be replaced.

When stainless steel is welded, it can create hexavalent chromium. Be sure to test welds on stainless steel for chromium before any surface preparation work is performed. When chromium is detected, be certain that steps are taken to protect workers and the environment from chromium dust.

3.2.0 Copper and Aluminum

Aluminum and copper don't need to be coated to protect them from atmospheric corrosion. The surfaces of both oxidize over time when exposed to air to form visible oxide protective films. The whitish film that forms on aluminum is not very attractive, so a clear or colored coating is usually applied to the metal. Uncoated aluminum is highly reactive. It will corrode when it contacts copper or copper alloys, concrete, certain acids, and strong bases. This is another reason that aluminum is normally coated.

WARNING

Aluminum powder must be carefully handled and kept away from ignition sources. If ignited, it can burn violently or explode.

Copper forms a brown/black oxide coating over time and eventually forms a green copper carbonate coating called verdigris that is very impervious to corrosion. Many roofs and other structures are allowed to remain uncoated because the verdigris is considered attractive. However, copper in contact with iron and steel will corrode them. Hydrogen sulfide and other sulfide solutions will corrode copper, and copper will react with hydrochloric acid to form copper chlorides.

On Site

Copper

The skin of the Statue of Liberty is made of copper. The verdigris on the copper is what gives the statue its greenish-blue color. When the statue was restored in the 1980s, most of the copper skin was intact—that was after almost 100 years standing on an island in New York Harbor.

3.3.0 Masonry and Concrete

Concrete is a widely used material in construction. Depending on its application, concrete may or may not require coating; most often, it does. Concrete is a very porous material. Uncoated concrete readily passes salts or acids, which can cause corrosion of any metal reinforcing material in the concrete (*Figure 14*). In cold climates, moisture in the concrete can expand as it freezes. This will break apart the concrete from internal pressure. Coating concrete that is exposed to moisture will extend its life.

Freshly poured concrete is usually allowed to go through a 28-day curing stage before it is coated; however, recent research has indicated that the concrete may be safely coated after 14 days if permitted by specification. After the concrete has been placed, inspect the surface for voids, and repair them as soon as possible to allow the repaired areas to cure while the poured concrete is curing. Projections and rough spots in the concrete surface can be ground down before or after curing, but it is better to do it before curing, because repairs may require curing.

Small blemishes in concrete can be repaired by stoning or sacking, which means scrubbing mortar over the concrete surface with an abrasive block or a gunnysack. This will fill voids and cracks with mortar. Be aware that mortar repairs may also need to cure before they can be coated, so it is best to make repairs on new concrete as soon as possible after pouring. Be sure to remove any form release agents or hardeners.

One method of repairing voids and cracks in concrete is to inject epoxies or resins into them. These products usually require that the surface under repair be dry. Follow the manufacturer's instructions when using these products.

205F14.EPS

Figure 14 Deteriorating concrete.

After the concrete has cured, it must be etched to remove laitance and roughen the surface. Concrete is etched by applying a diluted solution of hydrochloric acid to the surface and allowing the solution to sit on the surface until it stops bubbling, usually about 5 to 10 minutes. Afterwards, the surface is thoroughly washed to remove the acid solution. Sometimes several applications of etching solution are needed to achieve the desired surface cleanliness and profile. Acid etching is covered in *NACE No. 6/SSPC-SP 13* and *ASTM D4260.*

WARNING

When mixing acid and water, always add the acid to the water—never the reverse. Always wear the proper PPE.

Assessing existing coated concrete can be difficult. It may appear to be sound, but underneath, the concrete could be deteriorating. Deteriorating concrete can be removed with hand or power impact tools. Unfortunately, impact tools can loosen the remaining layer of concrete so that it may appear to be intact, but actually be detached from the concrete substrate. To remove the loosened layer, use wet abrasive or high-pressure waterjetting. *NACE No. 6/SSPC-SP 13* and *ASTM D4259* cover abrading of concrete. When sound concrete is exposed, the surface can be repaired with the specified repair material.

During the pre-preparation inspection, look for some of the following conditions:

- Pits, crack, gouges, and voids.
- Fluffy white deposit on the surface, which is called efflorescence and is caused by water reacting with salt in the concrete and rising to the surface. It must be removed by a dry method (mechanical or dry abrasive blasting) before the surface is cleaned by water.
- Oils or grease.
- Visible dirt or other foreign matter.
- Water, ice, or ice crystals on the surface, which can indicate excessive moisture content.

NOTE

There are standards for water quality when performing waterjetting and cleaning to prevent chloride or other contamination. Refer to www.advisorycouncil.org for additional information.

Chemical contaminants, oil, and grease can usually be removed by either steam cleaning or cleaning with chemicals or detergents. When the contamination is severe, the concrete in the area may need to be removed and the spot repaired. Note that solvents, detergents, and steam cleaning as described in *ASTM D4258* can dissolve contaminants and allow them to drain deep into the concrete. The surface appears to be clean, but when the temperature increases and the concrete breathes, the contaminant can bubble to the surface and cause a coating to fail.

Abrasive blast cleaning will provide a rough, irregular surface (profile) and will remove oil and grease. It will also remove laitance. Abrasive blasting opens holes and voids so they can be sealed. Concrete surfaces cleaned in this way are rougher than the surface of abrasive-blasted steel, so more coating is needed to cover the same area.

Lower-pressure water cleaning at pressures of 3,000 to 4,500 psi (246 to 316 kg/cm^2) is often used on poured concrete surfaces. This range of pressures doesn't usually open voids and pockets, and water cleaning won't achieve as good a surface profile on concrete as abrasive blasting. High-pressure waterjetting can open voids and pockets and also remove deteriorating concrete. Wet abrasive blasting also may be used.

Before concrete is coated, it should be carefully swept, vacuumed, or blown off with clean compressed air to remove dust, especially after abrasive blasting. Concrete surfaces can be water washed to remove gross contamination of dust, but must still be vacuumed or blown off when they are dry.

3.4.0 Synthetic Substrates

Synthetic (manmade) materials, such as polyvinyl chloride (PVC) or fiber-reinforced plastic (FRP), are widely used in commercial and industrial projects. In general, synthetic surfaces are prepared for coating by washing with mild detergents or degreaser, followed by a light sanding or abrasive blasting to achieve the desired surface profile. Blemishes and voids may be repaired by using a similar or compatible material or filler. Be certain that no fibers are visible on FRP before any coating takes place.

3.5.0 Existing Coatings

One of the challenges with overcoating operations is determining the nature of the existing coating. This is important because the new coating must securely bond to the existing coating. Some coatings are incompatible, which means that they will not bond properly. If the existing coating cannot be identified, it will be necessary to perform a **patch test** on a small area to determine whether the two coatings will bind. Two incompatible coatings can still be used, but a **tie coat**, which is a coating that is compatible with both the existing coating and the new coating, must be applied to the prepared surface before the top coat can be applied.

The condition of the coating on a surface may range from intact to completely failed. The surface may even be badly deteriorated. Good practice in overcoating operations follows the same general steps used for coating new surfaces:

- Pre-inspection of the structure to be coated
- Surface preparation
- Coating selection

If the existing coating is intact and the entire area is to be recoated without removing the existing coating, the surface will need to be treated to attain a surface profile. A solvent may be used to soften the surface, or abrasive-coated paper may be used to roughen the surface. A light abrasive blast may also be performed to create the desired surface profile. When the surface preparations are completed, use a patch test on a small area to determine whether the coatings are compatible to each other.

Once the surface and overcoat have been tested, the entire surface should be inspected to locate and mark failed areas. These may include areas with blistering, loss of adhesion, under-coating corrosion, and chalking. Some areas may be contaminated by grease, chemical salts, dirt, or other substances.

Once the areas to be repaired have been located and marked, surface-preparation operations may begin. The same tools, equipment, and techniques used in surface preparation of new construction may be used in overcoating operations. Solvent cleaning and/or waterjetting or water blasting may be used to remove chalky, crumbly coatings and grease, as well as dirt, chemical salts, and other gross contaminants. Hand or power tool cleaning and/or abrasive blasting may be used to open blisters, chip cracked and peeling paint, remove tightly adhered mill scale, and provide an acceptable surface profile.

One unique technique used during overcoating operations is feathering. Feathering takes a light touch. The repaired area is worked normally, but the edges are worked lightly so there is a smooth transition from the repair area to the sound coating.

4.0.0 Dry Abrasive Cleaning

Dry abrasive blasting is the most common method of surface preparation. It is commonly referred to as sandblasting, although sand is seldom used as an abrasive media today. Dry abrasive blasting is typically used on metals, but can be used on other surfaces, too. Dry abrasive blasting applies a stream of small abrasive media, under pressure, to a surface. It can be used to remove old coatings, mill scale, rust, and unsound concrete.

In addition to removing old coatings and contaminants, dry abrasive blasting produces a good surface profile on all substrates that allows for good coating adhesion. There are a number of industry standards for abrasive blast cleaning. They are as follows, listed from the most stringent to the least stringent:

- *NACE No. 1/SSPC-SP 5, White Metal Blast Cleaning* – This standard requires that after cleaning, the surface (when viewed without magnification) will be free of all visible oil, grease, dust, dirt, mill scale, rust, coating, oxides, corrosion products, and other foreign matter.
- *NACE No. 2/SSPC-SP 10, Near-White Metal Blast Cleaning* – This standard requires that after cleaning, the surface (when viewed without magnification) will be free of all visible oil, grease, dust, dirt, mill scale, rust, coating, oxides, corrosion products, and other foreign matter, except that random staining may be present on not more than 5 percent of any unit area of the surface. The staining may consist of light shadows, slight streaks, or minor discolorations caused by stains of rust, mill scale, or previously applied coating.
- *NACE No. 3/SSPC-SP 6, Commercial Blast Cleaning* – This standard requires that after cleaning, the surface (when viewed without magnification) will be free of all visible oil, grease, dust, dirt, mill scale, rust, coating, oxides, corrosion products, and other foreign matter, except that random staining may be present on not more than 33 percent of the surface. The staining may consist of light shadows, slight streaks, or minor discolorations caused by stains of rust, mill scale, or of previously applied coating.
- *NACE No. 4/SSPC-SP 7, Brush-Off Blast Cleaning* – This standard requires that after the surface has been subjected to blasting, the surface (when viewed without magnification) will be free of all visible oil, grease, dirt, dust, loose mill scale, loose rust, and loose coating. Tightly adherent mill scale, rust, and coating that pass the dull putty knife test may remain.
- *NACE No. 8/SSPC-SP 14, Industrial Blast Cleaning* – This standard requires that after cleaning, the surface (when viewed without magnification) will be free of all visible oil, grease, dust, and dirt. Evenly distributed traces of tightly adhered mill scale, rust, and coating residue are permitted if they do not exceed 10 percent of the surface area.

Dry abrasive blasting is a fast and easy way to prepare a surface for coating. Unfortunately, it is also an easy way to over-prepare a surface. The operator needs to be able to identify the levels of preparation. The key to being a good blaster is to understand the desired outcome. When the project specification calls for the surface to be prepared to *NACE No. 3/SSPC-SP 6, Commercial Blast Cleaning* and it is prepared to *NACE No. 1/ SSPC-SP 5, White Metal Blast Cleaning*, time and resources are wasted.

Choice of abrasive is usually stated in the project specification and may be influenced by the coating manufacturer specifications. *Table 1* shows some abrasives that may be used to attain a specified surface profile.

> **NOTE**
> *Table 1* is a guide only. Specific abrasive information should be obtained from the manufacturer.

4.1.0 Dry Abrasive Media

A number of different materials can be used as abrasive media, such as steel, mineral, or copper slag. Some, such as slags, are byproducts of other manufacturing processes, so they are very inexpensive. Others are manufactured especially for their unique properties, such as hardness or angularity. Manufactured media are usually expensive, and can be recycled. Large, hard, or angular blast media produce a deeper surface profile than small, soft, or round media.

Media that have a smooth round shape produce a peened surface, while angular media produce a sharp surface profile. Always check the project specification to determine the required surface profile. A surface profile that's too deep cannot be corrected, and will require more coating to achieve adequate coverage. In addition, hard, angular media also wears blasting equipment more than smooth, round media.

Several variables affect the surface profile resulting from abrasive blasting. They include the following:

- Type of media (size, hardness, and shape)
- Pressure at which the medium is delivered
- Distance and angle of the blast nozzle in relation to the surface
- Dwell time
- Nozzle aperture size

Table 2 provides a summary of various abrasive media, including usage, size, hardness, shape, dwell time (cleaning speed), and blast pressures.

> **DID YOU KNOW?**
> ***Using Corn Cobs***
> Corn cobs are sometimes used for abrasive blasting. The fibrous parts of the cob are ground into a grit that can be used to remove contaminants without affecting the surface being cleaned. Corn cob grit is good for blasting fragile items and wood.

Table 1 Choosing Abrasives for a Specified Surface Profile

Surface Profile		
Mils	**Micrometers (μm)**	**Abrasive Media (Delivered at 90 to 100 psi)**
0.5	12.7	80/120-mesh silica sand, 100-mesh garnet, 120-grit aluminum oxide, or G-200 iron or steel grit
1	25.4	30/60-mesh silica sand, 80-mesh garnet, 100-grit aluminum oxide, or G-80 iron or steel grit, S-110 steel shot
1.5	38	20/50-mesh silica sand, 36-mesh garnet, 50-grit aluminum oxide, or G-50 chilled iron or steel grit, S-170 steel shot
2	50.8	16/40-mesh silica sand, 30-mesh garnet, 36-grit aluminum oxide, or G-40 chilled iron or steel grit, S-280 steel shot
2.5	63.5	12/30-mesh silica sand, 20-mesh garnet, 24-grit aluminum oxide, or G-25 iron or steel grit, S-280 steel shot
3	76.2	8/20-mesh silica sand, 16-mesh garnet, 16-grit aluminum oxide, or G-16 chilled iron or steel grit, S-330 steel shot

205T01.EPS

Table 2 A Summary of Various Abrasive Media

	Benefits				Usage Properties					Physical Properties			
	Finishing	Cleaning and Removal	Peening	Etching	Average Cleaning Speed	Reusability	Dust Level(s)	Chance of Metal Removal	Typical Blast Pressures (psi)	Hardness MOH Scale (Rockwell C)	Bulk Density (ft^3)	Mesh Sizes	Angular (^) Spherical (o)
Glass Beads	•	•	•		MED	HI-LO	LO-HI	Very LO	20–55	5.5	100	30–440	o
Ceramic Shot	•	•	•		MED	HI	LO	Very LO	20–90	7 (57–63)	150	6–46	o
S.S. Cut Wire	•	•	•	•	MED	HI	LO	MED-HI	20–90	6–7.5 (35–55)	280	20–62	^
Steel Shot	•	•	•		MED	Very HI	LO	Very LO	20–90	6–7.5 (20–66)	280	8–200	o
Steel Grit	•	•		•	MED-HI	Very HI	LO	MED	20–90	6–7.5 (40–66)	230	10–325	^
Aluminum Oxide	•	•		•	HI	MED-HI	LO-MED	MED-HI	20–90	8–9	125	12–325	^
Silicon Carbide	•	•		•	Very HI	LO-MED	MED-HI	MED-HI	20–90	9	95	36–220	^
Crushed Glass	•	•		•	HI	LO-MED	MED-HI	LO-MED	20–50	5.5	100	30–400	^
Plastic Media		•		•	MED-HI	MED	MED	Very LO	20–60	3–4	45–60	12–80	o
Wheat Starch		•			MED	MED	MED	Very LO	20–35	3	40	12–100	^
Walnut Sholl		•			LO-HI	LO	MED-HI	Very LO	10–40	3	40–80	Many	^
Corn Cob		•			LO-MED	LO	MED-HI	Very LO	20–60	1	30	8–40	^
Sodium Bicarb.[1]		•			MED-HI	NR[3]	HI	Very LO	20–90	3	62	54–150	^
Silica Sand[2]	•	•		•	HI	NR[3]	HI	HI	30–90	7.5	100	12–100	^

NOTES:

•Above information intended as a general reference guide.

[1]Sodium bicarbonate must be treated with a flow agent.

[2]Do not use silica sand in a hand cabinet.

[3]Not recommended.

205T02.EPS

Although you may be able to select the type of media used during surface preparation, the decision may be determined by the project specification. The other variables that affect surface cleaning (over which you do have control) are the pressure at which the media is delivered to the surface, the distance the nozzle is held from the surface, the angle that the nozzle is held to the surface, and the dwell time of the sweep.

High delivery pressure, a small aperture, holding the nozzle close to the surface, and a slow sweep results in a greater surface cleaning. It can also result in a surface profile that is deeper than intended, burns from dwelling too long, and damage to thin metal structures. The nozzle should be held at about a 45- to 80-degree angle to the surface for the best cutting power.

When abrasive blasting, the operator must ensure the following:

- The abrasive type and size are as specified in the project specification.
- Abrasives are clean of contaminants, including fines, dust, paint, and mill scale, and also free of moisture and oil.
- The specified recycling procedure is followed.

Abrasives are sized using a sieve. The number of the sieve tells how many holes there are per inch of screen. A No. 10 sieve has 10 holes per linear inch (2.54 cm), so particles smaller than a tenth of an inch (0.254 cm) can pass. A No. 200 sieve has 200 holes per inch, so it passes only very small particles. The larger the number of the sieve, the smaller the particle it passes. Abrasives are often referred to by size, so 30/60-mesh means that all of the particles need to pass through a No. 30 but not a No. 60 sieve. New shipments of abrasive go through a **sieve test** to determine whether they are according to specification.

Some abrasives are recycled for reuse. Abrasive media reclamation units vary. Simple units may consist of only a cyclone separator (*Figure 15*) which removes dust and fine particles but passes the abrasive and any heavy debris, such as coatings, rust, and mill scale. Other units are much more sophisticated and pass the discharge from the cyclone separator through a scalping drum and a magnetic drum separator to remove steel shot or grit from nonferrous material. This may be followed by passing the media through a series of vibrating sieves, sometimes called a classifier, to separate oversize and undersize media and debris. The media is finally passed through waterfall air-wash media curtains created by channel baffles to separate very fine debris.

All of these techniques are shown in *Figure 16*. However, only one or two of them are used in any one recycling unit depending on the type of media being recycled.

As abrasives are used, they break down, so recycled abrasives are typically smaller than new abrasives. Check recycled abrasives daily to be certain that the size is adequate for the job.

As abrasives are used, they may become contaminated with coating material, dirt, salts, oils, or grease. Using contaminated abrasives may remove coatings, but it also embeds salts, oil, grease, and other contaminants into the surface.

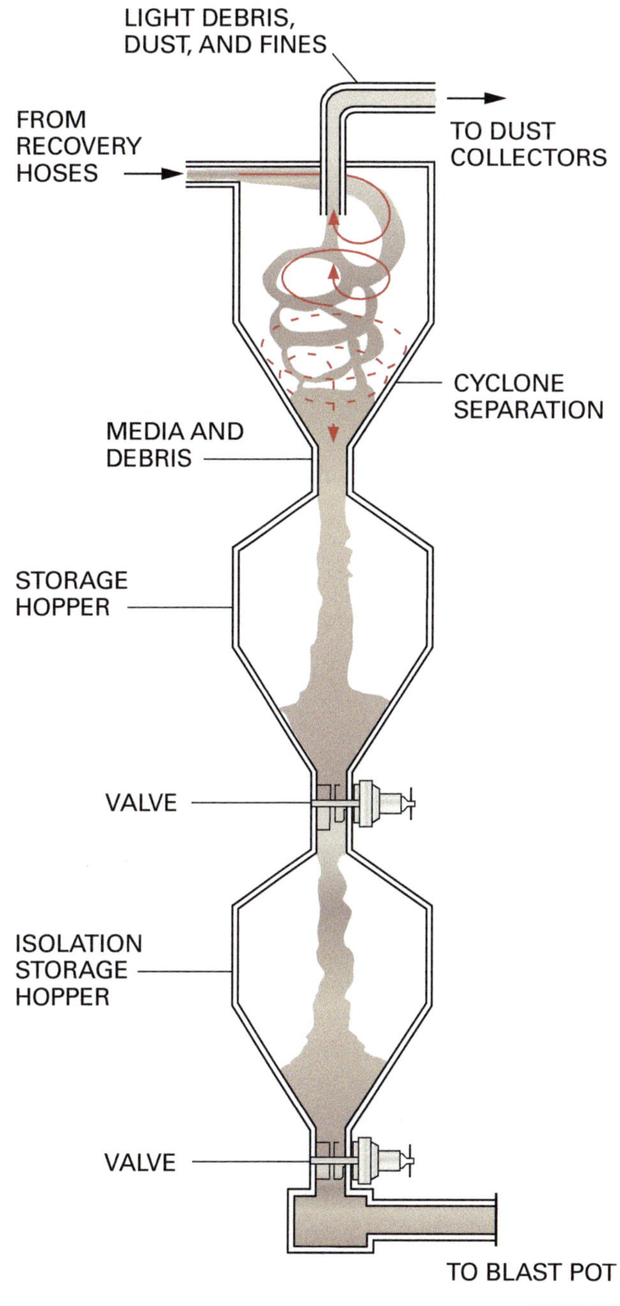

Figure 15 Cyclone separator.

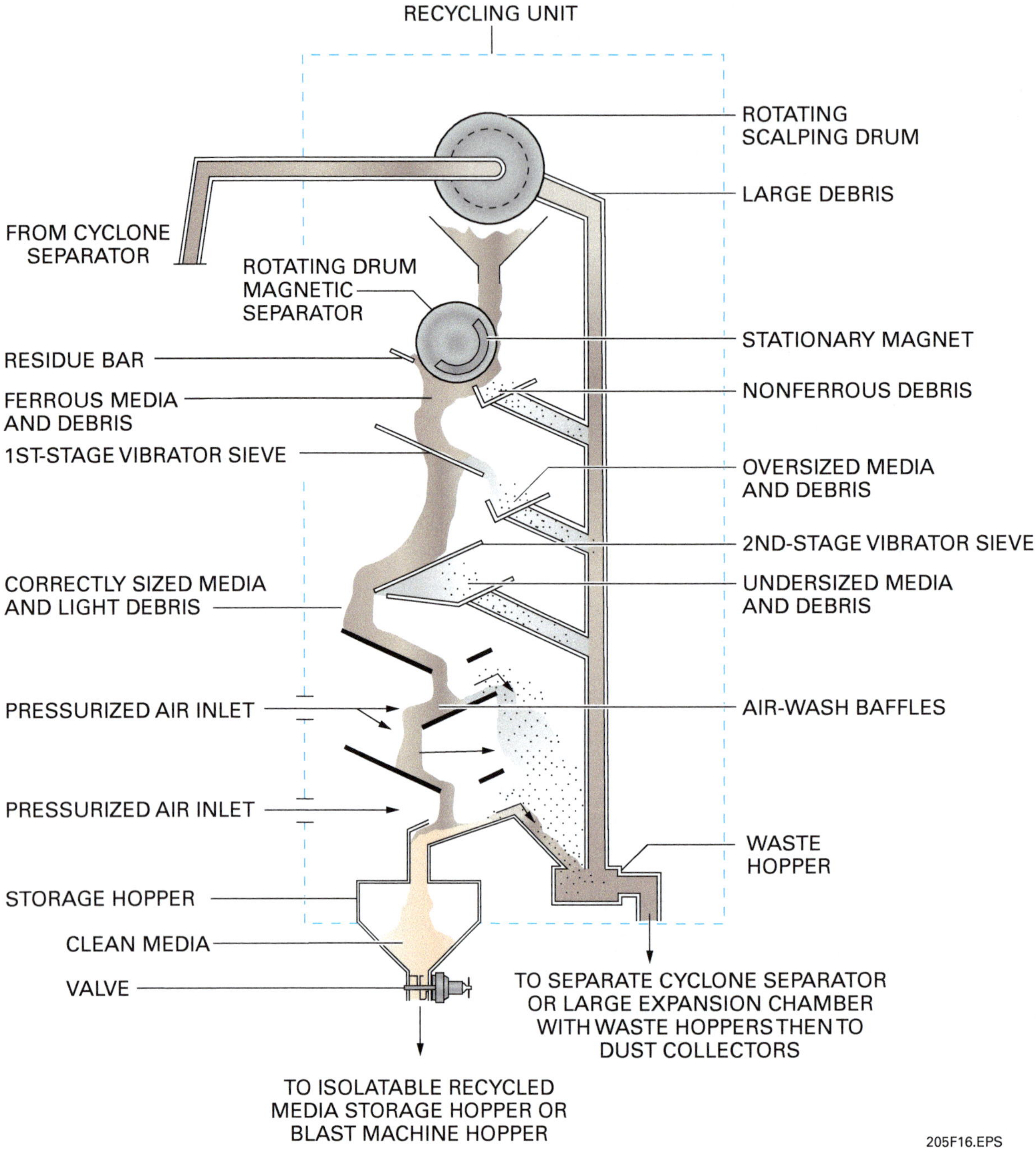

Figure 16 Recycling unit.

> **NOTE**
>
> *SSPC-AB2 Cleanliness of Recycled Ferrous Metal Abrasives* covers the requirements for cleanliness of recycled ferrous metal blast cleaning abrasives used for the removal of coatings, paints, scale, rust, and other foreign matter from steel or other surfaces.

Per work plan requirements, mineral and slag abrasives can be tested for cleanliness using a vial test. A bulk sample may also be required.

During a vial test, some of the abrasive is dropped into a small vial of distilled or deionized water of known pH (distilled or deionized water has a pH of 7) and shaken. Typically, the ratio should be one part abrasive to two parts water. The vial is allowed to sit undisturbed, until the abrasives settle to the bottom, and then the top of the water is inspected for a film of grease or oil. The water can be visually checked for turbidity (cloudiness due to sediment), which is usually a sign of excess dirt, dust, or clay in the abrasive.

To test for water-soluble contaminants, such as salts, use litmus or pH paper to test the water in the vial. Litmus paper cannot identify the contaminants, but it can tell whether the abrasive is acid or alkaline. If the result is neutral (pH 7.0), add a drop of 5 percent silver nitrate solution to the water. If a white solid forms, the abrasive is contaminated with chlorides. If the abrasive is contaminated, dirty, acidic, or alkaline, it should be replaced.

The water-soluble contaminants can also be checked by the water's conductivity to electricity, in accordance with *ASTM 4940, Standard Test Method for Conductimetric Analysis of Water Soluble Ionic Contamination of Blasting Abrasives*. This test measures the contamination level of the abrasive, but it cannot identify the contaminants. Conductivity results are measured in **microsiemens**. Clean deionized water has a low conductivity of 0.012 to 5.5 microsiemens, depending on how much oxygen is dissolved in the water. As contamination increases, conductivity also increases. The project specification will state at which conductivity level the abrasive is too dirty to use. The requirement is usually between 250 and 1,000 microsiemens.

There are a number of products that can be added to abrasives to improve the abrasive characteristics if specifications call for them. For example, a dust suppressor additive helps to reduce the amount of dust generated, and anti-static additives help to keep plastics from adhering to metal or fiberglass surfaces. Other products, such as BlastTox® and LeadX®, are added to abrasives during lead-based paint removal to make the debris disposable as non-hazardous.

Compressed air may be contaminated with water or oil from the compressor. Allow the compressor to run for 15 minutes. Always drain the water and oil traps on the air compressor at the start of the shift. Test the airstream for water and oil by performing a blotter test per *ASTM D4285* at least daily after the equipment is set up for an actual blasting operation, or whenever you notice the abrasives clumping or glistening. A blotter test is performed as follows:

Step 1 Set up blasting equipment, being certain to drain the oil and water traps.

Step 2 Allow the compressed air system to reach operating conditions.

Step 3 Discharge compressed air at operating conditions to remove accumulated condensation in the system.

Step 4 Using white absorbent paper as a collector, fasten the collector material to a rigid backing, and center the collector about 24 inches (61 cm) from the discharge point of the nozzle. Avoid contact with the air stream.

Step 5 Discharge air toward the collector for one minute.

Step 6 Visually inspect the collector for moisture or oil.

Any indication of water or oil contamination on the collector may require compressor maintenance. Consult the manufacturer's documentation for information.

4.2.0 Maintaining Dry Abrasive Equipment

There are a number of designs for dry abrasive blasting, but the two that will most likely be used are the suction system and the pressurized system (*Figure 17*). In the suction dry abrasive system, compressed air is delivered to the blast nozzle via a hose. A second hose is run from the blast pot to the nozzle. As the compressed air is released from

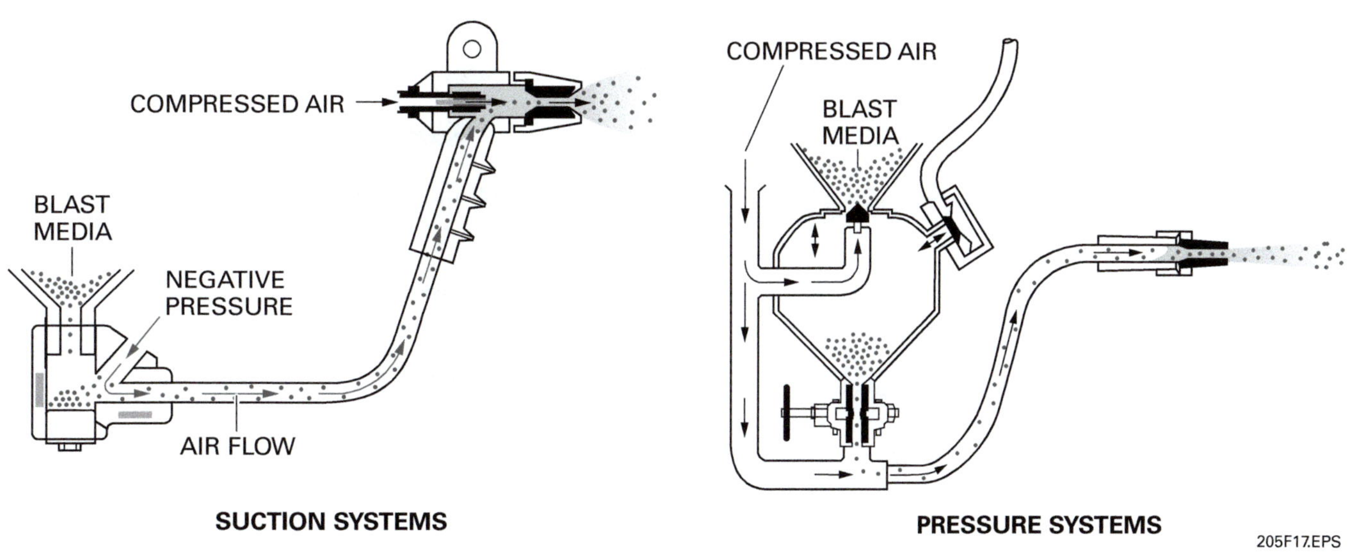

Figure 17 Dry abrasive blast systems.

the nozzle, a suction is created, which pulls blast media through the nozzle. In the pressurized dry abrasive system, the compressed air is run into the blast pot so abrasive media and compressed air are pushed through the hose together. Two types of control systems (*Figure 18*) are used on pressured blast machines. They are known as tank-under-pressure control and tank-pressure-release or tank-blowdown control. The control system valves may be pneumatically operated by twin-line systems as shown or, for non-explosive areas, by 12V or 24V control systems. Single-line control systems may also be encountered, but they are not OSHA-approved. Some systems also have a separate control at the nozzle handle to regulate and/or shut off abrasive flow from the blast machine, but leave the air flow on. This provides a convenient way to clean the hoses and work surfaces.

Troubleshooting problems in the system depends on the exact type of system in use. Always consult the manufacturer's instructions before using or troubleshooting a system. Some general problems and causes are listed in *Table 3*.

It is important to maintain pressure at a blast nozzle. As shown in *Table 4*, as the blast nozzle wears, the area increases according to the formula $A = \pi r^2$ where $\pi = 3.14$ and r is the radius of the blast nozzle.

Nozzle size is measured in gradients of 1⁄16 of an inch; therefore going from a #4 nozzle to a #6 nozzle increases the diameter of the nozzle by 1⁄8 of an inch and increases the volume cubic foot per minute (CFM) of air passing through the nozzle by two (*Table 5*). This has a direct result on blasting efficiency. As the area increases by the square of the radius, the volume of air passing through the nozzle increases, and as a direct result, the pressure decreases. Each 1 psi (7 kPa) drop in air pressure reduces efficiency by about 1.5 percent. Thus, a drop of 10 psi (kPa) equals a 15 percent loss of efficiency. *Table 6* shows metric air requirements for various nozzle sizes.

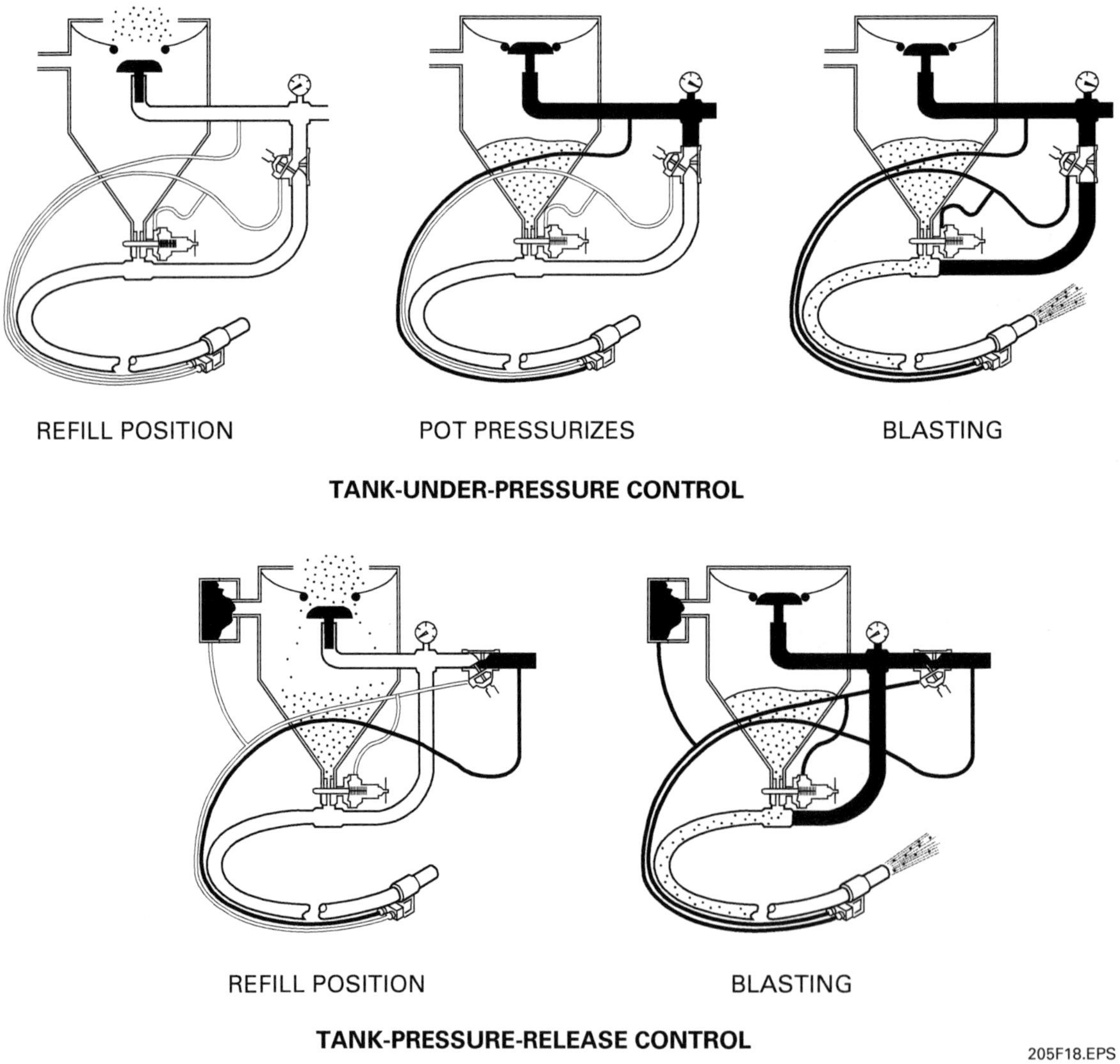

Figure 18 Blast machine pneumatic control systems.

A 1⁄16 inch (1.6 mm) increase in nozzle orifice size (going from a #3 to a #4 nozzle) can increase the volume of air required to maintain the same pressure by 25 percent. To ensure enough CFM to the blast nozzle, the blast hose diameter should be three to four times the diameter of the nozzle being used.

Three things can adversely affect the pressure at the nozzle:

- Length of the air hose
- Number of bends in the air hose
- Elevation of the blast nozzle above the compressor

Table 3 Dry Abrasive Blasting Troubleshooting

Symptom	Suction System	Pressurized System
Lower than expected air flow.	Check air compressor operation. Check hose connections.	Check air compressor operation. Check hose connections. Check for a clogged outlet valve at the blast pot.
No air flow.	Check air compressor operation. Check hose connections.	Check air compressor operation. Check hose connections. Check for a clogged outlet valve at the blast pot.
Lower than expected abrasive flow.	Check abrasive hose connection. Check for a clogged outlet valve at the blast pot.	With good airflow, the blast pot is likely empty or the blast pot pressure is too low. With poor airflow, the blast pot outlet valve or the nozzle is likely clogged.
No abrasive flow.	Check for a clogged outlet valve at the blast pot. Check for abrasive hose connection. Check for an empty blast pot.	With good airflow, the blast pot is likely empty. With poor airflow, the blast pot outlet valve or the nozzle is likely clogged.
Abrasive flows in a pulsating fashion.	Not applicable.	Pressure at the blast pot is too high.
Abrasive media is expelled in clumps.	Check for moisture or oil in the compressed air.	Check for moisture or oil in the compressed air or in the blast pot.

205T03.EPS

Table 4 Relationship Between Nozzle Size and Area

Nozzle Size	Diameter (in)	Radius (in)	Area (sq in)
#4	0.25	0.125	0.05
#5	0.31	0.155	0.075
#6	0.38	0.19	0.113
#7	0.44	0.22	0.15
#8	0.5	0.25	0.2

205T04.EPS

Table 5 Air Requirements for Various Nozzle Sizes (Standard)

Blast Nozzle Orifice		Volume of Air Required (ft³/min)				
Size	Diameter (in)	60 psi	70 psi	80 psi	90 psi	100 psi
#4	1⁄4	67	76	85	94	103
#6	3⁄8	151	171	191	211	232
#8	1⁄2	268	304	340	376	413

205T05.EPS

Table 6 Air Requirements for Various Nozzle Sizes (Metric)

Blast Nozzle Orifice		Volume of Air Required (L/min)				
Size	**Diameter (mm)**	**4.1 bar**	**4.8 bar**	**5.5 bar**	**6.2 bar**	**6.9 bar**
#4	6.3	1,900	2,150	2,400	2,600	2,920
#6	9.45	4,280	4,840	5,410	5,980	6,570
#8	12.6	7,900	8,610	9,630	10,650	11,700

205T06.EPS

To maintain efficient blast operations, the blast nozzle orifice and the nozzle pressure should be checked regularly.

4.2.1 Nozzle Orifice Check

Nozzle orifices are measured in 1/16-inch increments, so a No. 3 nozzle has a 3/16-inch (4.8 mm) orifice. Nozzle orifices should be checked regularly with a pressure blast analyzer gauge (*Figure 19*), commonly called a nozzle gauge, as follows:

Step 1 Mark the rounded side of the gauge using a grease pencil for about 5/8" centered at the approximate size of the nozzle.

Step 2 Insert the gauge into the back of the nozzle (*Figure 20*) and twist it one full turn (360 degrees).

Step 3 Withdraw the gauge and examine the black mark. The inner diameter of the nozzle is indicated by where the grease pencil mark is rubbed off.

Replace worn nozzles in accordance with company policy or the manufacturer's recommended practice—whichever is more stringent. Some workplaces replace nozzles when they are one size larger (1/16-inch or 1.7 mm) than their original size, but others may require that a nozzle be replaced when the amount of wear reaches 50 percent of the original size. For example, a No. 3 nozzle has a 3/16-inch (4.8 mm) aperture, and 50 percent of 3/16 inch is 3/32 inch (3/16 = 6/32, and 50 percent of 6/32 = 3/32). When the nozzle aperture is 9/32 inch in diameter (6/32 + 3/32 = 9/32), replace it.

4.2.2 Nozzle Operating Pressure Check

Nozzle pressure must be tested when all of the nozzles are operating, as well as the entire system. This is done using a hypodermic needle pressure gauge (*Figure 21*) during the blasting operation. The needle of the gauge is inserted into the hose in the direction of abrasive flow, as close as possible behind the blast nozzle when the hose is blasting with abrasive. The expected pressure is 90 to 100 psi (6.33 kg/cm² to 7.03 kg/cm²). When

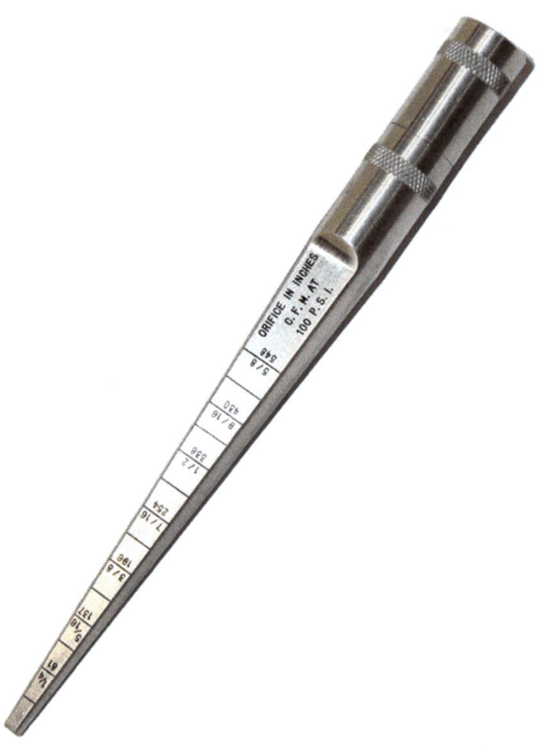

205F19.EPS

Figure 19 Pressure blast analyzer gauge.

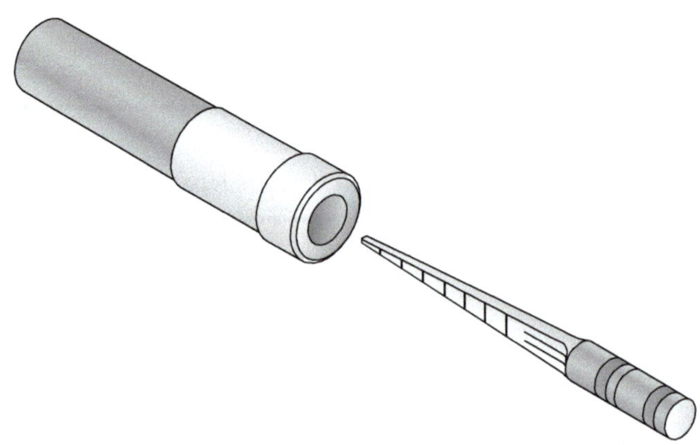

205F20.EPS

Figure 20 Inserting gauge at back of nozzle.

205F21.EPS

Figure 21 Needle pressure gauge.

inserting the needle, it should be put in at approximately a 30-degree angle. Hoses are self-healing.

Abrasive media, especially steel media, must be stored in bags in clean locations under low humidity conditions. If temporarily stored outdoors for very short periods prior to use, the bags must be raised off the ground and covered.

4.3.0 Low-Volume Water Abrasive Blasting

Low-volume water abrasive blasting is a cleaning method where a small amount of water is injected into the abrasive stream at the nozzle of a conventional pressurized blast machine. This type of blasting may also be referred to as air/water/abrasive blasting and includes water-shroud blasting, wet-head blasting, wet blasting, water induction nozzle (WIN) blasting, and slurry blasting. The major advantage of water abrasive blasting is that the water helps remove contaminants from the substrate and substantially reduces the dust normally generated by dry blasting. After blasting, the substrate must be rinsed using low-pressure water cleaning methods to remove any residual abrasive and contaminants. The disadvantage is that collection and disposal or recycling of the water may be a problem, along with flash rusting of steel surfaces. Clean potable water must be used to prevent contamination of the substrate during blasting and subsequent rinsing. An inhibitor, if compatible with the base coating, can be used to reduce flash rusting.

Various types of equipment are used for water abrasive blasting. The water ring, also called a radial water injector, is an accessory that clamps on the end of a dry abrasive blast nozzle (*Figure 22*). The ring has a number of outlet jets at the edge of the ring that provide a water shroud around the abrasive blast stream. A needle valve in the water line near the nozzle allows adjustment of the amount of water shrouding for the abrasive blast stream. The wet-blast injector is an accessory that fits between the abrasive blast hose and the nozzle. Water is injected via jets inside the adapter and mixes with the abrasive in the nozzle. As with the water ring, a needle valve allows water adjustment for the abrasive blast stream. If the blast machine is equipped with an abrasive cutoff capability at the nozzle, the air and water can be used to clear abrasive and rinse the substrate.

A slurry blast machine operates differently. In this machine, pressurized water, abrasive, and air are very precisely mixed to form slurry, usually at the machine. The slurry is supplied through a blast hose to a nozzle. Nozzle pressures range from 20 to 80 psi (138 to 551 kPa). These types of units can also be used for underwater abrasive blast cleaning up to depths of 490 ft (149 m).

As with dry abrasive blasting, the profiles obtained using these types of equipment meet the requirements of white metal, near white, commercial, and brush-off blast cleaning. However, the surfaces look different than dry abrasive cleaned surfaces. They normally appear darker and duller than the same substrates cleaned with the same

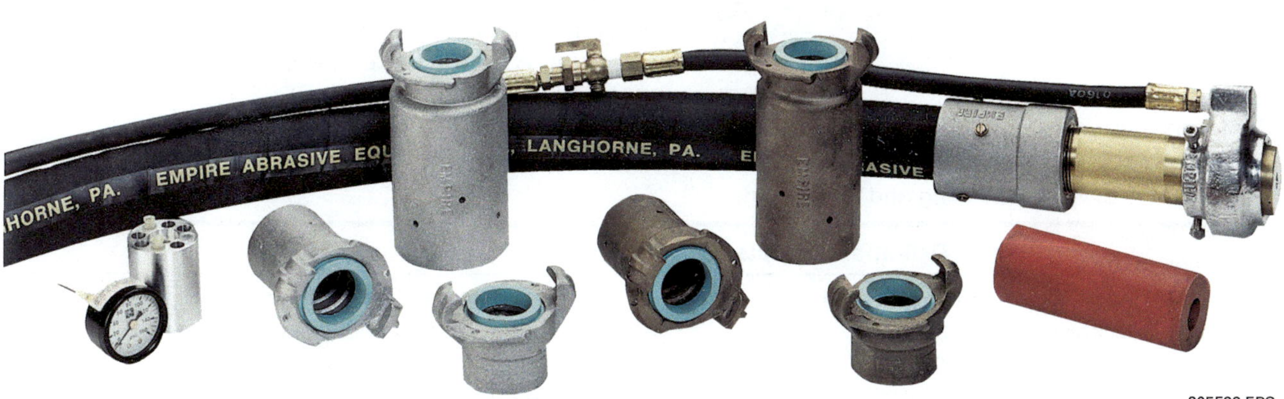

205F22.EPS

Figure 22 A water ring accessory.

abrasive using dry blasting. Additionally, variations caused by the type of steel, original surface condition, fabrication marks, and blasting techniques may occur. When the surfaces are wet or damp, defects, variations, and shading are magnified. After drying, streaks will form that may not be seen in small comparison photographs, but are visible over larger areas. Comparison photographs for wet abrasive blast cleaned steel surfaces are provided in *NACE VIS 9/SSPC-VIS 5*.

5.0.0 WATERJETTING

Pressurized water cleaning of surfaces without an abrasive is usually referred to as waterjetting (*Figure 23*). The pressures (low, high, and ultrahigh) that are used to distinguish water cleaning and waterjetting are defined in *NACE No. 5/SSPC-SP 12 Surface Preparation and Cleaning of Metals by Waterjetting Prior to Coating*. These definitions are shown in *Table 7*; however, other trades or organizations may define these pressures over different ranges. An example is the *ASTM 1575* standard that defines high as anything over 1,000 psi (6.9 MPa).

Water cleaning and waterjetting at various pressures can clean, but cannot abrade a surface to provide a profile for coating adhesion. If only water cleaning or waterjetting is used, any coatings applied must rely on the existing surface profile for adhesion. However, when abrasive is added to an HP or UHP waterjetting process, a surface can be cleaned and profiled or cut. This process is usually referred to as pressurized water/abrasive blasting or hydroblasting in the trade. Cutting that also uses abrasives is called water cutting or hydrocutting.

205F23.EPS

Figure 23 Waterjetting.

For manufacturing purposes, highly controlled and contained commercial waterjetting applications use water pressures up to 100,000 psi (700 MPa), with or without abrasives, for machining metals and other materials. However, waterjetting equipment manufactured for field use is currently limited to a maximum of 50,000 psi (340 MPa). A typical diesel-powered, wheel-mounted field unit for waterjetting is shown in *Figure 24*. This type of equipment can also be skid-mounted and powered by gasoline engines or electric motors. Besides the power source, the equipment is usually comprised of pumps, hoses, and various tools such as guns, nozzles, lances, and in some cases, hand-guided machines or remotely controlled robotic crawler units. A process called hydrodemolition, using a type of robotic crawler unit with pressures up to 50,000 psi (340 MPa) and with or without abrasives, can be used to remove and scarify damaged concrete. Concrete layers in floors and structures down to various rebar levels can also be removed without damaging the rebar. It is also used to remove entire concrete bridge decks without damaging the rebar.

To be effective, the water used for waterjetting should be clean so it does not contaminate the surface being prepared. The water should be free of solid particles and other contaminants. This also avoids damage to the equipment. This water must be tested to make sure it meets specified requirements.

UHP waterjetting can remove tightly adherent mill scale and coatings, but this is not always necessary. The dull putty knife test still applies. If mill scale, rust, and existing coatings cannot be removed by lifting with a dull putty knife, they may not need to be removed. The project specification will indicate the degree of cleanliness needed.

Table 7 Definitions of Water Cleaning and Waterjetting

Definitions of Water Cleaning and Waterjetting	
Low-pressure water cleaning (LP WC)	Less than 5,000 psi (34 MPa)
High-pressure water cleaning (HP WC)	5,000 to 10,000 psi (34 to MPa)
HIgh-pressure waterjetting (HP WJ)	10,000 to 30,000 psi (70 to 210 MPa)
Ultrahigh-pressure waterjetting (UHP WJ)	Over 30,000 psi (210 MPa)

205T07.EPS

Figure 24 Diesel-powered, wheel-mounted waterjetting unit.

Waterjetting can also remove the following:

- Surface oil and grease
- Rust
- Salts
- Concrete spatter
- Water-soluble contaminants

The specifications may require testing of the prepared surface for a variety of contaminants that can prevent coatings from adhering to it. These include salts such as chlorides, nitrates, and sulfate contaminants. Surface cleanliness after waterjetting is defined as follows:

- *Nonvisible contamination (NV)* – The presence of invisible organic matter, such as oil and grease, and/or salts on the surface after cleaning.
- *Visible surface cleanliness (VC)* – When viewed without magnification, the surface is free of all contaminants after cleaning.

The project specification should state the method to be used for determining the amount of nonvisible contaminants allowed on the surface after cleaning. The joint NACE-SSPC standard for surface preparation with waterjetting is *NACE No. 5/SSPC-SP 12, Surface Preparation and Cleaning of Metals by Waterjetting Prior to Recoating*. It specifies surface cleanliness in terms of WJ-1 to WJ-4 for visual cleanliness. These terms are defined in *Table 8*, which is reproduced from the joint standard *NACE No. 5/SSPC-SP 12*. *Table 9* from the standard provides flash rust definitions.

5.1.0 Safety Guidelines

The following safety rules apply whenever waterjetting operations take place:

WARNING

Only thoroughly trained personnel should ever be allowed to operate a waterjetting gun. This equipment can cut through heavy timbers or concrete block. Accidental contact with the jet is always a medical emergency. It is possible for skin penetration to take place at 100 psi. Even the low-pressure systems running below 1,000 psi can seriously injure a person struck by the jet.

- Never point a waterjetting nozzle at or yourself or anyone else
- Remember, the water that comes out of the nozzle is capable of removing skin. If it is close enough, it may cut through flesh. Even with personal protective clothing, the jet will cut away the armor more quickly than expected.
- Even the best armor is designed to only resist a short swipe. It cannot stand up to a constant stream of high-pressure water.
- If the water does not cut off when the dump valve is released, stop working immediately. Shut off pressure at the pump, drain the system, and have the valve repaired or replaced.
- Never try to repair the system while it is pressurized. Always stop the operation, shut down the pump, and release pressure before repairing leaks or failures.
- Never leave the pump unattended while it is running.

Table 8 Visual Surface Definitions

Term	Description of Surface
WJ-1	*Clean to bare substrate* – A WJ-1 surface shall be cleaned to a finish which, when viewed without magnification, is free of all visible rust, dirt, previous coatings, mill scale, and foreign matter. Discoloration of the surface may be present.
WJ-2	*Very thorough or substantial cleaning* – A WJ-2 surface shall be cleaned to a matte (dull, mottled) finish which, when viewed without magnification, is free of all visible oil, grease, dirt, and rust except for randomly dispersed stains of rust, tightly adherent thin coatings, and other tightly adherent foreign matter. The staining or tightly adherent matter is limited to a maximum of 5 percent of the surface.
WJ-3	*Thorough cleaning* – A WJ-3 surface shall be cleaned to a matte (dull, mottled) finish which, when viewed without magnification, is free of all visible oil, grease, dirt, and rust except for randomly dispersed stains of rust, tightly adherent thin coatings, and other tightly adherent foreign matter. The staining or tightly adherent matter is limited to a maximum of 33 percent of the surface.
WJ-4	*Light cleaning* – A WJ-4 surface shall be cleaned to a finish which, when viewed without magnification, is free of all visible oil, grease, dirt, dust, loose mill scale, loose rust, and loose coating. Any residual material shall be tightly adherent.

205T08.EPS

Table 9 Flash Rust Definitions

Term	Description of Surface
No flash rust	A steel surface which, when viewed without magnification, exhibits no visible flash rust.
Light (L)	A surface which, when viewed without magnification, exhibits small quantities of yellow-brown rust layer through which the steel substrate may be observed. The rust or discoloration may be evenly distributed or present in patches, but it is tightly adherent and not easily removed by lightly wiping with a cloth.
Moderate (M)	A surface which, when viewed without magnification, exhibits a layer of yellow-brown rust that obscures the original steel surface. The rust layer may be evenly distributed or present in patches, but it is reasonably well adherent and leaves light marks on a cloth that is lightly wiped over the surface.
Heavy (H)	A surface which, when viewed without magnification, exhibits a layer of heavy red-brown rust that hides the initial surface condition completely. The rust may be evenly distributed or present in patches, but the rust is loosely adherent, easily comes off, and leaves significant marks on a cloth that is lightly wiped over the surface.

205T09.EPS

- In addition to the standard equipment, all workers in the area of a waterjetting operation must wear waterproof suits, head protection, and protective goggles or a full-face shield. Shotgun operators must wear protective toe and shin guards, gloves, and a full face shield. Since water and surface debris is sprayed around the area, it is a good idea for operators to pause at intervals and wipe the face shield. This way they can see hazards, as well as the work itself. Lance operators are not required to wear shin and toe guards, only rubber boots. The waterproof clothing worn for this work is not enough to stop a direct swipe with a jet. Armor will protect a person from only a very brief contact with the jet. The jet will deliver a very heavy physical blow, even through the armor.
- Be very careful to make sure of proper footing. Back pressure from the lance can cause a fall if a misstep occurs.
- The sound level of a waterjetting operation runs in excess of 90 decibels. Hearing protectors or earplugs are required. In addition, if an operator is removing possibly toxic substances from surfaces, breathing equipment or masks may be necessary.

All blast accidents must be reported immediately. Even when the blast does not visibly penetrate the skin, any contact can force water, air bubbles, and toxins under the skin. This is true even if there is no visible blood. There can be a serious risk of infection and permanent damage. Take the injured person to a doctor, hospital, or emergency room immediately. The Water Jet Technology Association says, in its Recommended Practices for the Use of Manually Operated High-Pressure Waterjetting Equipment, that all operators should carry a waterproof card to show to physicians that reads:

> This man has been involved with high-pressure waterjetting at pressures up to 14,500 lb/in^2 (1,019.5 kg/cm^2) with a jet velocity of 900 miles per hour (1,448 km/hr). Please take this into account when making your diagnosis. Unusual infections with micro-aerophilic organisms occurring at lower temperatures have been reported. These may be gram-negative pathogens such as are found in sewage. Bacterial swab and blood cultures may therefore be helpful.

If the incident takes place in a remote area, the wound should be covered. The worker should remain quiet until medical care can be provided.

One of the injuries caused by water jets is called channeling. This is an injury where the jet enters the body through a small entrance wound and travels along some channel inside to produce trauma or infection at some other place. Take no chances with injuries. Check the company's rules on accident reporting; if any accident occurs, follow company procedures. Report all accidents, no matter how small, to the appropriate supervisor.

5.2.0 Environmental Issues

While waterjetting is a relatively clean technology and uses water efficiently, it can produce a lot of wastewater. On many sites, clean water may be channeled to storm drain systems. In some places, water is recovered and recycled, using either filtration systems or a series of settling tanks, or both. Be sure to check local codes and company requirements affecting discharge of cleaning water before setting up equipment for a project.

It is very important to know what is to be blasted because it may be toxic. If so, the project must capture wastewater. Many times, the waterjetting operator will be working to remove potentially toxic materials from a surface. The materials may be blasted loose and either thrown into the air as tiny particles or as vapors mixed with water. If it is possible that toxic airborne contaminants will be released, wear a breathing mask capable of protecting from such hazards.

5.3.0 Waterjetting Equipment

Besides the waterjetting unit, various tools such as guns, nozzles, and lances are used. The high-pressure hose, hose connections, and all other equipment, including the control valve (dump valve), lance, and nozzle, should have minimum burst strength of 2½ times the capability of its maximum-rated operating pressure. *Figure 25* shows a shotgun with the various components identified.

The hose connectors are fitted with a safety device known as a whiplock or whip check. A whiplock is a short length of cable or wire looped over each end of two hoses connected by a coupling. The whip check (*Figure 26*) prevents the ends of the hose from whipping around if the coupling breaks.

The section of hose next to the gun is fitted with a hose shroud (see *Figure 25*). The shroud is a short length of heavy-duty hose placed over the high-pressure hose. It provides protection if the hose bursts. A hose shroud also can be used over the hose connections. The shroud is designed to protect the operator from a high-pressure stream of water. It is not designed to allow the user to keep operating with a damaged hose or broken connection.

Waterjets are produced by orifices or tips (*Figure 27*) that come in different forms. Straight jets are most commonly used, but other shapes are available. The straight jets (*Figure 28*) are spot removers. The fan jets (*Figure 29*) are sweepers and scrapers and/or pushers. Interchangeable nozzle tips normally are used to produce the desired streams.

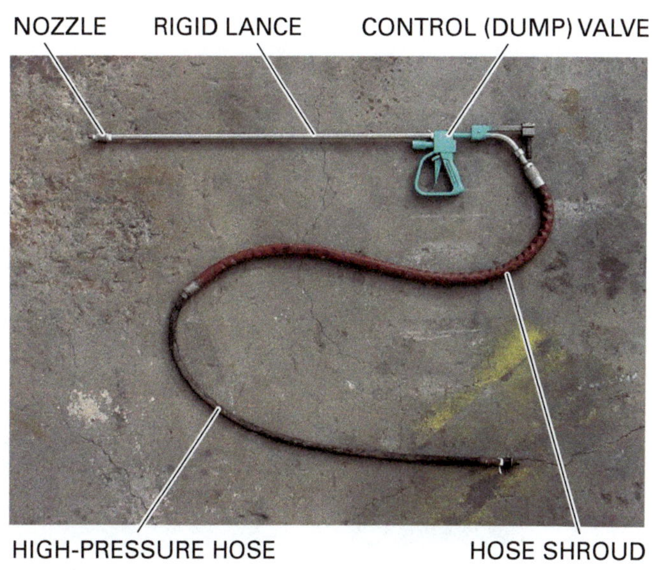

Figure 25 A shotgun.

Typical water flow rates are 1 to 14 gallons per minute (3.8 to 53 lpm). Water can be propelled through a single jet, a fan jet, or multiple rotating jets. Rotating jets are turned by small air, electric, or hydraulic motors. They can also be rotated by water pressure when the orifice is slightly tilted in a multiple-orifice nozzle. Tips can be designed to produce multiple jets of water that normally are rotated to achieve higher removal rates.

The tools can be handheld or mounted on a robot. When working with pressures of 10,000 psi (69 MPa) or higher, the cleaning team consists of a gun operator, a pump operator, and other workers as needed. The gun operator controls the cleaning operation by holding the gun and delivery hose and controlling the motion and direction of the jet(s). The pump operator monitors and controls the pressurizing pump during the jetting operation. The pump operator watches the gun operator and needs to be able to react if any difficulty arises, or if the operator begins to show signs of fatigue. The pump operator also monitors the working area and its surroundings in case anyone tries to enter the area, or should a potentially hazardous condition occur. If the pump is located at some distance and out of sight of the nozzle operator, a team member may be required to monitor the jetting operation and to communicate with the gun operator and pump operator.

One element of operator fatigue is from the back thrust from the high-pressure water (*Figure 30*). The operator should not be required to withstand a back thrust of more than one-third of his or her body weight for an extended period. Back pressure can be calculated using the following formula:

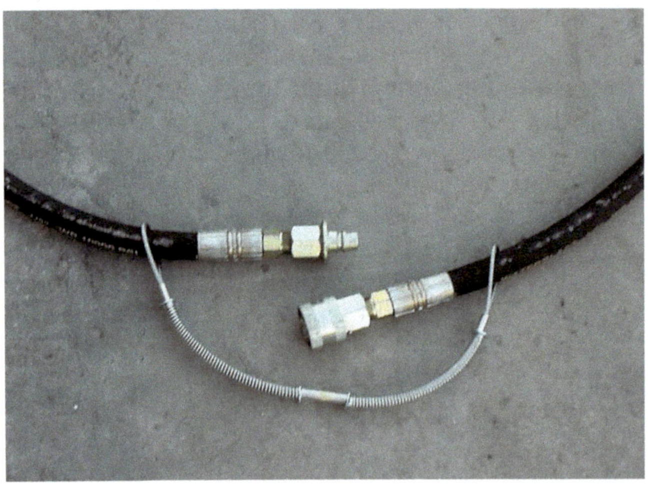

205F26.EPS

Figure 26 A hose whip check.

205F27.EPS

Figure 27 Typical waterjetting tip.

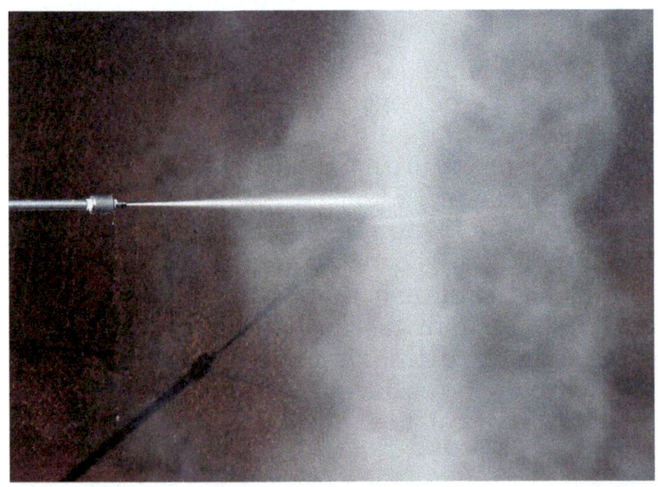

205F28.EPS

Figure 28 Straight jet.

205F29.EPS

Figure 29 Fan jet.

Back pressure (in pounds) equals 0.052 times the flow in gallons per minute (gpm) times the square root of the pressure in pounds per square inch.

For kilograms, the formula is 0.02364 times the flow in liters per minute, times the square root of the pressure in kilograms per centimeter squared.

Using this formula, it is easy to calculate the back pressure from a shotgun. For example, if you are using 10,000 psi and a flow of 12 gpm, the calculation is as follows:

$$B = 0.052 \times F \times P$$
$$B = 0.052 \times 12 \times \sqrt{10{,}000}$$
$$B = 0.052 \times 12 \times 100 = 62.4 \text{ pounds}$$

An operator working with a jet flowing at 10,000 psi (703.07 kg/cm^2) and 12 gpm (45.4 lpm) will experience a back-thrust force of 62.4 pounds (28.4 kg). The operator should weigh at least 187 pounds (85 kg) to operate the nozzle. Newer units operate with less back thrust than some of the earlier units.

When the pump operator needs to reduce the pressure in the supply hose for any reason, it must be done slowly. Rapidly reducing the system pressure can cause the gun operator to lose his or her footing due to the sudden change in back pressure.

Depending upon the size and scope of the project, other operators or workers may be required to assist in handling the jetting gun. This is especially true if it has more than one jetting extension or if the hose must be fed to the workpiece. To decrease operator fatigue and to ensure a safe operation, workers can alternate the gun operator position at designated intervals.

5.4.0 Waterjetting Preparation

Before starting the job, the surface preparation team should ensure the following:

- The work area is properly barricaded and appropriate warning signs are installed.
- Electrical equipment is properly covered and protected from the water.
- Electrical connections are not allowed to sit in water.
- One member of the team should be assigned the responsibility of checking that all necessary permits are in place. This includes any confined space permits, as well as those that may be required by local, state, and federal agencies.
- All fittings and hoses are in good condition (not worn or damaged). They must be properly pressure rated for the working pressure to be used.
- Nozzles are open and free of obstructions.
- The complete hook-up is flushed clean and air removed from the system before installing the nozzle.
- The dump system and all control systems are operational.
- All relevant moving equipment, such as conveyors, mixers, etc., is mechanically or electrically disabled. The appropriate lockout devices must be in place.

5.5.0 Dump Valve

Waterjetting devices are equipped with a dump valve (*Figure 31*). Activating the dump valve allows the water to pour out of the system at a safe pressure. Never operate a system if the dump valve does not work. Always stop work, turn off the pump, and repair or replace the valve. Do not

205F30.EPS

Figure 30 Braced for back pressure.

205F31.EPS

Figure 31 Dump valve.

service or adjust any components when the pressure is on. Always shut down first.

5.6.0 Waterjetting Procedure

When setup and testing is completed, the operations stage begins. It is not in the scope of this text to describe the specific operation of the equipment. However, a few general observations may be useful.

Before bringing the pump up to high pressure for cleaning, start at a low pressure, without a nozzle on the lance. Flush the system to remove air bubbles and particles. Then, check the operation of the dump system and look for leaks anywhere in the system.

If any component is malfunctioning or leaking, do not attempt to repair the connection under pressure. Shut down the pump, and find out why. Do not ignore leaks or other problems. Catastrophic failure of components could result, with possible serious consequences. If the hose begins to pulsate when running, the pump may not be receiving enough water at the suction side. Turn it off and check for problems. Another indication of insufficient water supply might be a banging sound from the pump. This is called cavitation.

After flushing the system, turn it off and install the nozzles. Now test the system at low pressure and gradually bring the system up to working pressure. Run the pressure up slowly to one and one-half times the working pressure to see if any leaks appear.

If there are any leaks, shut down the equipment and repair them. Never try to repair or tighten connections under pressure.

There is a distinct jolt when the dump valve or dry shutoff valve is operated. This is called a line pulse and any backpressure at the gun will immediately disappear. Be prepared for it.

5.7.0 Shotgunning

The most common waterjetting technique is shotgunning. When using a shotgun to clean large surface areas, be continually conscious of where the tip is pointed. The effectiveness of the cleaning depends on how close the tip is to the surface being cleaned. With a fan jet, the wider the area being covered in a sweep, the less force is available. The closer the tip is to the surface, the greater the backsplash. Be sure to get accustomed to the back pressure gradually. Loss of control of the gun or falling may occur due to sudden exposure to the kick. Remember to check footing constantly, as the combination of water and oil can produce a very slippery surface. If loss of control feels imminent, activate the dump valve immediately.

5.8.0 Tanks

Waterjetting tanks, boilers, cooling towers, and other enclosures require specific techniques. The use of a telescoping lance and a rotating T-head with two spinning nozzles (*Figure 32*) allows the operator to work inside without being inside. It is still very critical that the nozzles be kept from coming out of the opening while under pressure.

Mark the lance at least 24 inches (61 cm) from the head to avoid accidentally pulling the head clear of the tank. It is best to lower the nozzles into the tank from above to prevent the wash material from interfering with control.

5.9.0 Cleanup

Cleanup after surface preparation is an important part of a job. Always meet or exceed cleanup requirements. Coil and store all hoses and check them in the process. Also check the valves and fittings. Examine the pump and driver so that there are no delays in using them on the next job.

5.10.0 Other Waterjetting Equipment

In addition to waterjetting guns, there are carriage-mounted surface cleaners, used for

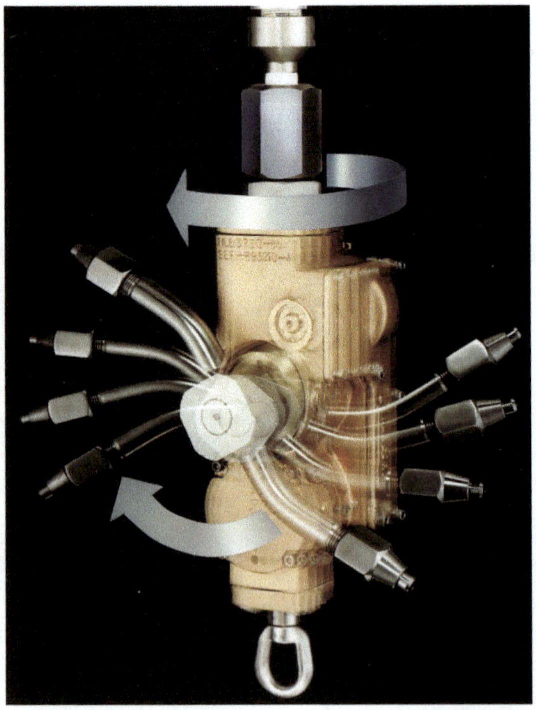

205F32.EPS

Figure 32 Rotating T-head with two spinning nozzles.

cleaning large floor areas (*Figure 33*). These are operated like a floor scrubber, cleaning a path on each pass, with the operator walking behind. Pressures of up to 40,000 psi (276 MPa) are used. Spin nozzles under the housing are completely protected by the housing, reducing the hazard for the operator.

For the outside of large pipes, there are automatic rigs on tracks. For large wall areas, including the exterior of tanks, robotic crawlers are available. Hydrodemolition of deteriorating concrete is accomplished at 20,000 psi (138 MPa) using remotely controlled equipment. These and most other waterjetting machines are equipped with vacuum shrouds to collect and remove the spent blast water, any abrasives, and debris. For hydroblasting surfaces to obtain cleaning as well as profile, abrasive injection waterjetting systems operating at up to 40,000 psi (276 MPa) are available along with various guns.

205F33.EPS

Figure 33 Waterjetting floor cleaners.

SUMMARY

Surface preparation can be the most time consuming and expensive part of a coating operation. The goal of surface preparation is always the same—to remove grease, oils, salts, and other contaminants that will prevent coatings from adhering to a surface. It requires experience and knowledge to properly prepare a surface for coating.

During surface preparation, document all aspects of the operation. Perform tests to determine the cleanliness of a surface including using indicator paper to test for pH, commercial kits to determine soluble salt contamination, heat lamps and ultraviolet lamps to identify hydrocarbon contamination, and solutions that identify hydrocarbon and salt contamination, as well as hexavalent chromium.

Specialized materials and equipment used for wet and dry abrasive blasting were also discussed, as well as safety precautions and environmental considerations.

Review Questions

1. The standard for evaluating the degree of rusting on painted steel surfaces is contained in _____.
   a. *SSPC-VIS 1*
   b. *SSPC-VIS 2*
   c. *SSPC-VIS 3*
   d. *NACE VIS 7/SSPC-VIS 4*
2. A dull putty knife is used to _____.
   a. determine whether old coatings are tightly adhered to a surface
   b. remove rust and mill scale from weathered steel surfaces
   c. repair voids and cracks in the surface of concrete
   d. restore the surface profile after grinding
3. When water beads up on a steel surface, it is most likely caused by _____.
   a. grease or oil
   b. salt
   c. dirt
   d. acid
4. To test for oil and grease on a concrete surface, use _____.
   a. chloride
   b. hydrochloric acid
   c. chromate
   d. lead acetate
5. To test for salts on metal, start with _____.
   a. a darkened area
   b. lead acetate
   c. a magnifier
   d. deionized water
6. A calcium chloride test determines _____.
   a. salt contamination on steel
   b. the moisture level of concrete
   c. hydrocarbon content of concrete
   d. presence of chromate on steel
7. Flash rust can be caused by _____.
   a. excessive dehumidification
   b. the lack of surface salts
   c. the use of an inhibitor
   d. surface contamination
8. A relative humidity meter measures humidity relative to _____.
   a. air temperature
   b. moisture
   c. a variable
   d. a constant
9. Abrasive blasting can achieve a surface profile on _____.
   a. steel
   b. concrete
   c. steel and concrete
   d. neither steel nor concrete
10. Hydrochloric acid can achieve a surface profile on _____.
    a. steel
    b. concrete
    c. steel and concrete
    d. neither steel nor concrete
11. The surface profile obtained using replica tape is measured with a(n) _____.
    a. thickness gauge
    b. hydrometer
    c. sling psychrometer
    d. RH meter
12. Environmental conditions that directly affect surface preparation operations are _____.
    a. temperature and humidity
    b. dust and temperature
    c. visibility and humidity
    d. ventilation and temperature
13. To prevent condensation on a surface, the surface must be held at a temperature that is _____ the air.
    a. higher than
    b. lower than
    c. the same as
    d. the same as or lower than
14. To measure air temperature, use a _____.
    a. manometer
    b. hygrometer
    c. micrometer
    d. thermometer

15. Of the following statements recorded in project documentation, which best describes why an action was or was not performed?
    a. I performed a heat lamp test on the concrete surface on September 18, 2008 with Dan Smith and Tom Jones.
    b. One ounce of hydrochloric acid solution bubbled for 10 minutes when applied to the surface of the concrete.
    c. Could not evaluate surface due to a uniform layer of mud on the surface. Will evaluate after power washing.
    d. See attached pictures.

16. Liquid-applied coatings tend to thin over _____ as they dry.
    a. gouges
    b. pinholes
    c. angles
    d. rivets

17. When testing for chromate with lead acetate, the metal is chromate-free when the test area _____.
    a. darkens
    b. lightens
    c. turns red
    d. turns green

18. On weathered steel, adherent and non-adherent corrosion _____.
    a. are easily distinguished by texture
    b. have the same color and texture
    c. should both be removed
    d. should be left in place

19. To determine whether corrosion is non-adherent on weathered steel, use a _____.
    a. wire brush
    b. power grinder
    c. sharp putty knife
    d. dry abrasive blasting machine

20. Copper surfaces _____.
    a. oxidize, so they do not need coating
    b. corrode quickly so they need to be coated
    c. never corrode
    d. are sacrificial, so they are supposed to corrode

21. Resin repairs should be made _____.
    a. before concrete cures
    b. while concrete is curing
    c. on wet surfaces
    d. on dry surfaces

22. Concrete is etched with _____ acid.
    a. hydrochloric acid
    b. nitric acid
    c. hydrocarbonic acid
    d. hydrogen peroxide

23. Acid etching is covered in _____.
    a. *ASTM D4360*
    b. *ASTM D4250*
    c. *NACE No. 6/SSPC-SP 13*
    d. *NACE No. 3/SSPC-SP 6*

24. Efflorescence occurs on _____.
    a. concrete
    b. steel
    c. copper
    d. plastics

25. A patch test is performed to determine _____.
    a. if new abrasives are in specifications
    b. if two coatings are compatible
    c. that compressed air is oil-free
    d. surface profile

26. A tie coat is compatible with _____.
    a. both the new and existing coatings
    b. neither the new nor existing coatings
    c. the new but not existing coating
    d. the existing but not new coating

27. To detect oil in compressed air, you need to _____.
    a. discharge air toward blotter paper for one minute
    b. drain the compressor oil trap
    c. use an ultraviolet light
    d. use Blastox®

28. A suction-type dry abrasive blast system requires two hoses to the nozzle, one for the media and one for the compressed air.
    a. True
    b. False

29. Blast nozzle aperture is determined using a _____.

    a. spring-loaded internal nozzle gauge
    b. hypodermic-type needle gauge
    c. simple internally inserted scale
    d. pressure blast analyzer gauge

30. Pressures that distinguish water cleaning and waterjetting are defined in _____.

    a. *NACE No. 6/SSPC-SP 13*
    b. *NACE No. 3/SSPC-SP 6*
    c. *NACE No. 5/SSPC-SP 12*
    d. *NACE No. 2/SSPC-SP 10*

Trade Terms Introduced in This Module

Feathering: Working the edges of the repaired area back, so there is a fairly smooth transition from the repair area to the sound coating.

Hydrocarbons: Petroleum products such as oil or grease.

Hygrometer: Humidity measuring device.

Microsiemens: Units of electrical conductivity.

Patch test: A test used in overcoating operations in which the selected coating is applied over a small area of the existing coating to see if they are compatible.

Sieve test: A test performed on abrasives as they are received to see if they are sized to specification.

Tie coat: A coating that is compatible with both the existing coating and the new coating.

Additional Resources

This module presents thorough resources for task training. The following resource material is suggested for further study.

Coating Inspector Program Level 1 © NACE International, 2010.

Coating Inspector Program Level 2 © NACE International, 2010.

"In Pursuit of Precision Blasting." J.C. Carson. *Products Finishing Magazine,* August, 2007. Cincinnati, OH: Gardner Publications, Inc. www.pfonline.com

ISO 8501-1 through ISO 8501-4, Preparation of Steel Substrates Before Application of Paints and Related Products – Visual Assessment of Surface Cleanliness.

ISO 8502-3, Preparation of Steel Substrates Before Application of Paints and Related Products – Tests for the Assessment of Surface Cleanliness.

Monitoring Environmental Conditions. David Beamish, DeFelsko Corporation. NACE Resource Center website. Coating Instrument Series from *Materials Performance 2003.* http://events.nace.org

NACE VIS 7/SSPC-VIS 4, Guide and Visual Reference Photographs for Steel Cleaned by Waterjetting.

NACE VIS 9/SSPC-VIS 5, Guide and Reference Photographs for Steel Surfaces Prepared by Wet Abrasive Blast Cleaning.

Painting: New Construction and Maintenance. EM 1110-2-3400. Department of the Army, April 30, 1995.

Reclothing the First Lady of Metals – Repair Concerns. Copper Development Association. www.copper.org

SSPC-VIS 1, Guide and Reference Photographs for Steel Surfaces Prepared by Dry Abrasive Blast Cleaning.

SSPC-VIS 2, Standard Method of Evaluating Degree of Rusting on Painted Steel Surfaces.

SSPC-VIS 3, Guide and Reference Photographs for Steel Surfaces Prepared by Hand and Power Tool Cleaning.

Surface Preparation of the Concrete Substrate. N. P. Mailvaganam, G. B. Pye, and M. R. Arnott. Construction Technology Update No. 24, December 1998.

Zinc Coatings. Centennial, CO: American Galvanizers Association, 2006. www.galvanizeit.org

Figure Credits

Rogers & Associates Consulting Services, Module opener

Photos courtesy of Paul N. Gardner Company Inc., 205F02, 205F04

DeFelsko Corporation, 205F03

Elcometer, Inc., 800-521-0635, www.elcometer.com, 205F05–205F07

O.T.B. Technology, 205F08

ARID-Dry by CDIMS, 205F09

Extech Instruments, 205F10 (electronic hygrometer), 205F13

Topaz Publications, Inc., 205F10 (electronic thermometer), 205F23, 205F26, 205F30

General Tools & Instrument Co., 205F11

Carrier Corporation, 205F12

Rogers & Associates Consulting Services, 205F14

NACE International, 205T01, 205T03-205T09

Clemtex, Inc., 205F19, 205F20

Mike O'Brien, 205F21

Empire Abrasive Equipment Company, 205F22

Photo courtesy of NLB Corp., 205F24, 205F32, 205F33

Ancon Marine, 205F25, 205F27–205F29, 205F31

NCCER CURRICULA — USER UPDATE

NCCER makes every effort to keep its textbooks up-to-date and free of technical errors. We appreciate your help in this process. If you find an error, a typographical mistake, or an inaccuracy in NCCER's curricula, please fill out this form (or a photocopy), or complete the online form at **www.nccer.org/olf**. Be sure to include the exact module ID number, page number, a detailed description, and your recommended correction. Your input will be brought to the attention of the Authoring Team. Thank you for your assistance.

Instructors – If you have an idea for improving this textbook, or have found that additional materials were necessary to teach this module effectively, please let us know so that we may present your suggestions to the Authoring Team.

NCCER Product Development and Revision

13614 Progress Blvd., Alachua, FL 32615

Email: curriculum@nccer.org

Online: www.nccer.org/olf

❏ Trainee Guide ❏ Lesson Plans ❏ Exam ❏ PowerPoints Other ____________

Craft / Level: Copyright Date:

Module ID Number / Title:

Section Number(s):

Description:

Recommended Correction:

Your Name:

Address:

Email: Phone:

Industrial Coatings Two

69206-10

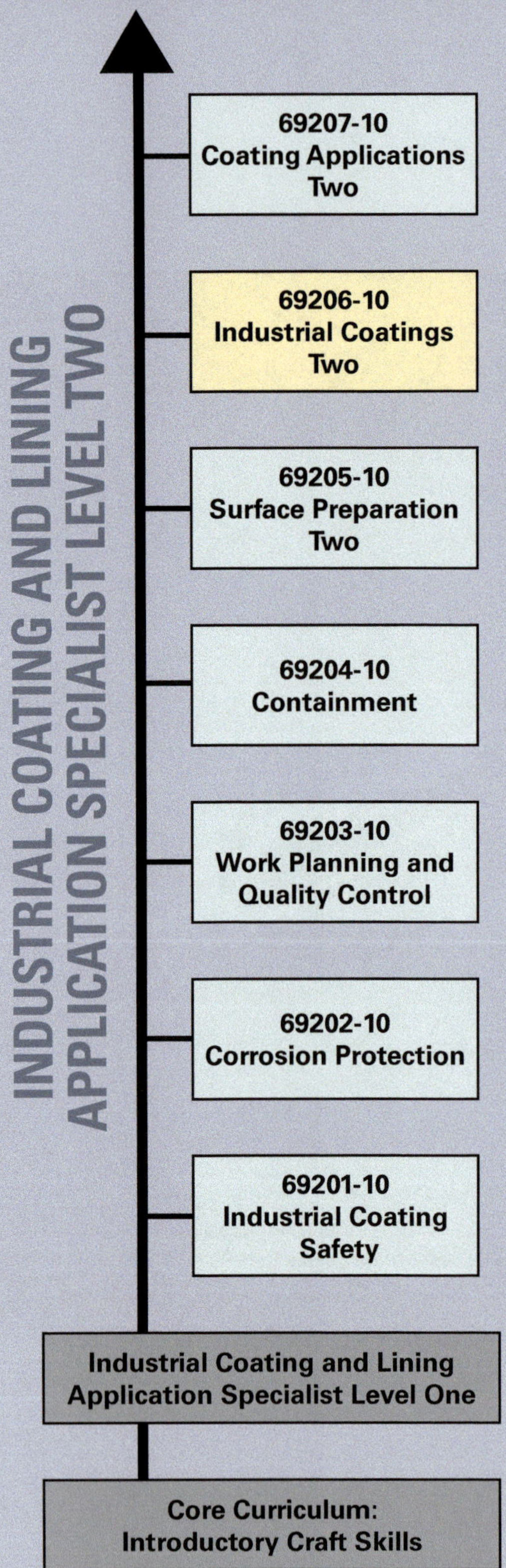

This course map shows all of the modules in *Industrial Coating and Lining Application Specialist Level Two.* The suggested training order begins at the bottom and proceeds up. Skill levels increase as you advance on the course map. The local Training Program Sponsor may adjust the training order.

Objectives

When you have completed this module, you will be able to do the following:

1. Describe physical coating properties.
2. Describe the characteristics of basic nonconvertible and convertible coatings.
3. Describe basic curing mechanisms and methods of film formation.
4. Describe basic pipeline coatings along with their advantages, disadvantages, and application methods.

Trade Terms

Barrier coating
Condensation-cure coating
Coreaction-cure coating
Hydration-cure coating
Inhibitive coating
Opacity
Oxidation-cure coating
Surfactants

Prerequisites

Before you begin this module, it is recommended that you successfully complete *Industrial Coating and Lining Application Specialist Level One* and *Industrial Coating and Lining Application Specialist Level Two,* Modules 69201-10 through 69205-10.

Contents

Topics to be presented in this module include:

Contents *(continued)*

Figures and Tables

1.0.0 Introduction

A fundamental method of corrosion control involves placing a protective coating or lining between a corrosive environment and the substrate to be protected. It is the most widely used method of protecting metals and other substrates. A protective coating is any material applied to a surface that will resist the environment and prevent serious breakdown of the substrate (*Figure 1*). This module provides additional details about the industrial protective coatings and linings covered in the Level One module, *Industrial Coatings*. It also provides detailed coverage for pipeline coatings, fire-resistant/retardant coatings, and powder coatings.

In general, protective coatings are defined as a film of material placed on the exterior of structures to protect them from the effects of the environment, but not necessarily from immersion. Linings and/or barriers are materials placed on the interior of pipes, tanks, containers, process equipment, and structures that are subject to immersion. The primary purpose of a lining is to protect the commodity from the substrate.

1.1.0 Industrial Protective Coating and Lining Uses

Industrial protective coatings and linings may be used for a variety of reasons. These uses are described in the following list:

- *Corrosion control* – Protective coatings are used to retard or eliminate the corrosion of a substrate.
- *Waterproofing* – Coatings or linings can prevent or reduce the transmission of a fluid such as water vapor.
- *Weatherproofing* – Besides waterproofing, some coatings also provide resistance to the ultraviolet rays of the sun or wind corrosion.
- *Biocides* – In the marine industry, tin-free antifouling coatings are used to reduce the attachment of marine organisms to the hulls of ships below the waterline and to other submerged structures when such protection is required.
- *Fire resistance/fire retardancy* – Fire-resistant and fire-retardant coatings are available. *Figure 2* shows shotcreted (slurry) fire-retardant material being applied to metal framing materials of a commercial building.
- *Appearance* – Coatings are used to improve and maintain the appearance of industrial and commercial locations and manufactured goods.
- *Color coding* – Substrates that might not be coated for corrosion-control reasons, such as stainless steel or plastic piping, can be coated as part of a color-code scheme.
- *Sanitation/decontamination* – Coatings may be applied to porous substrates, such as concrete, wood, and plaster, to allow for easier cleaning. Sanitation is important in food processing plants. Decontamination is important in nuclear power plants, where there are millions of square yards of surface area that must be coated to provide a readily decontaminable surface in case of an accident. (This is required by law in *10 CFR 50*.)

206F01.EPS

Figure 1 Coating a building.

206F02.EPS

Figure 2 Applying fire-retardant material.

- *Safety* – Coatings may be applied to make hazards more visible; to symbolize the presence of certain types of hazardous substances, processes, or equipment; and because of safety-marking and color-coding requirements mandated by public law and administered by OSHA under the ANSI program, and in other countries by similar regulatory bodies.
- *Prevention of product contamination* – Linings may be applied to the interior of containers, pipes, or vessels to prevent their deterioration and/or the contamination of a contained product.
- *Friction reduction* – A substrate may be coated to reduce friction; for example, lining the interior of pipelines to decrease resistance to flow.
- *Wear resistance* – High-wear substrates, such as the interiors of slurry pipelines or hoppers that are subject to abrasive materials, may be coated with abrasion-resistant coatings and linings to reduce wear of the substrate.
- *Heat transfer* – Coatings may be used to increase heat transfer (black paints in solar heating applications) or to cut heat transfer (white and reflective paints on buildings and other structures).
- *Electrical insulation* – Certain coatings may be used to provide electrical insulation for wire used in applications such as transformer or motor windings.
- *Sound deadening* – Foam insulation and fireproofing are examples of a coating material that may also be used for sound deadening.

1.2.0 Protection Mechanisms of Industrial Protective Coatings

A coating can provide corrosion protection by one or a combination of three basic mechanisms. These include barrier coatings, inhibitive coatings, and sacrificial coatings.

1.2.1 Barrier Coatings

This type of coating acts as a physical barrier between a substrate and its environment. **Barrier coatings** keep moisture and oxygen away from a substrate (*Figure 3*). They are often used as immersion coatings. Examples of barrier coatings include certain epoxies, coal-tar enamels, and many inert vinyl coatings.

1.2.2 Inhibitive Coatings

An **inhibitive coating** contains pigments that react with absorbed moisture within the coating and then react with the steel surface, passivating the metal and interfering with the corrosion process. Inhibitive coatings are normally used for structures exposed to atmospheric conditions, but not to immersion because many pigments will cause water to be drawn into the coating, causing blistering.

1.2.3 Sacrificial Coatings

Sacrificial coatings use a metal (usually zinc) that provides galvanic protection for steel. In the presence of an electrolyte, the zinc will corrode instead of the steel. In most instances, the zinc reaction products will fill in a scratched or damaged area to further protect it from atmospheric conditions. There are several types of sacrificial coatings, including zinc-rich, hot-dip galvanized, and spray-metallized coatings.

Zinc-rich coatings include the following four types:

206F03.EPS

Figure 3 Application of a barrier coating to a pipe.

- *Inorganic post-cure, water-based coatings* – These coatings are water soluble and need a separate wash coat to complete the curing process.
- *Inorganic self-cure, water-based coatings* – These coatings can be thinned with water and cure as a result of a chemical reaction that takes place as water evaporates from the coating.
- *Inorganic self-cure, solvent-based coatings* – This type of zinc-rich coating can be thinned with a solvent. The curing process is dependent on chemical reactions that are influenced by moisture in the atmosphere.
- *Organic coatings* – Organic zinc-rich coatings are compatible with both organic and steel surfaces. They can adhere to bare steel as well as various primers and top coats; however, they will only perform as a sacrificial coating when in direct contact with a steel substrate. Urethanes, chlorinated rubber, vinyl, and catalyzed epoxy are good vehicles for organic zinc-rich coatings.

Hot-dip galvanized sacrificial coatings involve the application of metals (usually zinc) to steel by dipping the steel into a molten metal. Hot-dip galvanized coatings provide a long-lasting barrier against most atmospheric corrosion processes.

Hot spray-metallized coatings also involve the application of metals to a substrate. Zinc or other metals are melted using special spray equipment and sprayed onto a steel substrate after the surface has been abrasive blasted. The metal is sprayed as soon as possible after the substrate is prepared to reduce oxidation of the steel. Substrates other than steel can also be coated using spray-metallizing techniques.

2.0.0 Desirable Properties of Industrial Protective Coatings and Linings

Certain coatings have properties that are considered desirable. The important properties for high-performance coatings are described as follows:

- *Resistance to water* – Water resistance is a very important property of a coating because most coatings come into contact with water to varying degrees over their lifespan. The types of water that may be encountered include:
  - Swamp water, which is ordinarily acidic and may corrode both steel and concrete
  - Sulfide water, which is prevalent in many areas, and which reacts readily with most metals such as iron, steel, brass, and copper
  - High-conductivity water or sea water, which can lead to rapid formation of anode-cathode areas on steel, causing severe pitting
  - Pure water from snow, which will dissolve the calcium out of concrete at a rapid rate, leaving the aggregate exposed
  - Pure water from ion exchangers or demineralizers, which may permeate the coating system, resulting in blistering and/or attack of pigments
  - Water with a high oxygen content, which may create anode-cathode type corrosion areas

  For a coating to have excellent water resistance, it must be able to withstand continuous or near-continuous immersion in water without loss of adhesion, blistering, cracking, softening, or swelling. It must also withstand wet-to-dry cycling, which is a normal condition for coatings exposed to the atmosphere as dew condenses on the coating during the evening and night and evaporates during the day. Not all high-performance coatings have excellent water resistance. The two major aspects of water resistance are resistance to moisture absorption and resistance to moisture vapor transmission. A coating's water resistance is also referred to as its low-moisture permeability. Moisture absorption and moisture vapor transmission are entirely different processes.

  Moisture absorption is the amount of water held within the coating. A corrosion-resistant coating should have low water absorption, which is the amount of water picked up and retained within the molecular spaces of the coating. Once the coating has formed and is in use, the water content comes to a state of equilibrium with the atmosphere, desorbing or evaporating water in the dry stage and absorbing more water when subject to high humidity or immersion. Generally, the less water absorption, the better the coating.

  Moisture vapor transmission is the rate at which moisture vapor will transfer through the coating. When there is a difference in moisture vapor on one side of a coating compared to the other, moisture vapor will generally transfer through the coating. Each individual coating has its own moisture vapor transfer rate; generally, the lower the moisture vapor transfer rate, the better the protection provided by a corrosion-resistant coating.
- *Chemical resistance* – Chemical resistance is the ability of a coating to resist deterioration of its properties due to chemicals that come in contact with the coating in its intended environment. Certain types of high-performance coatings are formulated to resist specific types of chemical attack.

- *Ease of application* – Ease of application is an important property of a coating, especially for critical exposures where the structure may be intricate, with many corners, edges, recesses, and similar areas. Most industrial structures are difficult to coat.
- *Adhesion to substrate* – Adhesion is created by the interaction of physical and chemical forces at the interface of the coating and the substrate. Good adhesion to the substrate is attained by good surface preparation, good coating formulation, and proper application of the coating. Adhesion is the ability of dry paint to attach and remain fixed on the surface without cracking, blistering, flaking, or being removed by tape. Organic coatings vary in adhesion value from approximately 100 psi (7.03 kg/cm^2) for some epoxies to more than 1,000 psi (70.3 kg/cm^2). Thermal spray coatings may show approximate values of over 2,000 psi (140.6 kg/cm^2).
- *Cohesive strength* – Cohesive strength is the ability of a coating to resist internal stress due to curing, thermal expansion or contraction, and cold working (forming). It is a measure of the internal strength of a coating and is usually a good indicator of toughness.
- *Tensile strength* – Tensile strength is the ability of a coating to resist breaking and cracking under linear stress.
- *Flexibility/elongation* – A coating must be able to withstand deformation of the substrate.
- *Impact resistance* – Impact resistance refers to the ability of a coating to withstand sudden mechanical shock. Impact resistance is closely related to abrasion resistance.
- *Edge retention* – Edge retention is a coating's ability to maintain coating thickness on a sharp edge.
- *Abrasion resistance* – Coatings that are applied to the decks or hulls of ships, barges, offshore platforms, and similar areas are examples of coatings that must be abrasion resistant. Other areas needing protection by abrasion-resistant coatings are floors that are subject to the movement of heavy equipment, foot traffic, possible wheel traffic, and damage by tools and equipment. Abrasion resistance may also be required in industrial applications for internal equipment surfaces where agitation of a solution containing particles is occurring; surface areas in direct contact with high-velocity solutions; or inside hoppers handling dry abrasive particles. To withstand this type of service, the coating must be tough, extremely adhesive, and shock resistant. Unless the coating has these properties, it will soon cease to resist corrosion effectively.
- *Temperature resistance* – Temperature resistance is the ability of a coating to resist deterioration of its properties within the range of temperatures that may be expected to occur at the substrate.
- *Resistance to cold flow* – Resistance to creep, or cold flow, is an important property in high-build thermoplastic coatings. Cold flow is the tendency to flow down or slump with age. Glass is a thermoplastic material that will develop cold flow. Coal-tar enamels and pitch compositions may also be subject to cold flow.
- *Dielectric strength* – Dielectric strength is a reflection of a coating's ability to break the electrical circuit set up during a corrosion reaction. A coating must resist the passage of electrons and prevent metal from going into solution at the anode. If electrons cannot travel to the cathode, the corrosion mechanism is arrested. When used with cathodic protection, a coating that is sufficiently dielectric prevents the cathodic current flow in areas protected by the coating, reducing the amount of current required to protect the substrate.
- *Other properties* – Any corrosion-resistant coating should possess a majority of the properties just described. Additional general properties include weather resistance; resistance to bacteria and fungi; longevity; and the ability to retain its appearance.

2.1.0 Types of Coatings

Coatings can be classified in a variety of ways. One of these classifications is organic or inorganic. Most coatings are organic. Organic coatings contain carbon-to-carbon bonding. They are made from organic materials such as vegetable oil, animal fats and waxes, petroleum, and coal. Alkyds, epoxies, oil, and tars are examples of organic coatings.

Inorganic coatings do not contain carbon-to-carbon bonding. They are generally made up of salts of sodium, lithium, potassium, and silica. Inorganic coatings have very specific properties and are primarily used for specialized areas. Examples include phosphate coatings, metallics, silicates, cement, and glass.

Ethyl silicate is an organic compound. However, it is frequently included in the list of inorganic coatings because during the curing process of ethyl silicate zinc, ethyl alcohol is given off and the remaining zinc ferrous-silicate matrix is inorganic.

2.2.0 Composition of Liquid/ Liquefiable Coatings

Conventional coatings have three major components: pigment, resin, and solvent or water (*Figure 4*). The combination of resin and solvent or water is referred to as a vehicle. Additives, such as driers, suspension agents, anti-sag agents, flow-control agents, and gloss-control agents, may also be included in the coating. The resin and the pigment make up the solids content for the coating and are described in literature as solids by volume or weight. Volume solids are a better indicator of coating coverage than are weight solids.

2.2.1 Pigment

A pigment is a discrete particulate solid used to impart specific protective or decorative qualities to the coating. The pigment remains as solid individual particles whether in the liquid coating or the solid film. The term *pigment* may also be used to describe inert fillers such as chalk, talc, or mica that are used to add bulk to the film. Various pigments are added to coatings for the following reasons:

- Inhibit rusting
- Decrease film permeability
- Hide the surface
- Provide color
- Protect the film from the effects of ultraviolet light and weather
- Decrease film flexibility
- Provide strength
- Provide self-cleaning and controlled chalking
- Increase abrasion resistance
- Enhance gloss control (decrease gloss and increase adhesion of subsequent coats)
- Assist in drying the coating binder
- Provide certain desired finishes, such as metallic luster or various glosses
- Aid in storage properties of paint

Pigments can be white or colored, opaque, or translucent. A pigment imparts a coating's color, level of sheen (gloss), and **opacity**. Opacity is a coating's ability to cover or hide the substrate. A coating's hiding power and dried film smoothness is generally dependent on how fine a pigment is ground. Most pigments in commercial coatings are ground to diameters ranging from 0.0001 mm to 0.006 mm. Pigments usually consist of two or more elements combined to form an organic or inorganic compound. Inorganic pigments are made from minerals that contain no carbon. Organic pigments are derived from chemical compounds based on carbon, including coal tar and petroleum. Pigments may be categorized by their primary function (*Figure 5*):

- *Color pigments* – These pigments can be used alone or in combination with one another. They are normally used in combination with white pigments for tinting purposes. Color pigments can be grouped into two general categories: earth pigments and precipitated pigments. Earth pigments are natural or mineral pigments, such as ochres, red iron oxides, siennas, and umbers. Precipitated pigments are made using a process that separates (precipitates) them from a chemical solution. They can be organic or inorganic. Chrome yellow and iron blue pigments are examples of inorganic precipitated pigments. Organic precipitated pigments include phthalocyanine blue and green pigments.

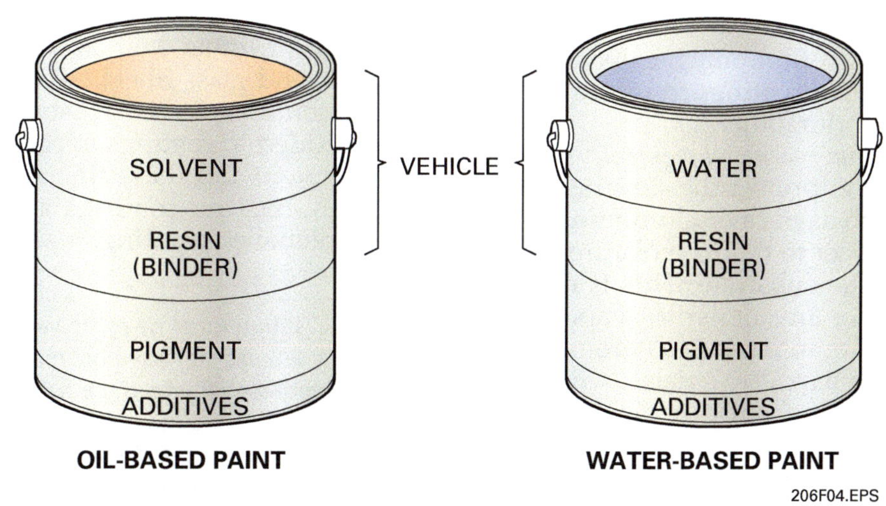

Figure 4 Basic components of conventional liquid coatings.

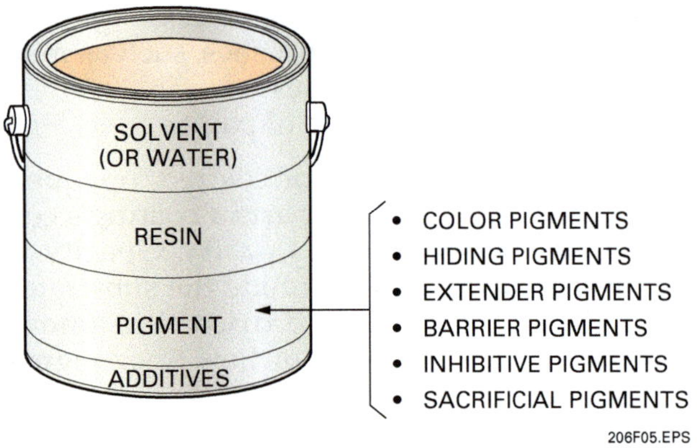

Figure 5 Categories of pigments used in coatings.

A color pigment gives the appearance of color because it absorbs a portion of the visible light spectrum that falls on it and reflects only the light that represents the color. White, black, and gray are called colors; however, black is caused by the complete absorption of light with no reflection; white is caused by complete reflection of all light; and gray is caused by a mixture of white and black that results in imperfect reflection of all light.

- *Hiding pigments* – These pigments have a high light-refractive index, typically between 1.94 and 2.7 (the higher the index, the greater the opacity). Inorganic titanium dioxide is a common hiding pigment that is white and is used as the base pigment in many coatings. Rutile titanium dioxide has the highest refractive index and the best hiding and coloring power of all pigments. Rutile is one of the naturally occurring crystal forms of titanium dioxide.
- *Extender pigments* – These pigments reinforce the coating film and make the wet coating thicker (more viscous). Among other things, they stabilize pigment suspension, improve leveling, control flow, and lower gloss. They also tend to have low refractive indices and reduce opacity. Use of too much extender pigment in some cheaper coatings results in excess chalking and poor durability.
- *Barrier, inhibitive, and sacrificial pigments* – These pigments all help to protect the substrate. Barrier pigments, such as mica or aluminum flake, are plate-like and act to retard moisture vapor transmission to the substrate. This is accomplished by forcing any moisture vapor that enters the paint film to work itself around and between the large barrier particles, which increases the distance the moisture must travel to reach the substrate.

  Pigments of chromate, molybdate, and borate are inhibitors that provide corrosion protection for metal substrates, usually steel or aluminum. These pigments are slightly water soluble and when moisture vapor enters the coating film, the pigments dissolve and are carried to the surface of the substrate. The dissolved pigment reacts with the metal to form a film that helps make the surface of the metal chemically inert and corrosion resistant. Chromate pigments in coatings have been severely limited in the United States because they have been identified as a health hazard.

  Sacrificial pigments provide corrosion protection for steel substrates. Zinc dust pigments are used as a sacrificial metal layer that must be in direct contact with a steel substrate. Zinc is more active than steel, and exposure to chemicals or corrosive agents causes the zinc to corrode (be sacrificed) by galvanic action to protect the steel substrate. This occurs for other types of zinc coatings, including those deposited by hot-dip galvanizing or thermal spraying.

Hundreds of pigments are available, and the properties of the different pigments vary with the manufacturer. Commercially available pigments are classified and listed by the National Paint, Varnish, and Lacquer Manufacturers Association in its Pigment Index.

Proper safety precautions should be taken to avoid the ingestion or inhalation of pigment. For example, appropriate respirators should be worn when blending zinc dust into the vehicle of an organic zinc-rich coating.

The shape of a pigment can influence the physical properties of the coating. Pigments are classified according to shape by the following types:

- *Nodular (lump-shaped) pigments* – Nodular pigments generally add color, such as titanium dioxide or are sacrificial, such as metallic zinc. Most commonly used pigments are nodular.
- *Acicular (needle-shaped) pigments* – Acicular pigments are used to reinforce and strengthen the coating film as well as to provide color. Zinc oxide and glass fibers are examples of acicular pigments. In the past, asbestos fibers were often added to coatings to provide reinforcement (cohesion) and these fibers may be discovered when existing coatings are removed during maintenance painting.

CAUTION

Caution must be taken when removing coatings that have been reinforced with asbestos fibers. If the asbestos is disturbed, these fibers may get into the air, where they can be inhaled. Exposure to airborne asbestos can cause lung cancer and other potential health risks.

- *Laminar (plate-like) pigments* – Laminar pigments overlap when the coating dries, increasing the degree to which the pigment is impermeable to moisture. Mica, aluminum flake, glass flake, and micaceous iron oxide are examples of laminar pigments.

2.2.2 Solvents

Most conventional coatings contain solvents. They are sometimes called a carrier or a volatile vehicle because they do not remain as part of the dry coating film. Solvents can be grouped into five broad classes, as shown in *Figure 6*.

Solvents are added to coatings for a number of reasons. One reason is to dissolve the base resin. A solvent is needed to make a solution of the resin to allow its use as a coating vehicle. Many synthetic resins, such as alkyd resin and epoxy resin, are solids. A solvent is also used to make the coating easier to apply. In such cases, the solvent is usually referred to as a thinner. The solvent also controls the evaporation rate and improves the quality of the final film.

Types of coating solvents are described in the following list:

- *Aliphatic hydrocarbon solvents* – Solvents in this group are also called paraffins. Chemically, they are open-chain (also called straight-chain) compounds. Mineral spirits and VMP (varnish makers' and painters') naphtha are common aliphatic hydrocarbon solvents. Mineral spirits, also called painter's naphtha, is a high-boiling-point petroleum product used to dissolve oils, asphalts, and alkyds.
- *Aromatic hydrocarbon solvents* – These solvents have a closed-chain, six-carbon group (often called the benzene-ring structure) as a principal part of the molecule. The simplest chemical of this family is benzene. Other family members include toluene (toluol) and xylene (xylol). Aromatic hydrocarbon solvents are active solvents for chlorinated rubber, coal tar, and certain alkyds, and are used as diluents in combination with other solvents for epoxies, vinyls, and polyurethane materials.

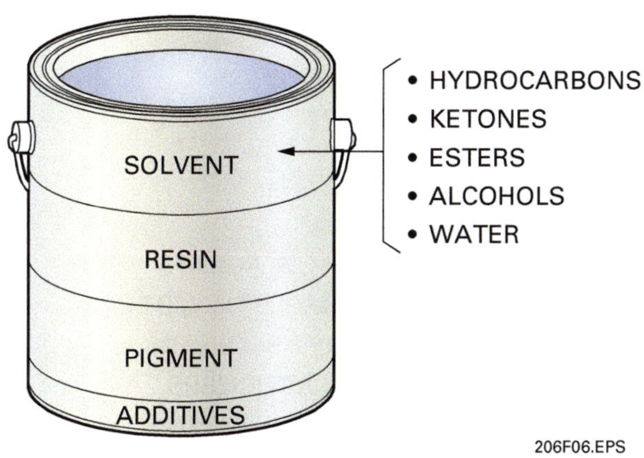

Figure 6 Classes of solvents used in coatings.

> **WARNING**
>
> Benzene is recognized by OSHA as a hazardous material. Breathing in benzene can cause drowsiness, dizziness, and unconsciousness. Because of health risks, OSHA has imposed benzene exposure limits in the workplace.

- *Ketone solvents* – These solvents are oxygenated hydrocarbons of the acetone family and include methyl ethyl ketone (MEK). They are the most effective solvents for vinyls and are often used in epoxies and other resin formulations.
- *Glycol ethers* – The usual form of ether used in protective coatings is glycol ether (alcohol ether), such as ethylene glycol monoethyl ether, commonly known as Cellosolve® (a trade name of Union Carbide). A number of glycol ethers are sold under this designation. Cellosolve® is a good solvent for many oils, gums, natural resins, and synthetic resins, such as alkyds, ethyl cellulose, nitro-cellulose, polyvinyl acetate (PVA), polyvinyl butyral, and phenolics. It is a slow solvent and is used in many lacquers to improve flow-out and gloss. These materials are combined with other solvents to help achieve solvent formulations that have higher flash points.
- *Esters* – These solvents are alkyd acetates and glycol ether acetates. They evaporate slowly and can be used as retarder solvents to slow the curing process. Esters are excellent solvents but are not as strong as ketones. They have a pleasant odor.
- *Alcohol* – Nonionic **surfactants** are important components in many applications, including industrial and architectural paints. Alcohol ethoxylates make up a large portion of the nonionic surfactants used in these applications, along with alkyl phenol ethoxylates. Water-dispersible chain-stopped alkyd industrial coatings that use Propasol and butyl alcohol thinner are available and exhibit both fast air drying and good corrosion resistance.
- *Water* – As environmental regulations become more stringent, waterborne latexes with self-crosslinking capability have gained popularity in the industrial coatings market due to their high performance capability. These products have a variety of applications, including industrial maintenance primers, clear top coats and sealers for wood and cement, high-temperature-tolerant factory-applied coatings for cement fiberboard, and as interior and exte-

rior concrete floor coatings, providing fire and abrasion resistance.

2.2.3 Additives

Additives are special-purpose ingredients added to a coating in small quantities in order to enhance its performance. Additives are classified by the function they perform. These functions are shown in *Figure 7* and are explained as follows:

- *Driers* – Driers are added to an oil-based coating to speed up the drying time by boosting the ability of the coating to absorb oxygen. Driers are compounds made of oil and certain metals. There are two classes of driers: surface driers and through driers. Normally, these are used together. Surface driers are compounds that cause the surface of a drying oil coating to set rapidly and turn almost solid while the underlying coating remains softer. Surface driers are compounds typically made of cobalt and manganese. Through driers are more active than surface driers. When used in conjunction with surface driers, they act to achieve drying throughout the coating film. Through driers are compounds typically made from cadmium, zinc, or zirconium naphthenate.
- *Antiskinning agents* – Antiskinning agents are compounds used to prevent a tough layer or skin from forming on the surface of coatings in containers. This skin forms as a result of the coatings being exposed to air.
- *Fungicides* – Fungicides, also known as mildewcides, are chemical agents added to coatings to prevent or retard the growth of fungi, such as mold and mildew. Fungicides are often used in water-based coatings that are especially susceptible to mold and mildew in damp conditions.
- *Thixotropes* – Thixotropes are additives that make coatings become liquid or fluid when stirred or otherwise agitated, but quickly thicken again when left undisturbed. This property allows the coating to be free flowing during application but quickly set to a gel once applied. The advantage of this is that the coating is less likely to drip from the brush and/or will have less roller splatter. Another advantage is that it can be applied in thicker films without running or sagging.
- *Flow agents* – Flow agents improve the flow of the coating after application. This helps eliminate brush marks, roller patterns, orange peel from spraying, and other film unevenness and defects.
- *Emulsion aids* – An emulsion is a liquid preparation in which tiny, immiscible (nonsoluble) particles of one liquid are suspended in another liquid, such as oil in water or vice versa. Emulsion aids, sometimes called surfactants, are additives used in oil-and-water emulsions that break down the surface tension of the liquid. This allows ordinarily dissimilar and immiscible materials like oil and water to readily mix and disperse in order to produce a stable emulsion. They also enable a water-based emulsion to remain in suspension to help keep ingredients in the emulsion from settling and caking at the bottom of a container during storage.
- *UV absorbers* – Ultraviolet light (UV) absorbers are added to exterior coatings to help prevent deterioration caused by the sun's UV rays. UV rays are invisible, short-length light rays that have an active chemical effect on paints and other materials. Excess exposure to UV light normally results in the coating chalking or becoming brittle.

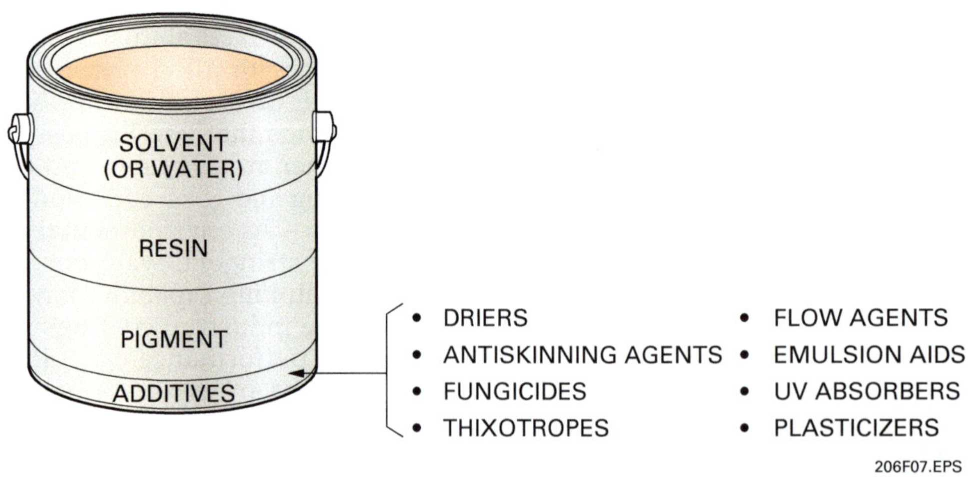

***Figure** 7* Functional categories of additives used in conventional coatings.

- *Plasticizers* – Plasticizers are additives used in certain protective coatings to make them flexible and to improve their adhesion. They also act to decrease the material's brittleness, hardness, and glass transition temperature. The transition temperature is the temperature at which a material changes from a flexible, rubber-like material to one that is hard and brittle.

2.2.4 Resins

Resins or resin blends (*Figure 8*) are often called the binder or non-volatile vehicle of a coating. Along with pigments, they constitute the primary film-forming component of the coating.

To create a protective coating film on a substrate, the resins must convert from a pliant liquid form (which allows application) to a cohesive solid form that adheres to and protects the surface. It is this capability to change from one form to the other that identifies resins suitable for use in coatings.

Generally, the name given to a coating material is that of its principal resin. Resins may be natural organic materials or synthetic materials. Some common resins that are described in the following sections include:

- Acrylic
- Alkyd
- Asphalt
- Chlorinated rubber
- Coal tar
- Drying oil
- Epoxy
- Phenolic
- Polyvinylchloride
- Silicone
- Urethane
- Vinyl
- Siloxirane

2.3.0 Coating Curing Mechanisms

The curing mechanism is the method by which a coating changes from a liquid to a solid state. It is the most important reaction that takes place in the formation of a coating. As shown in *Table 1*, the types of curing mechanisms include convertible coatings and nonconvertible coatings.

2.3.1 Convertible Coatings

Convertible coatings cure by one of several types of polymerization reactions. Polymerization is the joining together of small molecules to make larger molecules. Convertible coatings are also referred to as thermoset materials. Thermoset materials undergo a chemical reaction by a variety of mechanisms. Some examples of thermoset mechanisms are oxidation-cure, coreaction-cure, condensation-cure, and hydration-cure.

Oxidation-cure coatings cure by oxidative crosslinking or polymerization. They use atmospheric oxygen in this process. Heat is sometimes applied to speed up the curing process. The main ingredient in oxidation-cure coatings is vegetable oil. Fish oils are sometimes added as modifiers in alkyd coatings because they are semi-drying. The oil reacts with atmospheric oxygen to form the film. Driers, such as cobalt, tin, and manganese compounds, may be added to speed up the reaction between the oil and oxygen. The wet film thickness (WFT) per coat must be limited because the oxygen can only enter the film at the surface. If the WFT is thicker than recommended, the oxygen may not be able to completely penetrate through the entire film. If this occurs, the film skins over and does not cure thoroughly, causing puckering and wrinkling.

Oxidation-cure coatings can take a considerable amount of time to cure, ranging from two hours to several days. The speed of the reaction

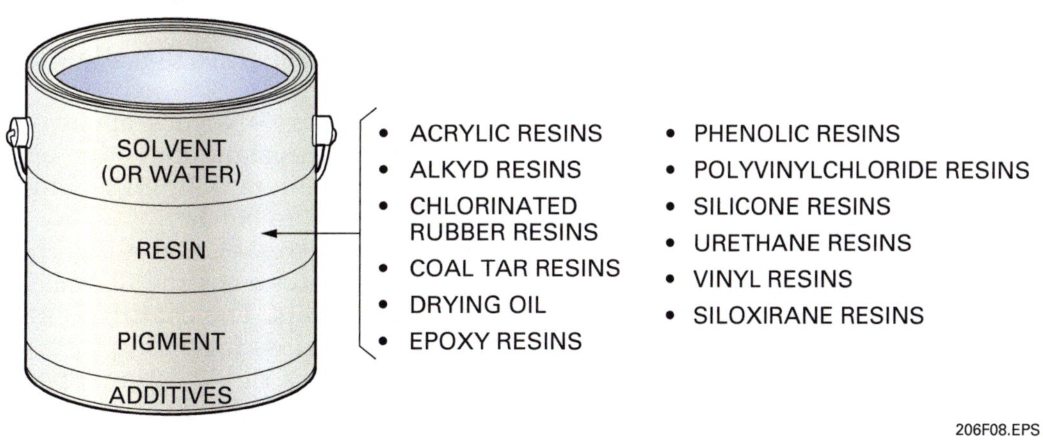

Figure 8 Common resins used in coatings.

can be increased with an increase in oxygen. A slow change of air, which increases the available oxygen over the surface, speeds up the reaction.

These types of coatings may saponify in contact with alkaline surfaces such as concrete or zinc. Saponification is the process that occurs when oils in oxidation-type coatings react with alkali and moisture to form soap.

Once dry, oxidation-cure coatings should not be overcoated with a coating that contains strong solvents, such as the vinyls or epoxies, unless specific tie coats are used. Strong solvents cause the film to swell and wrinkle.

Oxidation-cure coatings include the following types:

- Alkyds

Table 1 Types of Coatings by Curing Mechanisms

Convertible Oxidation Cure	**Convertible Coreaction Cure**
Alkyd	Epoxy
Drying oil	Coal tar epoxy
Epoxy ester	Zinc-rich epoxy
Phenolic (oil-modified)	Epoxy emulsion
Silicone alkyd	Rubber sheet lining
Urethane (oil modified)	Phenolic (epoxy-modified)
	Urethane (two-component)
	Inorganic zinc, post-cure type 1-A (waterbase)
Convertible Condensation Cure	**Convertible Hydration Cure**
Vinyl wash primer	Cementitious materials
Furan	Inorganic zinc, self-cure type 1-C (solvent reducible)
Phenolic (highbake)	Urethane (moisture-cure)
Inorganic zinc, self-cure type 1-B (water reducible)	
Nonconvertible Evaporation (Lacquer Dry)	**Nonconvertible Coalescence (Latex Dry)**
Asphalt/coal tar	Acrylic/latex emulsion
Chlorinated rubber	
Polyethylene	
Vinyl	
Vinyl acetate	
Cooling and Plating Vitreous	**Cooling and Plating Metallic**
Porcelain/glass enamels	Cementation
Ceramics	Cladding/electroplating
	Condensation of metal vapor
	Metal spraying
	Hot dipping

206T01.EPS

- Drying oils
- Epoxy ester
- Phenolic (oil-modified)
- Silicone alkyd
- Urethane (oil-modified)

Coreaction, or chemical-induced polymerization, is achieved when a curing agent is added to the base material so that there is an end-to-end addition of monomers to form long-chain molecules. **Coreaction-cure coatings**, also called plural-component or multi-component coatings, require the addition of an activator or catalyst to the base material. Once mixed, many have a specified wait time, called a sweat-in time or induction time, to allow a chemical reaction to take place. All have a pot life beginning as soon as the coating is mixed that limits the time allowed for the application of the coating. Plural- or multi-component coatings include the following:

- *Epoxy* – Epoxy (two-component) coatings are produced by adding a curing agent to an epoxy resin. Two common types of curing agents are generally used. The first is polyamine. Polyamine converters have good chemical resistance. However, when polyamine epoxy coatings are applied in cool, damp conditions, amine blush can occur. The second curing agent is polyamide. Polyamide converters generally have better wetting characteristics. They have less of a tendency to bloom and/or blush than polyamine converters. Epoxy coatings are resistant to chemicals, solvents, and water, offer high film build per coat, and demonstrate good exterior durability, but will chalk over time.

  Inert pigments, such as silica, mica, talc, and titanium oxide, are sometimes added to epoxy coatings to produce high-performance coatings. These modified epoxy high-performance coatings are generally used in severe corrosion conditions. Penetrating sealers are used as tie coats and are typically applied at 1 to 2 mils. Penetrating sealer is a thin, low-viscosity material with properties that allow it to wet the substrate.
- *Novolac epoxy* – Novolac epoxy is more heat- and chemical-resistant than other epoxies. They are often used in tank linings. Some varieties have a tendency to blush.
- *Siloxane epoxy* – Combining epoxy with siloxane produces a coating that resists acids, bases, and solvents. The coating is also flexible and holds its gloss for a long time. In general, epoxies form a rigid coating. Epoxy and siloxane combinations also result in good weatherability without the presence of toxic isocyanates.

- *Coal-tar epoxy* – Coal-tar epoxy is made with coking byproducts. Selected coal tars are combined with certain epoxies, reducing the permeability of the epoxy films. The advantages of coal-tar epoxy include heavy-duty protection and high resistance to water, chemicals, and solvents; however, it is difficult to repair or recoat and generally has a short pot life. Coal-tar epoxy is used in a wide variety of applications, including oil tanks, pipelines, river and ocean docks, sewage and water treatment plants, oil refineries, and the marine industry.
- *Zinc-rich epoxy* – Zinc-rich epoxy contains high proportions of zinc in the cured film. It is available as two- or three-package products. A two-package product consists of the epoxy resin premixed with zinc and a curing agent (hardener). The three-part package consists of the epoxy resin, the curing agent, and zinc dust with color or filler pigments.
- *Epoxy emulsion* – Epoxy emulsion systems are sometimes referred to as acrylic epoxy. They are water-based rather than solvent-based. Epoxy emulsion coatings almost equal the performance of solvent-based epoxies in terms of durability. They also have the added advantage of low solvent odor.
- *Polyester epoxy* – Polyester epoxy is formed by the combination of a polyester resin and an epoxy resin. They are generally two-part reactive coatings. Polyester epoxy coatings can be formulated to produce high-gloss, tile-like surfaces that can be easily cleaned. These coatings have better weather resistance than polyester resins but are less resistant to chemicals and solvents than pure epoxy resins.
- *Epoxy-modified phenolic* – Epoxy-modified phenolic coatings must have a converter added to the base before application. The converter is normally a primary amine or a modified primary amine and is more sensitive to low temperature conditions during application and cure than other epoxy converters. Epoxy-modified phenolic coatings are often used as a lining material on tank interiors and pipeline interiors. These coatings contain a significant amount of phenolic resin, which contributes to water and solvent resistance but detracts from sunlight resistance.
- *Catalyzed (activated) urethane* – There are two varieties of activated urethanes, also called polyurethanes. They are aliphatic- and aromatic-activated materials. Aliphatic-activated urethanes are two-component systems that have a limited pot life. They are extremely abrasion resistant and flexible with excellent color and gloss retention. Aliphatic-activated urethane coatings are used in most heavy industrial applications and environments. Aromatic-activated urethanes have some of the properties of the aliphatic materials but can also be used in some immersion service situations. They are also two-component systems with a limited pot life. Both two-component systems use an isocyanate hardener as the activator.

WARNING

The fumes given off by some activated urethane coatings are hazardous. A respirator, along with adequate ventilation, is required. Check the MSDS for coating-specific safety procedures. Hot-sprayed activated polyurethane and polyurea coatings are hazardous; a full body suit, heat-resistant gloves, and full face mask with a supplied-air respirator must be worn when working with these coatings.

On Site

Plural-Component Hot-Sprayed Elastomeric Coatings

The use of plural-component hot-sprayed elastomeric coatings for concrete and steel is becoming increasingly common because they contain no volatile organic compounds (VOCs). They are 100-percent-solids elastomeric aromatic polyurethane or aliphatic polyurea coating systems with an isocyanate hardener. They are applied in a 1:1 ratio at temperatures of 160°F (71.1°C) and thicknesses of 30 to 250 mils (0.78 mm to 8.35 mm) or greater in multiple passes. Mixing is done at the gun or in a manifold feeding the gun and, in both cases, the component barrels and lines are heated. The gel time ranges from several seconds to 45 seconds, and recoat times range from seconds to hours depending on the product. The coating exhibits extraordinary toughness, including impact and abrasion resistance, and will bridge gaps of up to ⅛ inch (0.25 cm). These coatings are normally used for non-immersive, aboveground applications that include food and beverage facilities and bridge coatings, as well as the interior and exterior walls, floors, and roofs of commercial and industrial structures. They can also be used on storage tanks, pipes, industrial machinery, and high-impact deck and rail areas of ships or offshore platforms.

- *Inorganic zinc* – Inorganic zinc coatings are composed of powdered metallic zinc mixed into a reactive silicate solution. When inorganic zinc coatings cure, the silicate zinc mixture becomes concentrated as the solvent evaporates. The solvent can be water or organic solvents. Once the initial drying has taken place, environmental reactions take over, and the chemical curing of the coating begins. Little shrinkage occurs during the curing process. Inorganic zinc coatings provide cathodic protection to the substrate, resist the environment, and are solvent and abrasion resistant, but they have limited chemical resistance.
- *Polyester* – Polyester coatings use polyester resins to produce coatings with a relatively high DFT per application of up to 80 mils (2 mm). These coatings have excellent chemical, solvent, water, and abrasion resistance. They exhibit slightly more color stability on exterior exposure than epoxy coatings and are slower to chalk.
- *Vinyl ester* – Vinyl esters use vinyl and epoxy-phenolic resins to produce coatings that are resistant to high temperatures (over 200°F or 93°C) and are effective in conditions of continuous immersion in strong acid. Vinyl ester coatings can be applied to alkaline surfaces such as concrete.
- *Epoxy-modified siloxirane* – Epoxy-modified siloxirane coatings are epoxy compositions that provide the properties of a high-performance epoxy and a polyurethane in one coat. These coatings offer a high degree of corrosion control coupled with long-term weather resistance with single-coat applications. In addition, they can be directly applied over inorganic primers, and they cure at low or room temperatures. They have a high gloss, are resistant to acid, and are environmentally safe. Epoxy-modified siloxiranes are typically used for structural steel, tanks, piping, industrial plants, and concrete walls and floors. This type of coating needs no thinning and can be applied by brush, roller, or spray.
- *Siloxirane* – Siloxirane coatings are ambient-temperature-cured coatings that can withstand continuous temperatures up to 2,000°F (1,093°C). Siloxiranes are used to protect both carbon steel and stainless steel under insulation. They are highly resistant to moisture, as well as to a wide range of chemicals, including acids, alkalis, solvents, and oxidizing agents. Siloxiranes offer good impermeability and flexibility. These coatings are environmentally safe and are used in power plants, refineries, chemical facilities, the exteriors of reactors, and under equipment and pipe exterior insulation. Siloxirane coatings are also repairable.

With **condensation-cure coatings**, a condensation reaction takes place when two or more organic molecules react, forming a new bond and giving off a molecule of water. The types of condensation-cure coatings include:

- *Vinyl wash primer* – A vinyl wash primer coating has a modified vinyl vehicle with a rust inhibitive pigment, such as zinc chromate. These coatings are used to provide adhesion between the substrate and top coat. Vinyl wash primers have a fast drying and curing time. They should be top-coated quickly, or a new coat of wash primer must be applied and then top-coated. The applied thickness is critical. When applied, the WFT should be approximately 2 mil to 3 mil (0.051 to 0.08 mm) to yield a DFT of no more than 0.5 mil (0.013 mm) but preferably 0.3 mil (0.008 mm).
- *Furan* – Furan coatings are made from resin formed by the reaction of furfural alcohol, which is derived from agricultural byproducts such as oat husks. They are widely used in the fabrication of corrosion-resistant processing equipment and in resinous cements. The coatings have good wetting or penetrating properties and chemical resistance, but they are brittle, subject to shrinkage, and have poor adhesion to nonporous surfaces.
- *Phenolic* – Phenolic (heat-converting/resole) coatings must be baked to cure. They are extremely brittle, have poor resistance to sunlight, and poor flexibility, which limits their total film thickness. However, they are resistant to a wide range of chemicals and solvents. Phenolic coatings are used primarily on the interiors of tanks and vessels, down-hole tubing, and certain types of drill pipe.
- *Inorganic zinc (Type 1-B)* – SSPC Paint 20 Type 1-B self-curing inorganic zinc coatings are water reducible. These coatings cure by crystallization after the evaporation of water from the coating. Once the water evaporates, a hard, metallic coating is left on the surface, and the film continues to cure by reacting with carbon dioxide and humidity from the air. Type 1-B inorganic zinc coatings are used on the interiors of tanks and tankers in warm or hot climates. These coatings do not work well in cold, high-humidity environments, which prevent water evaporation from the film.

2.3.2 Hydration-Cure Coatings

Hydration-cure coatings cure by a process in which water combines chemically with the coating and its constituents. Types of hydration-cure coatings include the following:

- *Concrete* – Concrete is widely used as a construction material and may provide a larger surface area for coatings than steel does. Even though there are a number of concrete surfaces that do not require a coating, some concrete may require a protective coating. In addition, concrete is considered to be a coating itself. It is applied over steel to prevent it from corroding. It provides a thick, dense, water-resistant barrier to corrosion. Lime, present in portland cement, causes high alkalinity that may passivate steel, provided that certain chemical salts (chlorides), which may promote corrosion, are not present. This high-alkaline condition can also cause problems for any coatings applied over a concrete surface.
- *Inorganic zinc (Type 1-C)* – SSPC Paint 20 Type 1-C inorganic zinc coatings are cured by reaction with moisture in the air, which is needed to complete the hydrolysis reaction. The rate of cure depends on the amount of moisture in the atmosphere. Generally, 50 to 90 percent relative humidity is required to complete the cure. The rate of cure is also influenced by the vehicle composition, catalyst, and temperature. Low temperatures and low humidity slow the curing rate. Type 1-C coatings are solvent reducible. Solvent-based inorganic zinc-rich coatings become water resistant more quickly than water-based inorganic zinc-rich coatings. The first step in film formation is the evaporation of the organic solvent. The primary form of this type of coating is the organic silicates, such as ethyl silicate. Type 1-C coatings are widely used throughout the industry. These products are often used as primers for organic top coats.
- *Urethane (moisture-cured)* – Urethane (moisture-cured) coatings require a relatively high ambient humidity (between 30 and 80 percent) to cure. They have a limited pot life and should not be used if left over from the previous day. Moisture-cured urethane is tough, flexible, and resistant to chipping, abrasion, dirt retention, solvents, acids, and alkalies. It is often used for bridges, tanks, structural steel, and over concrete.

2.3.3 Nonconvertible Coatings

Nonconvertible coatings include evaporative (lacquer dry) and coalescence (latex and water-based acrylic) coatings. They cure solely by evaporation of a solvent or dilutant. Some nonconvertible coatings are also referred to as thermoplastic or thermoset coatings. A thermoplastic material becomes soft or fluid when heated. Upon cooling, it regains its original physical and chemical properties. Heated types of thermoplastic or thermoset coatings that are applied in powder form are covered in more detail in later sections of this module.

Evaporative lacquer dry coatings cure by simple solvent evaporation. The base resin, such as chlorinated rubber, remains unchanged. The base resin is easily redissolved by any aromatic hydrocarbon solvent, no matter how long the film has cured. For this reason, an evaporation-type coating should not be top coated with a coating that contains a strong solvent. For example, an epoxy top coat should not be applied to a chlorinated rubber base coat. Evaporation-type coatings should not be applied too thickly, so that all the solvent can evaporate to complete the cure. Heat can be applied to speed up the evaporation of the solvent or to help fuse the resin particles together. Evaporative-cure coatings include the following types:

- *Asphalt coatings* – Asphalt is a dark-colored, petroleum-based thermoplastic material. The coatings are available as enamels that must be heated for application, as an emulsion in a water-based vehicle, and as a cutback solution using a volatile solvent. Hot-melt asphalt can be applied to tank bottom coverings or brick floors by pouring and leveling off. Liquid cutback and emulsion asphalt coatings can be applied by brush, mop, or spray. Multiple coats should be applied at right angles to each other to minimize pinholes and discontinuities.
- *Coal-tar (thermoplastic) coatings* – The main ingredient of thermoplastic coal-tar materials is coal-tar pitch. This pitch is produced from the coal tar recovered in the manufacture of coke. Hot-applied materials are often called coal-tar enamels and consist of pitch and filler. The material is heated before use and applied in molten form. These coatings are often used on the exterior and interior surfaces of pipe. A spinning method can be used to apply hot coal tar to interior lengths of pipe. A rolling rig, end feed machine, or line traveling machine can be used for application on exterior lengths of pipe. Cold-applied coal-tar materials are made by reducing the amount of pitch with a suitable solvent. If a filler is used, it is added after the solvent reduction is accomplished. These materials range from thin liquids to heavy pastes and can be applied by brush, spray, or roller. Curing time will vary according to the thickness of the coating.
- *Chlorinated rubber coatings* – Chlorinated rubber is a resin formed by the reaction of natural rubber with chlorine. Chlorinated rubber top coats, applied using a solvent vehicle, provide a good maintenance coating because they are

hard, tough, and chemical- and weather-resistant. They can be easily repaired and recoated and are resistant to alkalis/acids, chemicals, and water. However, they have poor resistance to solvents, animal and vegetable oils, and heat.

- *Polyethylene (PE) coatings* – Satisfactory liquefied PE coatings have so far proven to be impractical. Current low-density PE application methods include dipping a workpiece in a melted resin bath, immersing a hot workpiece in a fluidized bed of powdered resin, and flame spraying the resin directly onto a workpiece.
- *Vinyl coatings* – Various vinyl copolymers are used in vinyl coatings, including vinyl acetate, vinyl chloride/vinyl acetate, vinyl alcohol, and carboxyl-modified vinyl chloride/vinyl acetate. The coatings use a high percentage of solvent vehicle. Because of their age resistance, toughness, and water/chemical resistance, these coatings have been used in many applications, including the marine industry, paper plants, petrochemical plants, refineries, food plants, and offshore structures.

NOTE

Because some coatings, such as lacquer, vinyl, or chlorinated rubber coatings, normally use a high percentage of VOCs, such as aromatic hydrocarbon or ketone solvents, their use in the United States has been reduced.

- *Coalescence (latex and water-based acrylic) coatings* – These types of coatings employ a special evaporative drying. The coatings have tiny particles of resin dispersed in water (an emulsion). When the water evaporates, the resin particles fuse together and form highly cross-linked molecules that are insoluble in water after curing. These types of coatings are typically acrylic/latex emulsion coatings and include vinyl (acetate) acrylic, polyvinyl acetate, styrene butadiene, and styrene acrylic. The coatings are easy to apply, VOC-compliant, a minimal fire hazard, and have a fast dry/recoat time.

3.0.0 Coating Methods for Some Non-Steel Substrates

The following sections describe coating methods used for cementitious, aluminum, and stainless steel substrates.

3.1.0 Cementitious Substrates

While concrete and other cementitious substrates can be a coating in and of themselves, many are still coated for waterproofing, damp-proofing, decorative and/or protective barrier purposes. Waterproofing coatings can be decorative and/or protective and prevent pressurized water flow into or out of concrete. They are generally thick films of asphalt, coal-tar urethane, or epoxy. The coatings are typically applied by spray or in sheets to the positive water pressure side of the concrete structure. Most protective barrier coatings are also effective waterproofing systems.

Damp-proofing coatings seal the surface of the substrate and prevent absorption of water but are not effective against hydrostatic pressure. They are generally thin-film asphaltic emulsions, asphaltic or coal-tar cutbacks, concrete stains, or penetrating sealers. A common use for damp-proofing coatings is to protect concrete against freeze-thaw damage. Typically, they are used on one side of a concrete structure, and a decorative or protective coating is applied on the other side.

Protective barrier coatings are used to protect concrete from exposure to chemicals, such as acids, alkalies and sulfates, and from penetration of salt solutions or other corrosive chemicals. They are also used to protect a product, such as food or purified water, from contamination by direct contact with the concrete. Proper cleaning of concrete surfaces prior to coating is required. All traces of any form-release agents, curing agents, hardening agents, salts, or efflorescence must be removed before the coatings are applied.

3.1.1 Penetrating Primers

Due to the porous nature of concrete, penetrating primers are used to produce a nonporous surface and provide a strong bond within the concrete. They also provide a surface that will chemically bond with subsequent coating layers. Penetrating primers produce a surface that is less affected by changes in temperature, concrete moisture content, and environmental conditions than other types of primers. They also increase the strength of the concrete surface. Penetrating primers in common use include epoxy, acrylic, silane, silicone, siloxane, silicates, methyl methacrylate, asphalt emulsion, and polyurethane. The 100-percent-solids primers are the most effective for immersion service. Solvents should not be used to thin conventional coatings that are to be used as penetrating primers.

3.1.2 Thin-Film Coatings

Thin-film coatings (less than 20 mils, or 508 microns) are primarily chosen for decoration, cleanability, damp-proofing, corrosion protection, and barrier protection in mild chemical environments. These coatings do not provide a continuous film over irregular concrete surfaces and should not be used when concrete surfaces must be completely isolated. The following are a few of the thin-film coatings in common use:

- *Acrylic latex* – Provides a breathable film
- *Polyurethane and epoxy* – Offer good chemical and water resistance
- *Asphalt emulsions* – Often used as positive-side damp-proofing coatings
- *Polyvinyl butyral* – Used as a primer or tie coat

3.1.3 Liquid Rigid-Barrier Thick-Film Coatings

Liquid rigid-barrier thick-film coatings are normally 20 to 80 mils (508 to 2,032 microns) thick and can be used to form continuous films over properly prepared concrete surfaces. They are used in mild to moderate chemical exposures to waterproof a concrete surface or to line concrete storage vessels. Since most of these coatings are rigid, they are susceptible to cracking due to concrete movement. The cracks that develop in the concrete may extend through the coating and expose the concrete. Common liquid barrier-type coatings include 100-percent-solids epoxy, filler-reinforced epoxies, bituminous mastics, and filler-reinforced vinyl ester/polyester.

3.1.4 Elastomeric Membranes

Elastomeric membrane coatings are highly elastic, rubber-like materials that are used for waterproofing and as a protective barrier. In some applications, a cloth mesh or scrim-type material is used to reinforce the elastomeric coating system, providing a structural film strength that is independent of the substrate. Elastomeric coatings stretch and bridge over small cracks that can develop in concrete surfaces due to temperature changes or settling. They are also sometimes used as an intermediate coat for more rigid coating systems, absorbing movement between the concrete and the rigid coating. Elastomeric materials are usually applied over a primer for adhesion. They are also less chemically resistant and more prone to puncture than many of the rigid coatings, making them undesirable for some concrete coating applications. The most common generic elastomeric coatings include polyurethane, acrylic, butyl rubber, synthetic rubber, and polysulfide.

3.1.5 Protective Barrier Laminate and Bonded Thermoplastic Liners

Laminate liners are manually laid-up systems. Laminate liner systems, also called laminate fiber-reinforced plastic (FRP) systems, are a hand-applied polymeric resin reinforced with fiberglass cloth or a similar mesh material. They can be applied as a thick single layer to provide a thick barrier lining or multiple layers to build up a new surface within an existing structure. Common laminate systems include epoxy, polyester, and vinyl ester materials. FRP using chopped fiber and laminated FRP systems can also be applied to steel surfaces.

Bonded thermoplastic liners can be mechanically bonded by knob extensions on the back of the liner. The liner is placed in the concrete forms, and the extensions are embedded in concrete when the concrete is placed. Other bonded thermoplastic liners are plastic sheets that are attached by adhesive to existing concrete structures. Bonded bag thermoplastic liners are prefabricated for vessel interiors and are bonded to the interior at selected points, such as manholes or fill areas.

3.2.0 Aluminum Substrates

Aluminum and its alloys can be coated with various types of finishes to enhance appearance and/or improve the functional properties of the surface. Organic coatings, anodizing, conversion coatings, porcelain enamels, or electroplating may be used. In all cases, complete cleaning and preparation of the aluminum surfaces in accordance with specifications is required before the application of the coatings.

3.2.1 Organic Coatings

Aluminum is an excellent substrate for organic coatings. The aluminum surface must be properly cleaned and prepared. If a substantial oxide film forms, the film may be too thick to allow subsequent organic coatings to penetrate the aluminum surface. For indoor applications, an organic coating may be applied directly to a clean surface, but the use of a primer coat usually improves the performance of the organic coating. Wash primers and zinc chromate primers are commonly used. Outdoor applications and indoor applications subjected to impact or abrasion may benefit from a surface treatment, such as anodizing or chemical-conversion coating prior to applying the primer and finish coat. Conversion coatings provide a good base for organic coatings and improve the service life of the coating by retarding corrosion of the aluminum substrate.

3.2.2 Anodic and Conversion Coatings

These are mill coatings applied to bulk raw material or manufacturing process coatings applied to fabricated products. Anodizing converts the aluminum surface to aluminum oxide by electrolytic action. Chemical conversion coatings are adherent surface layers of low-solubility oxide, phosphate, or chromate compounds produced by the reaction of various reagents with the aluminum surface.

3.2.3 Porcelain Enamels and Electroplating

Like anodic and conversion coatings, porcelain enamel application and electroplating are mill coatings applied to bulk raw material or manufacturing process coatings applied to fabricated products as a final finish. Porcelain enamels are usually composed of lead silicate or cadmium silicate bases, but phosphate or barium bases may also be used. They are used to improve the appearance, chemical resistance, and weather resistance of aluminum alloys. Electroplating is also used for decorative and/or functional purposes. Chromium, nickel, cadmium, copper, tin, gold, zinc, and silver are metals that can be electroplated on aluminum. Zinc or tin plating, overplated by copper, is usually applied to aluminum prior to plating the desired metal.

3.3.0 Stainless Steel

Although stainless steel surfaces are frequently left bare, they may be coated for corrosion protection (that is, protection from chloride or sulfide attack), appearance, job-site markings, and safety. As with aluminum, stainless steel surfaces must be completely cleaned and prepared in accordance with specifications prior to coating.

The following coating systems are recommended for stainless steel:

- *Metal marking for identification during construction* – Clean using *SSPC-SP 1*. Apply one coat of industrial waterborne acrylic for metal that will not be exposed to temperatures above 300°F (149°C). Apply one coat of silicone acrylic for metal that may be exposed to temperatures of 301°F (149°C) and higher. These coatings may be used for stencils, striping, or complete painting. Not all colors are stable to the temperature limits of the silicone acrylic.
- *Color coding, safety colors, and permanent stencils* – Clean using *NACE No. 2, SSPC-SP 1*, or *SSPC-SP2* as permitted. Coat per *Metal Marking for Identification During Construction* for normal atmospheric exposure. Coat with polyamide or polyamine epoxy for corrosive environments. Top coat the epoxy with polyurethane for better color and gloss retention indoors or outdoors. Colors may change at temperatures above 200°F (93°C) with either epoxy or polyurethane.
- *Heavy-duty corrosion protection* – Clean using *SSPC-SP 10* as permitted. Existing coatings must be completely removed. For services from -50°F to 300°F (-45.5°C to 149°C), apply two coats of epoxy phenolic at 10 to 14 mils (254 to 355 microns) DFT. If an anchor profile in excess of 1 mil (25 microns) cannot be obtained, apply a DFT of only 5 to 8 mils (127 to 203 microns) for service below 40°F (4°C) or above 120°F (49°C). The thicker coating needs the deeper anchor profile to resist the differential thermal expansion between the coating and the metal at the extreme temperatures.
- *High-temperature coatings (301°F to 1,500°F)* – Clean using *NACE No. 1/SSPC-SP 5* as permitted. Existing coatings must be completely removed. The anchor profile may be 1 to 2 mils (25 to 51 microns). Higher or lower anchor profiles may cause premature failure.

Mill or manufacturing process electroplating is also used to coat stainless steel surfaces. Stainless steels may be plated with copper, chromium, nickel, cadmium, or precious metals. Electroplated coatings provide lubrication during cold heading, spring coiling, or wire drawing. They also reduce scaling at high temperatures, improve electrical conductance, and prevent rusting.

4.0.0 Pipeline Coatings

Coating and cathodic protection systems are the methods used to protect pipelines from external corrosion. A coating's success depends on its ability to provide a barrier from the environment and to work with any cathodic protection system. Pipelines may be buried in various types of soil, submerged in water, or installed above ground either on supports or directly on top of the ground. The selection of a pipeline coating and its application varies according to the pipe material and its location. NACE International has a number of standards that discuss pipeline coatings. Teaching the actual contents of these standards is beyond the scope of this course. However, the importance of following the project specification, especially on pipeline jobs, must be stressed.

The material used for pipelines is generally governed by the service or operating conditions of the pipeline. The following list briefly covers some of these materials:

- Carbon steel is widely used for pipe construction and is used with crude oil, natural gas, and refined petroleum products. Carbon steel is often used for slurry and raw water lines, and for recirculating water pipes in power plants.

- Fiber-reinforced plastic (FRP), a plastic reinforced with glass fibers, is used for pipes that carry some chemicals, acids, alkalis, and chemical wastes. It is primarily used in and around petrochemical complexes.
- Ductile iron is less brittle than regular cast iron, and it has greater strength when stressed. Ductile iron is very difficult to weld, and it must be joined with mechanical couplers or mechanical joints.
- Asbestos cement (AC) pipe is made from high-grade portland cement reinforced with asbestos fibers. This pipe is very dense with a smooth interior surface that permits good fluid flow through the pipe. Regardless of other asbestos regulations, asbestos cement pipe is approved for potable water use in the United States.
- Plastics are widely used in the pipeline industry to carry natural gas, water wastes, and chemicals. Depending on the service and operating conditions, polyethylene (PE), polypropylene (PP), or polyvinyl chloride (PVC) pipe may be used.
- Prestressed concrete forms pipe with a great deal of strength. It is manufactured by applying concrete to thin-wall steel pipe; a steel wire under controlled tension is spirally embedded into the concrete and then covered with more concrete. The pipe is then cured in a steam kiln.
- Steel cylinder concrete pipe is formed by lining and coating a steel pipe with concrete, which is then air cured or cured in a steam kiln. Both prestressed and steel cylinder pipe are used to transport water (untreated or treated), domestic and industrial wastes, and treated sewage. They are frequently coated to provide chemical and corrosion resistance to the waste.
- Alloy steel pipe is used in the petrochemical industry for the transportation of corrosive liquids or special fluids. The most common types of alloy steels used include stainless steel, titanium, Monel™, and Carpenter 2™ steel.

The primary purpose of coating the exterior of a pipe is to separate the pipe from its environment and to aid cathodic protection systems. Ideally, the coating should provide a barrier to moisture as well as to electron flow. However, although many coatings are very resistant to current flow, as they break down they allow current flow, thus accelerating the corrosion process. A good pipe coating should exhibit the following properties:

- *Durable bond to the pipe* – A well-bonded coating helps prevent coating film creep and cathodic disbonding. Coal-tar epoxy, hot-applied coal-tar enamel, fusion-bonded epoxy (FBE), and three-layer polyolefin exhibit strong, durable bonds to pipe.
- *Moisture resistance* – To isolate pipe from water, the coating must be resistant to moisture absorption and migration through the coating. Coal-tar epoxy and some polyethylene tapes are very moisture resistant.
- *Mechanical strength* – Because most pipeline coatings are applied in stationary plants and handled at least twice before installation, the coating must be able to resist mechanical damage. Once installed, the coating must resist stress loads resulting from soil movement caused by settling and cycles of wetting and drying of the soil. In known high-stress soil areas, an extra layer of outer wrap may be applied to supply additional strength to the coating.
- *Dielectric strength* – Pipe coatings should resist the passage of electrons to avoid loss of electrical potential from cathodic protection systems. Coatings initially provide 99 percent of the pipeline protection; the remaining 1 percent comes from cathodic protection. Over time, the coatings may deteriorate and cause the need for cathodic protection to increase.
- *Cathodic disbonding* – Cathodic protection systems are a backup to coatings that can prevent the corrosion of pipelines. Coating selection is critical when coatings are used in conjunction with cathodic protection.
- *Bacterial and chemical resistance* – Acidic or alkaline soils, chemicals from toxic waste disposal sites, products from nearby industrial or refining sites, and microbiologically influenced corrosion (MIC) can all dictate the type of coating applied to pipe. Bacteria- and chemical-resistant coatings must be used in these types of environments.
- *Durability and longevity* –The coating must be able to resist physical and chemical changes over the life of the pipeline. The design life of some pipelines may be as long as 75 years.

Some pipeline coatings are only applied in stationary plants or in the field. Others may be applied in a plant and in the field. *Table 2* summarizes a variety of coatings and their application methods. The ends of pipes are not welded until they are fit together in the field, so the coatings are not applied to the ends in the plant. The part of the coating that is placed in the plant is called the mainline coating. The portion that is coated in the field is the field joint only. Some coatings can be applied in the field as a mainline as well as a joint coating.

Table 2 Application Methods for Various Pipeline Coatings

Pipe Coating	Plant Application	Field Application
FBE	Yes	Yes (Field joints only)
Extruded polyethylene crosshead-extruded	Yes	No
Side-extruded	Yes	No
Hard adhesive-bonded	Yes	No
Hot asphalt mastic	Yes	Yes (Field joints only)
Coal-tar enamel	Yes	Yes (Mainline/ joints or rehab)
Coal-tar epoxy	Yes	Yes (Mainline/ joints or rehab)
Epoxy	Yes	Yes (Mainline/ joints or rehab)
Coal-tar urethane	Yes	Yes (Mainline/ joints or rehab)
Polyethylene tape wrap	Yes	Yes (Mainline/ joints or rehab)
Concrete weight coating	Yes	Yes
Coal-tar pitch tapes	No	Yes (Field joint/ rehab)
Petrolatum tapes	No	Yes (Field joint/ rehab)
Shrink-fit sleeves	No	

206T02.EPS

Some of the coatings covered in *Table 2*, including powder coatings, are briefly described in following sections. These coatings, which can be applied to pipe up to 60 inches (0.154 m) in diameter at a stationary plant, are generally applied in the following sequence:

Step 1 Bare pipe arrives. Pipe sections are logged in for storage and inspected for damage.

Step 2 The pipe is delivered to a stationary plant coating site and is inspected again for damage from delivery.

Step 3 The pipe is precleaned and preheated as necessary (*Figure 9*).

Step 4 The pipe is abrasive blast cleaned to a white metal or near-white metal surface as required by the coating manufacturer or coating specification.

Step 5 The pipe is finally blown clean inside and out and then inspected and repaired if necessary.

Step 6 The pipe is primed if required by the coating manufacturer or specifications.

Step 7 Pipeline coating is applied and cured as necessary.

Step 8 Pipeline coating is inspected. Inspection methods include visual inspection, holiday inspection (*Figure 10*), and DFT measurements. The pipe is wrapped with a protective buffer, usually rope, to prevent damage.

Step 9 Holidays are repaired. The pipe is loaded for storage or delivery to the pipeline installation site.

4.1.0 Fusion-Bonded Epoxy Pipeline Coatings

Fusion-bonded epoxy (FBE) coatings are the most widely used pipeline coatings in North America. FBE coatings are normally applied in thicknesses of 12 to 16 mils (305 to 406 microns). The coating is a one-part, heat-curable thermosetting powder that typically consists of epoxy resins, curing agents, pigments, flow and stability control agents, and other ingredients. When the powder is manufactured, it is heated to partially react and then chilled to prevent further reaction. FBE powder is ground very fine and packaged in weatherproof containers.

4.1.1 FBE Coating Advantages

FBE coatings have the advantage of offering the best bonding to steel of any coating. They also cure almost completely prior to cooling so that they can be handled immediately after coating application. In addition, they can be repaired easily with two-part epoxies or patch sticks. Because slight moisture within the coating allows some current flow through the film, corrosion and pitting are rarely encountered under disbonded FBE coatings, provided adequate cathodic protection is available. The coatings can be overwrapped, but the use of any overwrap material that blocks cathodic protection can result in a shielding situation and loss of protection.

4.1.2 FBE Coating Disadvantages

FBE coatings require careful surface preparation and application methods. Steel imperfections and contamination can cause holidays, provide poor moisture permeation resistance, and easily damage the coating during construction. One disadvantage of FBE is excess DFT that can result in cracking.

4.1.3 Plant-Applied FBE Coatings

Plant application of FBE coatings (*Figure 11)* is similar to other types of pipe coating applications, but it is more stringent in the cleaning, cur-

206F09.EPS

Figure 9 Pipe in an infrared preheat oven.

206F10.EPS

Figure 10 Holiday/pinhole inspection with a curved high-voltage detector.

ing, and inspection phases than other types of coatings. The coatings are applied to very clean, inductively-heated pipe (450°F to 500°F [232°C to 260°C]) by production electrostatic gun heads located in a closed or semi-closed powder chamber (*Figure 12*) as the pipe rotates and passes through the chamber. In some cases, the pipe may be powder coated as it passes through a cloud chamber. Once the powder is applied to the heated pipe, the powder goes through the following four phases:

1. *Melt* – During the melting phase, the powder particles begin to absorb heat and liquefy.
2. *Flow* – The flow phase occurs when the powder particles begin to flow, fully liquefying and wetting the pipe surface.

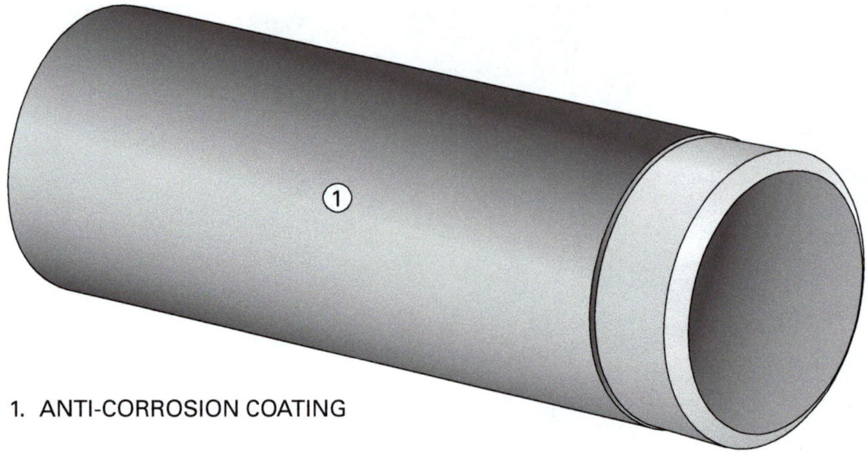

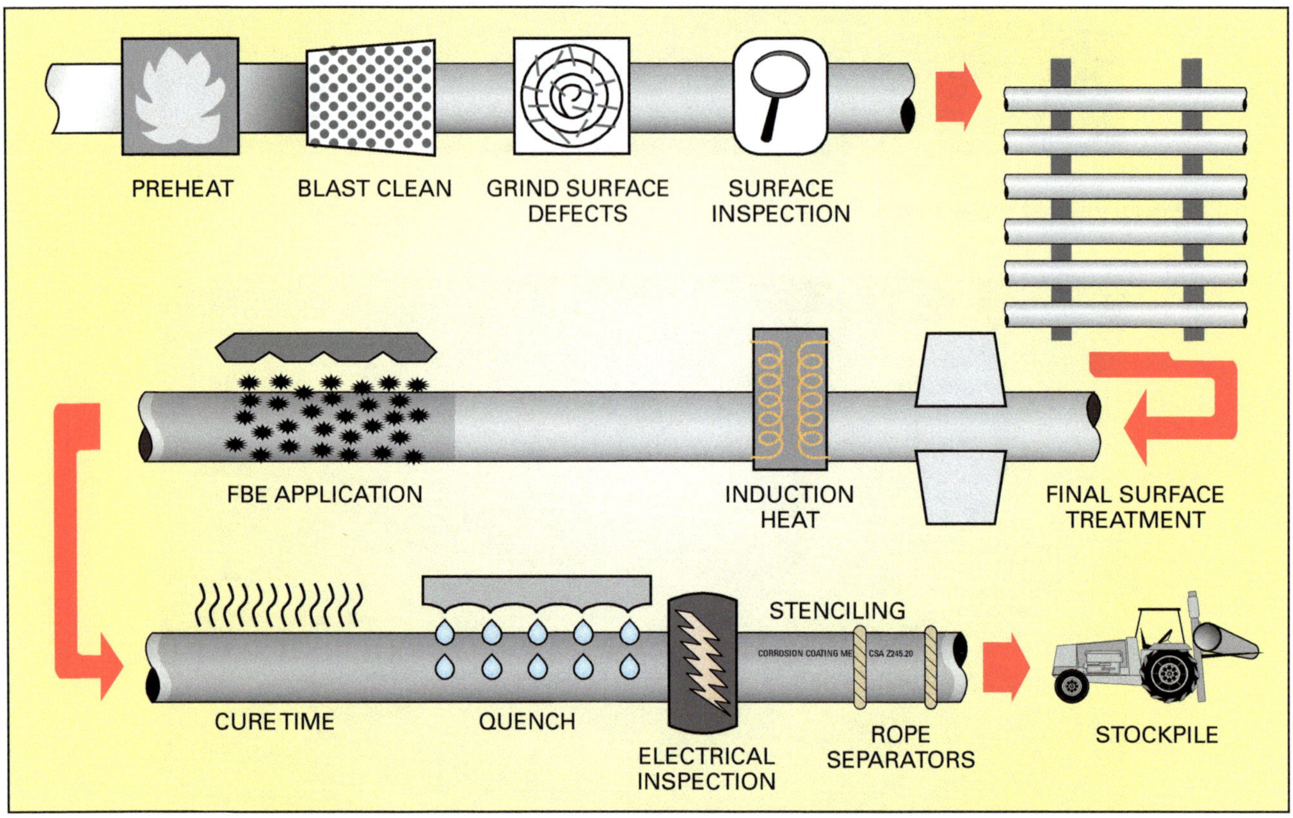

Figure 11 FBE application process in a plant.

3. *Gel* – The gel phase takes place when the particles of the liquefied material gel, converting to a solid.
4. *Cure* – During the curing phase, further changes occur that allow the FBE powder to cure fully.

The complete process occurs as the pipe rotates down the conveyor line and generally takes less than two or three minutes, which makes this an ideal process for production-line application. Some FBE powders may require a post-heating stage, called reduction heating, to ensure proper curing and to reduce thermal shock from sudden cooling. For most FBE resins that do not require post heating, the cure time varies depending on the powder. Once the FBE coating has cured or nearly cured, the pipe may be water quenched to allow immediate handling. Then the cutback at the ends of the pipe are ground (*Figure 13*) and the pipe is inspected, repaired as necessary, and stenciled (*Figure 14*).

The FBE application process is critical. The coating operation may be monitored to ensure the following:

Figure 12 Pipe emerging from a powder chamber.

- The specified powder is used and batch numbers are recorded.
- The powder is dry and has not picked up moisture during transit or storage.
- The powder mix (the original-to-recycled powder ratio) is properly maintained.
- Voltage settings are correctly maintained.
- The number of powder gun heads is matched to the pipe diameter, and the proper powder-delivery ratio is maintained.

4.1.4 Field Joint FBE Coating Application

Pipes coated in a plant have the coating removed and the pipe beveled at the ends to allow joining by welding in the field. The uncoated area is called a coating cutback and generally extends back about 2 inches (5.1 cm) from the ends of the pipe. After pipes are welded together in the field, the cutback area and 1 to 2 inches (2.4 to 5.08 cm) of the existing coating on both sides of the cutback area are abrasive blast cleaned to white metal. As shown in *Figure 15*, a portable electric induction coil heater is used to heat the pipe to the proper application temperature. The induction heater is then removed, and a portable automatic FBE powder sprayer is immediately moved into place and activated to rotate around and back and forth over the pipe to uniformly coat the entire joint area of the pipe. The powder coating fuses together just as it does in a plant operation.

Figure 13 Pipe cutback grinding.

206F14.EPS

Figure 14 Pipe stenciling.

206F15.EPS

Figure 15 Field joint induction heater and powder sprayer.

4.2.0 Coal-Tar Pipeline Coatings

Coal-tar pipe coatings were one of the first pipeline coatings to be used successfully in plant applications. They are formulated with coal-tar pitches mixed with combinations of fillers, extenders, and other ingredients in various coatings, including coal-tar enamels, epoxies, and tapes.

Hot-applied coal-tar enamel (CTE) is the oldest pipeline coating with the longest corrosion performance record of any coating in use and is the standard by which other pipeline coatings are measured and evaluated. Current CTE coating systems have evolved to a sophisticated application of primers, multiple grades of enamel, and high-strength resin-bonded fiberglass tapes. Five grades of enamel and several types of primer are available for various service environments. Similar to other thermoplastic coatings, high temperatures will cause the enamel to soften and sag. This may eventually cause it to disbond from the pipe. Low temperatures may cause the enamel to become brittle, crack, and disbond. Therefore, selection of the correct grade depends on the highest continuous operating temperature anticipated for the life of the pipeline, the ambient temperature anticipated during the coating installation of the pipeline, expected soil stress, and the condition of the backfill.

4.2.1 CTE Coating Advantages

The molecular structure of coal-tar pitch, from which coal-tar enamels are formulated, provides an excellent bond to a primer-coated steel pipe and prevents moisture absorption and penetration. It also provides the CTE with the ability to resist attack by hydrogen, water, soil chemicals, and soil microorganisms that can cause corrosion. CTE coatings are also resistant to cathodic disbondment and are easy to repair in the field with coal-tar enamel, coal-tar epoxy, or tape coatings.

4.2.2 CTE Coating Disadvantages

While many long-term CTE-coated pipelines are still in service, coating failures have occurred. The failures are usually the result of improper field or plant application techniques, improper surface preparation, application in improper weather conditions, or rough handling during storage, transportation, or installation. Soil stress is a major factor in many CTE coating failures and can cause the coating to wrinkle, crack, and disbond. As CTE coatings age, cathodic protection requirements normally increase. If water penetrates under a disbonded coating, current may not reach the pipe surface, resulting in shielding the cathodic protection. In some cases, CTE coatings on high-temperature pipelines can cold flow, leaving the top of the pipe without adequate coating. In addition, these types of coatings are not recommended for use in soil that can become contaminated by hydrocarbon products from petrochemical plants, refineries, or storage facilities for such products.

4.2.3 Plant-Applied CTE Coatings

After the inspected pipe to be coated has been precleaned, it is then abrasive blasted to a specified cleanliness and profile, typically 2 to 4 mils (51 to 102 microns). Immediately after blast cleaning, it is sprayed with a primer made from coal tar, a fast-drying synthetic primer, or a plural-component epoxy primer that is applied to preheated pipe. If coal tar or synthetic primers are used, they are allowed to dry or are force dried by heat before the CTE coating is applied. If an epoxy primer is used, the hot CTE coating is applied directly over the wet primer, and the latent heat of the CTE coating cures the epoxy primer.

The CTE to be applied is melted in kettles to a temperature between 450°F and 490°F (232°C to 254°C) and is then pumped to a coating station. As the rotating primed pipe passes through the station, the hot enamel is sprayed or flooded onto the pipe to a thickness of about 100 mils (2,540 microns). While the pipe is being coated, a band of 20-mil (508-micron) porous inner fiberglass mat is simultaneously pulled into the coating around the pipe. This continuously applied band is positioned in the middle of the coating with an overlap of about ½ inch (1.3 cm) and must be completely coated with hot enamel. The mat provides strength to the coating after cooling. At another station, an overlapping band of denser, heavier fiberglass that has been dip-coated with hot enamel is wrapped around the outer surface of the hot coated pipe. This overwrap provides additional strength to the coating and offers additional resistance to soil stress after the pipe is installed. A final wrap of Kraft paper (or whitewash) is applied to the pipe to allow visual inspection for subsequent damage in handling and to provide heat reflection and UV protection during storage. After the coating has cooled, sometimes by a water quench, the coating at the ends of the pipe is cut back for joining by field welding.

4.2.4 Field Joint CTE Coating Application

Molten coal-tar enamel is poured over the primed field joint and caught by a handheld nonporous wrapper. The hot enamel is applied to the bottom and sides of the pipe when the wrapper is moved upward in a see-saw manner. To complete the joint, the wrapper is folded over the top of the pipe with its end facing downward to prevent water from entering.

Coal-tar epoxies have been used for years in both plant and field applications. They have the advantages of high bond strength and moisture resistance, as well as better chemical resistance and higher temperature resistance than solvent cutback coatings. Unlike coal-tar enamel, which requires a primer in order to develop a reliable bond, coal-tar epoxies are self priming. Coal-tar epoxy coatings give excellent service for many years and are used extensively for pipeline reconditioning. They are applied as a plural-component spray.

Fabric-reinforced coal-tar pitch tapes consist of a cotton fabric saturated with coal-tar pitch that has a softening point of 250°F (121°C) and has no inert filler. A thin polyester film is used as a separator on one side between rolls of the tape and is removed as the tape is applied. The tapes are applied in the field as field joint coatings and can also be used for an entire pipe.

During application, the pipe or welded field joint is first cleaned, most often by wire brushing, and then primed, generally using coal-tar cutback or special chlorinated rubber primer. Finally, the pipe or field joint is wrapped with softened tape that has been heated with a portable, low-pressure gas (usually propane) torch. Once attached to the pipe, the tape is again heated to soften the tar and make it fluid enough to be spirally wrapped around the pipe. The tape is alternately heated and wrapped around the pipe. Only a short section of the tape is heated at a time, and it must not be overheated. Excessive heating causes the pitch to flow off the fabric and may cause the pitch to burn. Enough heat should be applied to melt the tar, yet retain it on the fabric as the tape is applied. When heated properly, the tape is pulled firmly and tightly onto the pipe surface. The softened tar is squeezed into all of the pipe surface irregularities and at the tape overlap step-down. The tape can also be wrapped lengthwise (like a sheet material).

4.3.0 Extruded Polyolefin Pipeline Coatings

Plant-applied polyolefin (PO) coatings have been used on pipelines for almost 40 years. Polyolefin coatings are made of polyethylene (PE) or polypropylene (PP). Because polyolefin coatings do not bond well to steel, they are normally applied over an adhesive or other material that furnishes the bond to steel. In many extruded two-layer polyolefin coating systems (2LPO), the bond to steel is provided by a rubber-modified asphalt adhesive or a butyl rubber compound. Three-layer extruded polyolefin systems (3LPO) use a FBE as the base coating, an ethylene copolymer or terpolymer adhesive as the tie layer, and a PO as the top coat (or a tie layer).

4.3.1 PO Coating Advantages

Plant-applied PO coatings have proven to be excellent coating systems, provided proper pipe preparation and application methods are employed. PO coatings are resistant to water permeation, many chemicals, many environments, and bacterial attack. They also provide a slick surface that is usually unaffected by soil stress. These coating systems also handle well during transportation and construction.

4.3.2 PO Coating Disadvantages

In PO coatings applied over an adhesive layer, the adhesive layer alone provides the bond to the steel. The adhesive can allow the PO to shrink laterally at the ends of the pipe due to temperature changes. When this occurs, the steel is exposed, making it susceptible to corrosive attack. Lateral shrinkage at the field joints can also expose the steel pipe surface. If the coating is damaged while in service, it can disbond and allow water penetration between the coating and the pipe. Because polyolefin coatings display high electrical resistance and shield the cathodic protection, underfilm corrosion can become a significant problem. Additionally, the primer may not offer much corrosion protection when the cathodic protection is shielded. External impact can also break the bond between the adhesive and the PO, especially when applied by a crosshead extrusion process.

The intermediate tie layer in three-layer PO systems allows the top layer to shrink and expand with temperature changes. If the FBE primer is applied as a thin layer or at a temperature that is too low, it will not properly bond with the steel. This can result in early disbondment and may not provide corrosion protection if the top layers are damaged.

A major disadvantage of all PO coating systems is determining the type of material to use on repairs, field joints, valves, and other components. Many field joint coatings (tapes and shrink sleeves) are affected by soil stress. They also do not bond well to PO coatings and may have application problems that can cause coating failure at the repair or field joint. In addition, liquid epoxies, mastics, and similar coating types do not bond well to PO coatings. As a result, water may penetrate at the junction of the two coatings. Better repair and field joint coating systems for the three-layer coating systems are currently being developed.

4.3.3 Plant-Applied Extruded PO Coatings

PE is the most common extruded polyolefin coating for service temperatures less than 150°F (65°C). PP coatings are used for higher pipe service temperatures of up to 265°F (129°C). Both PE and/or PP coatings may be applied by a number of methods, including crosshead extrusion, side extrusion, hard-adhesive extrusion, and sintering.

- *Crosshead extrusion* – See *Figure 16*. For pipes using only a sheath of PO coating, the pipe is preheated and shot blast cleaned to the proper profile. Note that a phosphoric acid wash and rinse is used before the coatings are applied. Then it passes through an annular (ring-shaped) extrusion head (die) in which hot mastic is applied. The mastic is a butyl-rubber/asphalt blend or a butyl rubber compound extruded uniformly on the pipe to a DFT of about 10 mils (254 microns). After being primed, the pipe enters a second annular extrusion head in which a seamless sleeve of polyolefin coating is extruded simultaneously around the pipe to a DFT of 40 mils (1,016 microns). The sleeve is larger in diameter than the pipe being coated. Just beyond the extrusion head, the pipe enters a water spray, which cools the PO coating and causes it to shrink onto the primed surface in a compression fit. Extruded PO coatings can also shrink from the ends, providing a cutback. PO coatings do not bond to the mastic primer. During the cleaning, coating, and extrusion process, the pipe travels longitudinally but does not rotate. Pipe up to 30 inches (76 cm) in diameter can be coated by this process. The PO pellets that are melted and used in the extrusion process must be checked for any sign of moisture. In addition, the temperature and WFT of the mastic and the DFT of the extruded coating must be closely monitored.
- *Side extrusion* – This method is commonly used for pipes up to 96 inches (244 cm) in diameter and for lap-welded pipe. This type of weld leaves a void that must be filled with an adhesive layer and PO coating. After pipe is abrasive blast cleaned and preheated, it rotates as it travels forward under a side extruder head. A base coat that can be a molten asphalt butyl-mastic adhesive or butyl-rubber compound is spirally applied by the side extruder. Immediately following the base coat, a film of melted PO coating is applied by the side extruder over the base coat in a spiral fashion, similar to a tape wrap. The film fuses together at the overlap due to the heat of the coating. Another side-extrusion method involves overcoating a previously cured epoxy with one or more layers of a PO coating. This is called a hard-adhesive extruded coating.

Three-layer polyolefin (3LPO) systems can use crosshead or side extrusion for coating application (*Figure 17*). These systems typically use a sprayed FBE coating as a primer with an extruded copolymer intermediate coating that is used as an adhesive for a side-extruded outer polyolefin. All three layers are applied in sequence as the pipe rotates and passes through an application station. Because of the FBE, 3LPO application requires very stringent preparation and application methods. First, the pipe must be abrasive blasted to a white or near-white metal surface with a profile of less than 50 microns; it must then be phosphoric acid cleaned and water rinsed. If a standard FBE is used, the

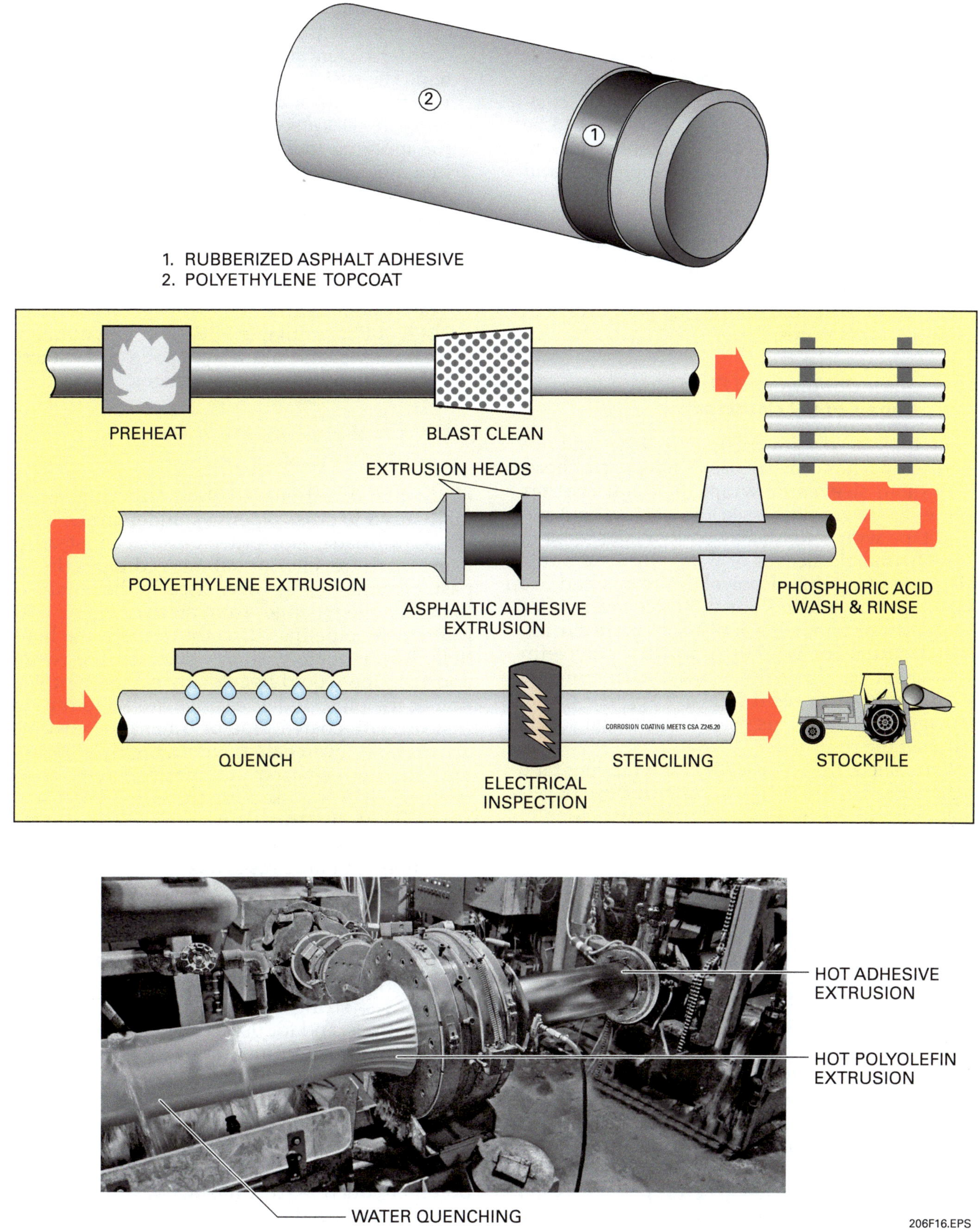

Figure 16 2LPO crosshead extrusion.

pipe must be heated to at least 450°F (232°C) for proper FBE bonding, and a high temperature copolymer intermediate coat must be used. If the intermediate coat cannot tolerate high temperatures, special low-temperature FBE coatings must be used. The FBE coating should be least 12 to 25 mils (305 to 635 microns) thick to allow a tapered cutback at the pipe ends and proper bonding of field joints to the FBE layer.

- *Sintering* – Fittings, such as valve bodies and bends, can be coated with PO coatings by a sintering process. Sintering is a method of applying polyolefin powders directly to a heated surface in the same manner as FBE, except at lower temperatures.

4.3.4 Joint Coating and Field Repair of PO Coatings

Commonly used joint coatings include cold-applied wraps, hot coal-tar-based wraps, and heat-shrinkable sleeves or wraps. The hot coal-tar tape wraps were discussed previously; cold-applied tape wraps are discussed in a later section. Heat-shrinkable sleeves or wraps for field joints are made of modified polyolefin resins and are coated on one side with an adhesive layer. After the sleeves or wraps are applied, they are shrink-fit by heating. For patching purposes, polyolefin resin sheet patches with a butyl base material and/or polyolefin resin rods are applied to the damaged areas. Heat is used to apply the sheets and melt the resin rods.

Automated, heat-assisted spiral-wrapping tape application machines (*Figure 18*) have been developed. With these systems, FBE is conventionally applied to the joint followed by hot-sprayed modified PP or PE dual base powdered layers. Using the machine, they are immediately followed with an application of single-layer PP or PE tape. This results in a nearly indiscernible interface between the factory and field coatings.

Other processes that have been refined or developed for coating field joints include the following:

- *Polypropylene flame spray* – This is a process for applying PP by flame spray (*Figure 19*) to field joints of 3LPP coated pipelines. Conventional equipment is used for application of FBE and modified PP dual-powder base layers. The overall thickness is then achieved by applying molten PP powder. The PP powder is fed directly into a flame-spray gun by a stream of nitrogen. The carrier gas feeds powder into the center of an annular combustion flame that heats the powder. A second annular gas nozzle feeds a gas stream around the powder stream to accelerate the spray towards the substrate and prevent the powder from burning. This coating system provides a full three-layer system. The system can also provide a useful service for tie-ins or as a backup for PP tape wrap.
- *Injection-molded polyurethane* – This is a process for applying thermally injected polyurethane in a mold clamped around a field joint (*Figure 19*). It can also be used with state-of-the art solid or syntactic polymeric-injected coatings (composed of very small spheres or bubbles of polymer in a matrix of another polymer) that can perform at higher operating temperatures. The system can be used for factory applied 3LPP, 3LPE, multilayer FBE/solid polyurethane elastomeric, and multilayer solid or syntactic polyurethane/PP coating systems.

4.4.0 Abrasion-Resistant Overlay (ARO) Pipeline Coatings

Horizontal directional drilling or boring is normally used to cross pipelines under rivers, highways, railroads, and other obstacles instead of using open trenches or submerged pipes. In the past, pipeline casings were used in the boreholes to protect the pipelines from abrasion rather than pulling the pipelines directly through the borehole. Unfortunately, shorts between the casing and the pipe would negate any cathodic protection of the pipe. Today, casings are seldom used. Instead, FBE-coated pipe or other pipe coatings are overlaid (overcoated) with a highly abrasion-resistant epoxy-based polymer concrete. These overlaid pipes are used for crossings. Because of their smooth surfaces, they can be drawn directly through a crossing in a slick-bore (slurry-assisted) operation with little resistance. The polymer concrete overlay on FBE-coated pipe can also be used as a rock shield in trenched applications.

Epoxy-based polymer concrete can be applied in a stationary plant or in the field. Several different formulations are available to accommodate more or less stringent pipe preparation, different base coatings, and for use as standalone corrosion coatings. Various formulations can be applied to bare steel pipe, FBE-coated pipe, or pipes coated with coal tar enamels. Polymer concrete chemically bonds to FBE.

When an FBE coating is new, little or no surface preparation is necessary before applying the polymer concrete. If it has aged, it must be removed or abraded and then cleaned, depending on the formulation of the polymer concrete being used. Since disbonding of an insulating overcoating could shield cathodic protection, the surfaces to which the polymer concrete is applied must be

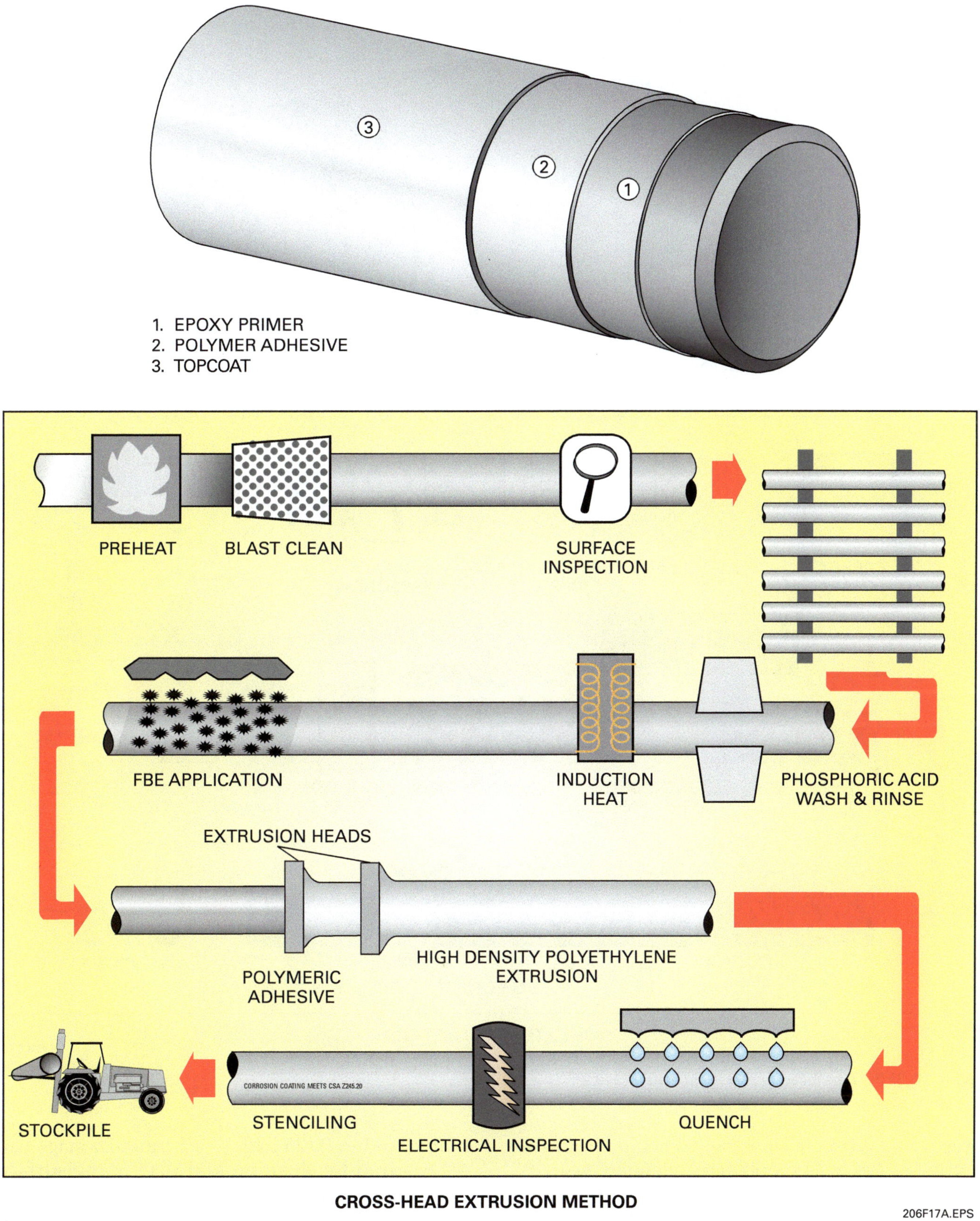

Figure 17 3LPO crosshead and side extrusion (1 of 2).

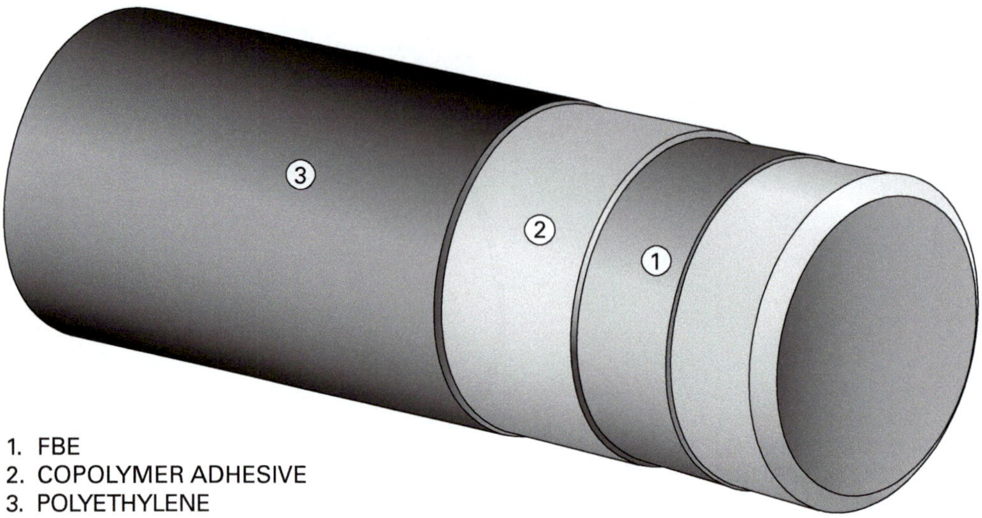

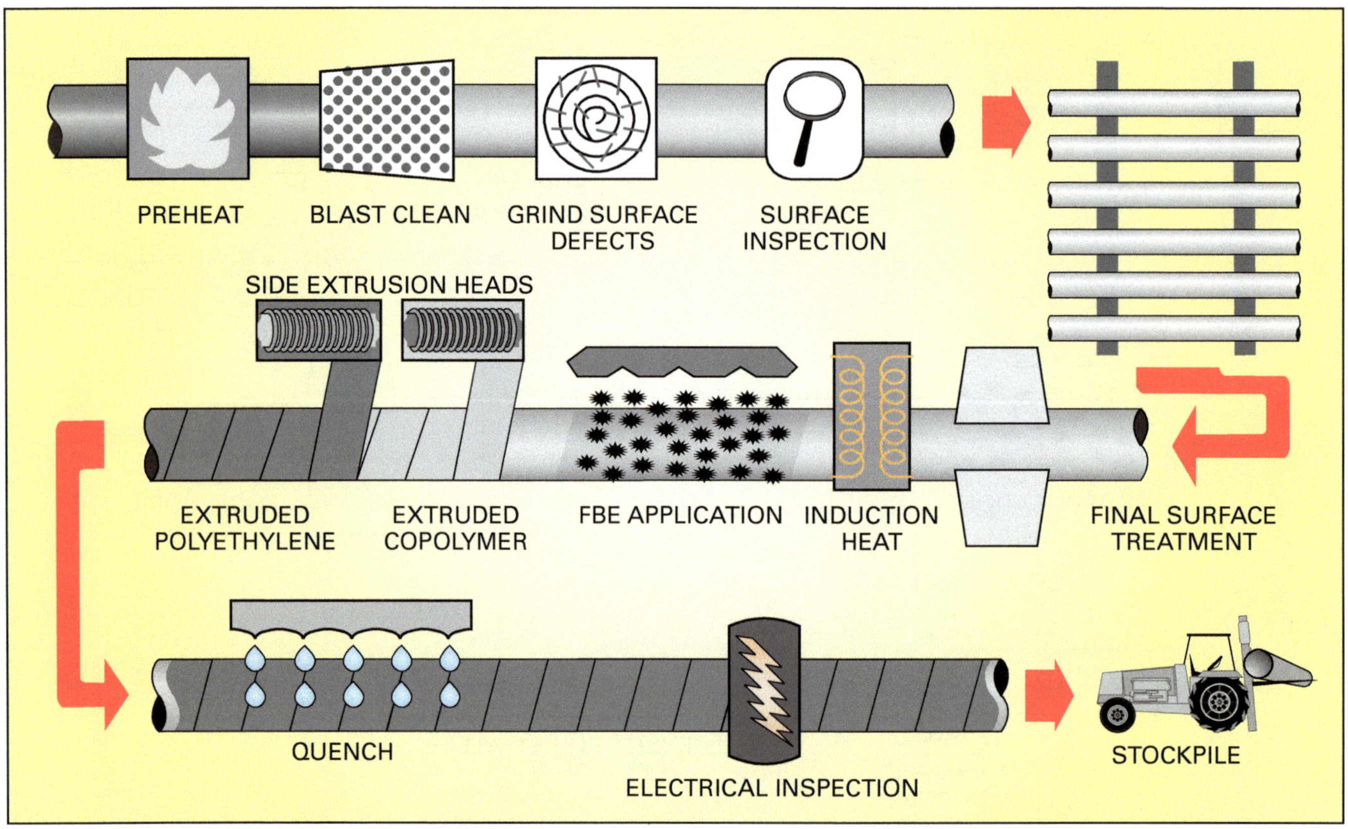

SIDE EXTRUSION METHOD

206F17B.EPS

Figure 17 3LPO crosshead and side extrusion (2 of 2).

prepared as directed for the particular formulation being used. The polymer concrete is usually sprayed, although some versions can also be applied manually. Each spray pass produces a thickness of nearly 0.030 inches (0.76 mm). Depending on the pipe diameter and/or intended pipe usage, multiple passes may be required to build up to the desired thickness. Maximum thickness is normally around 0.125 inch (3.175 mm). Damage can be repaired by cleaning and drying the surface and applying the same formulation of polymer concrete. At a temperature of 90°F (32°C), the concrete sets to a smear-free surface in 30 minutes. It then hardens after four hours and cures after seven days.

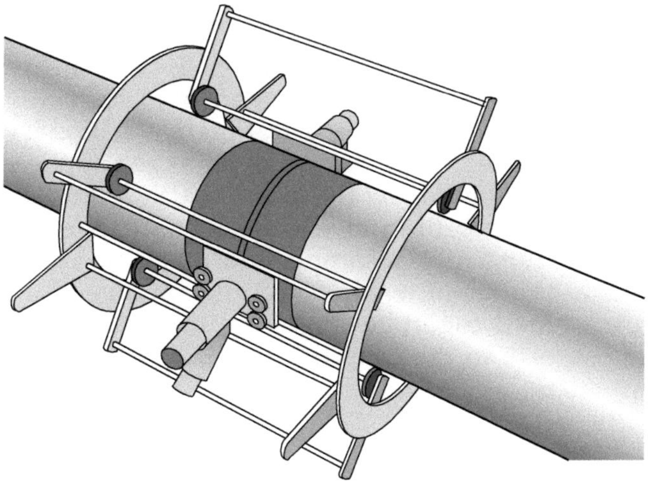

206F18.EPS

Figure 18 Automated tape wrap machine.

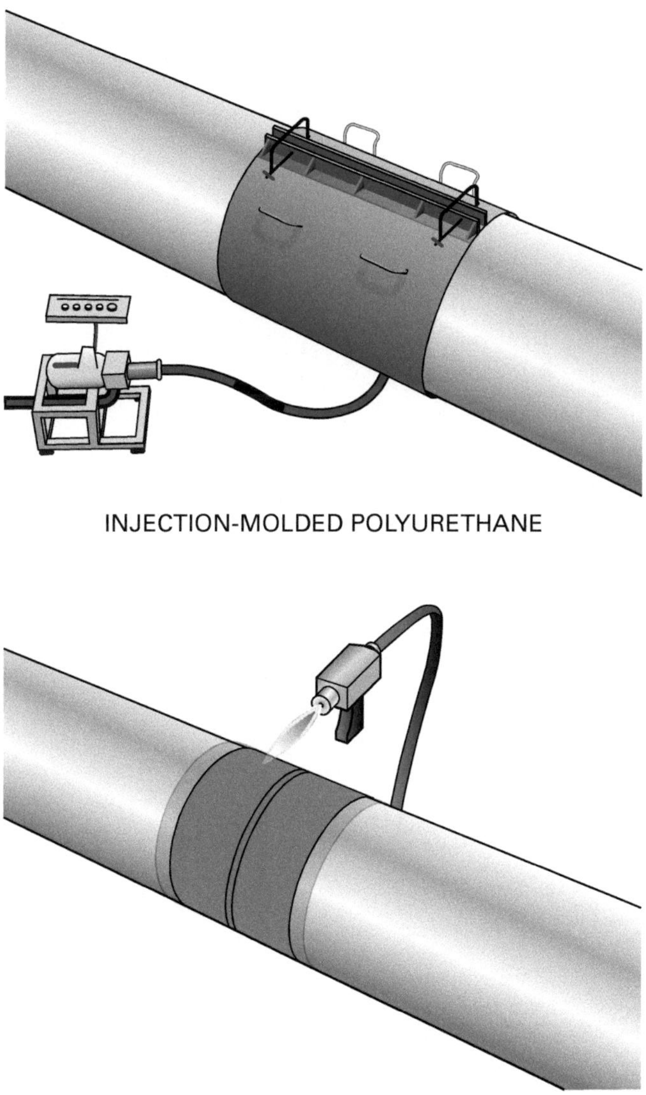

206F19.EPS

Figure 19 Injection-molded field joint and flame spray.

4.5.0 Cold-Applied Pipeline Tapes

PE and petrolatum tapes are applied cold; no heat is needed for the application process. Both types are used as coatings for mainline pipe construction and field joints, and both must be applied to pipe that has been coated with a compatible primer. The tapes are available in rolls with a thin plastic separator between the layers and in various widths. Petrolatum tapes are generally nonwoven fabric tapes impregnated with a neutral petrolatum compound that is a residue of paraffin-based hydrocarbons with added fillers. PE tapes are made of laminated or co-extruded materials. Laminated tapes are manufactured by applying a heated compatible adhesive to one side of a prefabricated plastic (PE) film. The tape can vary in total thickness from 15 to 50 mils (381 to 1,270 microns). The adhesive portion typically varies in thickness from 5 to 20 mils (127 to 508 microns). Co-extruded PE tape is manufactured by funneling three separately heated materials (backing, tie layer, and adhesive resins) through an extrusion die. The materials emerge as a homogeneous tape rather than a laminated tape. With laminated tape, the adhesive can be separated from the backing; however, co-extruded tape is almost impossible to pull apart. Co-extruded tapes have the appearance of a smooth and dense film on one side and an adhesive surface on the other.

4.5.1 Advantages of Cold Tape Coatings

Neither PE nor petrolatum tapes require particularly stringent surface preparation before application. Petrolatum tapes can be applied to wet surfaces. Both types are available in a variety of formulations to satisfy various service conditions. Both can be hand-applied and the PE tapes are normally machine-applied in the field by line traveling systems.

4.5.2 Disadvantages of Cold Tape Coatings

Both types of tapes have experienced disbondment due to the cathodic protection system. They have been subject to spiral corrosion because of a poor seal at the tape spiral overlap. Soil stress can displace portions of the tapes and cause disbondment, especially with the petrolatum tapes. Some of the tape adhesives can be degraded by bacterial attack, causing corrosion at the pipe/coating interface.

4.5.3 Petrolatum Cold Tape Application

Petrolatum tapes are manually applied and, for additional mechanical protection, can be overwrapped with a compatible protective tape, such as fiberglass or polyolefin. Bare pipe surfaces are normally prepared to an *SSPC-SP 2* or *SSPC-SP 3* finish and primed with a tape-compatible primer. The tape is applied in a spiral wrap using sections that are less than 6 feet long. The sections are overlapped at the ends by an amount equal to the tape width. For normal applications, the spiral overlap is usually 1 inch (2.5 cm). For severely corrosive environments, a 55 percent overlap is recommended. While wrapping, all air pockets must be pressed out, and all lap seams must be smoothed.

4.5.4 PE Cold Tape Application

Both laminated and co-extruded PE tape are applied under controlled tension over specially primed surfaces. Tape application in a stationary plant is done in the same manner as the application of polyolefins by the side-extrusion process. New or rehabilitation mainline coating in the field is normally accomplished by a line travel machine. For new construction, pipe sections are welded together above the ditch and bent to conform to the contour of the ditch. With the

On Site

Dual-Layer FBE Abrasion-Resistant Coatings

Plant-applied abrasion-resistant coatings are also available. The figure below shows the application of a dual-layer, plant-applied FBE coating. The coating provides protection for operating temperatures up to 221°F (105°C) and allows pipe bendability of 1.5 degree/pipe diameter length.

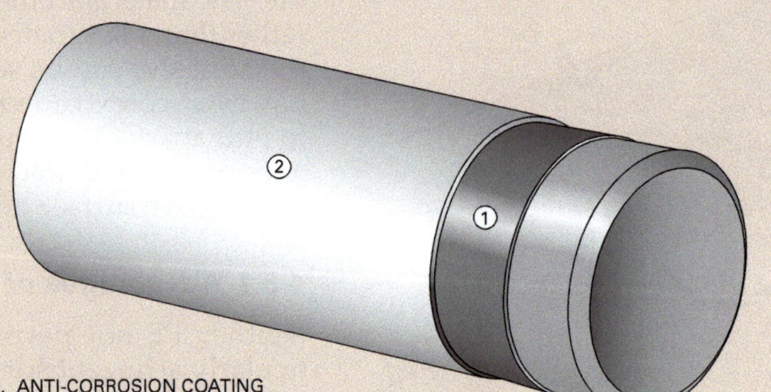

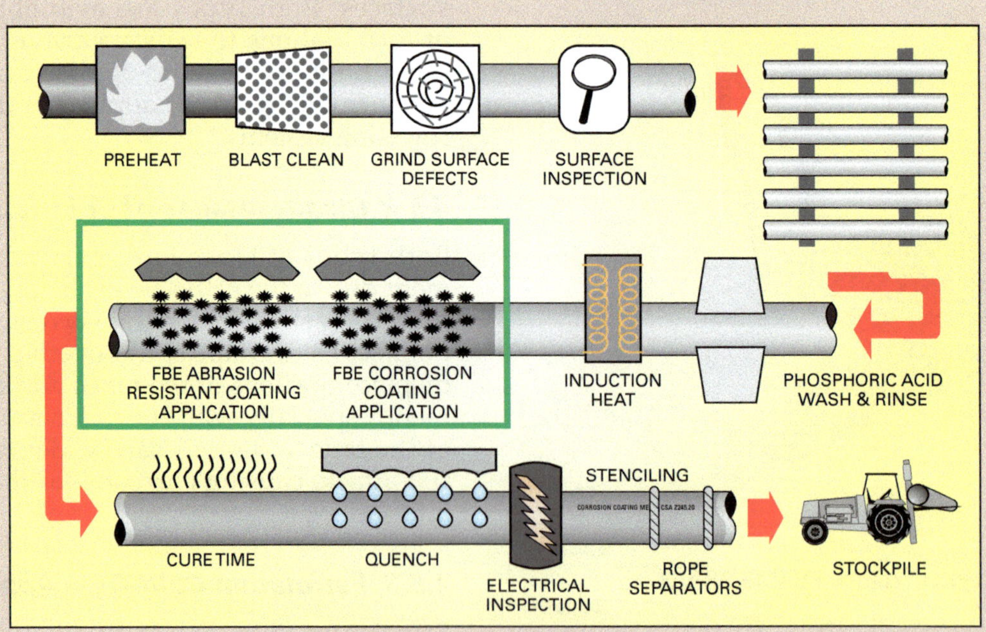

pipe above the ditch, a line travel machine, supported by a side-boom tractor, travels along the pipe, cleaning, priming, and wrapping the pipe in a continuous operation. The pipe surface is cleaned to a *SSPC-SP 2* or *SSPC-SP 3* finish and primed with a quick-dry primer to allow immediate wrapping by the machine. The amount of tape overlap depends on the anticipated service conditions. Rehabilitation coatings are accomplished on ditched pipe that has been exposed and is still in service. As a result, the machine must be frequently removed and reinstalled to avoid the necessary pipe supports. For cold tape that is factory applied or for other types of coatings, field joints can be hand-wrapped in either a spiral or lengthwise fashion.

On Site

Insulated Pipeline Coatings

Pipelines can be coated in a stationary plant with a polyurethane foam insulation and covered with an extruded adhesive and PE jacket. These insulated pipelines can be buried or placed above ground. They are used to reduce heat loss and prevent hydrate formation in gas pipelines, maintain viscosity of hot oil lines, or provide freeze protection to water and sewage lines. A variety of anti-corrosion undercoat systems can be overcoated with the foam insulation and PE jacket. The process is shown in this figure.

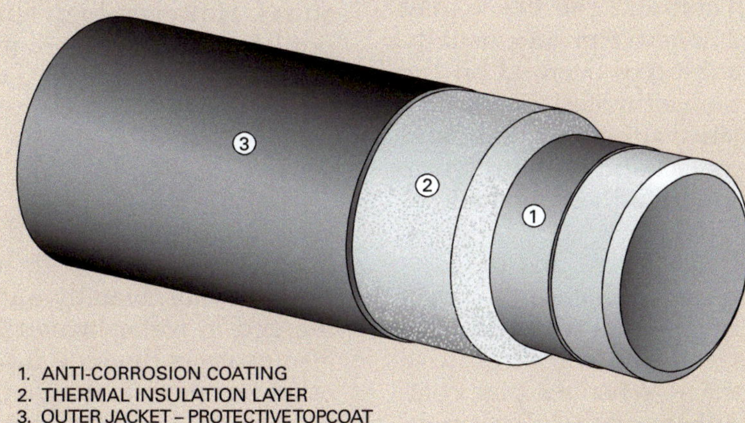

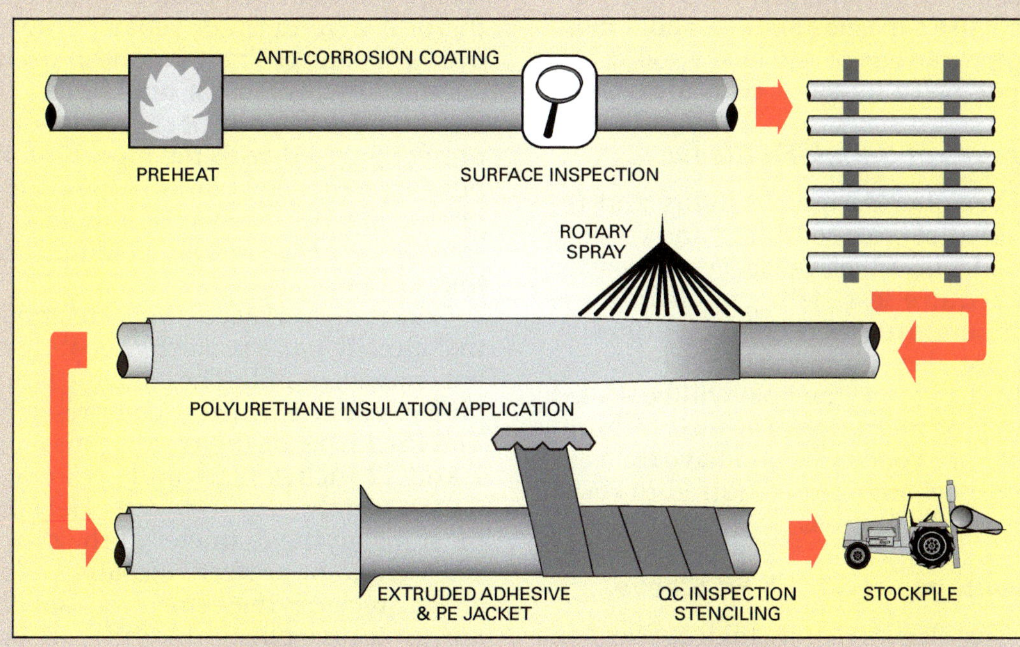

206SA02.EPS

4.6.0 Shrink-Fit Pipeline-Joint Sleeves

Shrink-fit sleeves are one of the most commonly used methods to protect pipe field joints from corrosion. Sleeves are made in sheets from the same composites as polyolefin tapes. The plastic sheet (backing) is chemically treated or irradiated to cause it to expand. When the material is heated, it shrinks to a tighter, denser film as it tries to return to its normal dimension. Some sleeves are tubular and must be slipped over one end of the pipe. Most sleeves are wraparound lengths of polyolefin backing that are overlaid with a special adhesive. The adhesive prevents moisture and air from contacting the pipe surface and holds the backing around the joint. The polyolefin backing also offers mechanical protection to the adhesive and the joint.

The sleeves are sized to match the pipe diameter with an allowance for overlap and are wide enough to cover the bare field joint and extend over the coating at each cutback. The sleeve may be applied over a special adhesive primer, usually an epoxy. The special adhesives overlaid on the sleeves are mastics or hot-melt materials. Mastics are tacky, pressure-sensitive adhesives that bond easily to substrates on contact and with pressure. Hot melts are non-tacky hard adhesives made of semi-crystalline resins.

Axial pipe movements, soil stress, and the type of mainline pipe coating each influence the selection of the adhesive. Most pipelines with relatively constant temperatures and low soil stress use sleeves with mastic adhesive. When the pipe coating has a mastic type of adhesive, the sleeve should also use a mastic adhesive. However, if the pipe coating does not use a mastic adhesive, and if there is significant pipe movement and heavy soil stress, a sleeve with hot-melt adhesive is typically used.

4.6.1 Advantages of Shrink-Fit Sleeve

Shrink-fit sleeves offer design flexibility and reliability, as well as service condition and coating compatibility. Mastic adhesive sleeves are easier to apply and require less stringent cleanliness and lower temperatures than hot-melt adhesive sleeves. They are also self-repairing if the backing is punctured, because the hole will be sealed by the mastic. Hot melts perform well in high-temperature service conditions and have the very high shear and peel strengths required to resist pipe movement and soil stress.

4.6.2 Disadvantages of Shrink-Fit Sleeve

Hot-melt adhesive sleeves require high application temperatures, as well as pipe preheating. They also require more stringent surface cleanliness than mastic adhesive sleeves. However, mastic adhesive sleeves have a much lower peel strength.

4.6.3 Shrink-Fit Sleeve Application

Wraparound sleeves are placed around a field joint with the overlap facing the bottom of the ditch. The sleeve is heated with a large propane torch that melts the adhesive and shrinks the sleeve to form a tight fit over the joint. After cooling, the adhesive solidifies and forms a tough bond to the pipe and coating. Pipe surface cleanliness for mastics can be *SSPC-SP 2* or *SSPC-SP 3* finishes, but hot melts usually require *NACE 3/ SSPC-SP 6* finishes. Hot melts also require pipe preheat temperatures in excess of 220°F (104°C) at installation. Certain hot melts can tolerate lower preheat temperatures by using an epoxy primer. A three-layer application using an epoxy primer layer provides maximum protection in most situations. However, high surface cleanliness, epoxy pot life, dust contamination, and cure temperature must be taken into consideration when applying a three-layer sleeve.

4.7.0 Pipeline Weight Coatings

Weight coatings are used to increase the strength of pipelines and provide negative buoyancy when the pipelines, usually 8 inches or larger, are submerged in water or marshes. A concrete coating also protects the pipe from impact, abrasion, and other mechanical damage. Heavyweight pipe can be coated with aggregate-fortified asphalt mastic that provides both a protective and a weight coating or a wire-mesh or rebar-reinforced concrete overlay coating. The wire mesh used in concrete overlay coatings must not contact the pipe surface in order to prevent shorting out any cathodic protection used with the pipe. Both types are applied in stationary plants. Aggregate-fortified asphaltic mastic is generally not used much today, so only concrete overlay coatings are covered in this section.

Concrete overlay coatings are applied over pipe that already has a protective coating of polyolefin, hot enamel, or FBE. The concrete is applied by impingement or extrusion/compression techniques. The thickness of the coating may vary between 1 and 12 inches (2.54 and 30.5 cm), depending on the weight requirements. Field-joint coatings must match the diameter of the concrete coating. The field joint protective coating, along with layers of quick-drying cement or asphalt mastic and any protective plastic covering, are used to fill the void at the field joints.

Impingement coating is accomplished by applying concrete using a throwing unit that consists of a conveyor belt and a high-speed circular wire brush cluster or slinger belt. A mechanical barrier coat must first be applied over the pipe's protective coating to prevent damage to the protective coating during application of the concrete. A stiff concrete mix is supplied from the conveyor belt to the high-speed wire brush. The brush or belt sprays the concrete at a high velocity onto the pipe surface. Controlling the pipe's rotation and forward speed produces the desired thickness. Simultaneous spiral wrapping of wire mesh at a controlled tension places the wire at the center of the concrete. If a rebar cage is used, the pipe is centered inside the rebar cage before it is moved into the throwing unit. A liquid sealer or steam is normally used during the curing period to assure complete hydration of the concrete. This method results in some amount of wasted concrete because of the high-velocity slinging.

Extrusion/compression coating (*Figure 20*) is accomplished in one method by spirally extruding stiff concrete through a feeder box along with wire-mesh reinforcement and a polyolefin outer wrap outside the concrete as the pipe rotates and moves forward past the feeder box. The feeder box controls the thickness of the coating applied

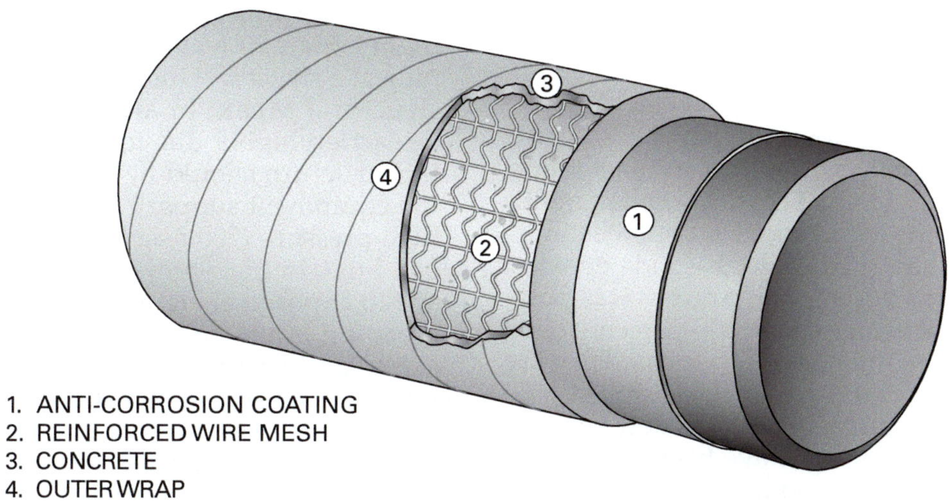

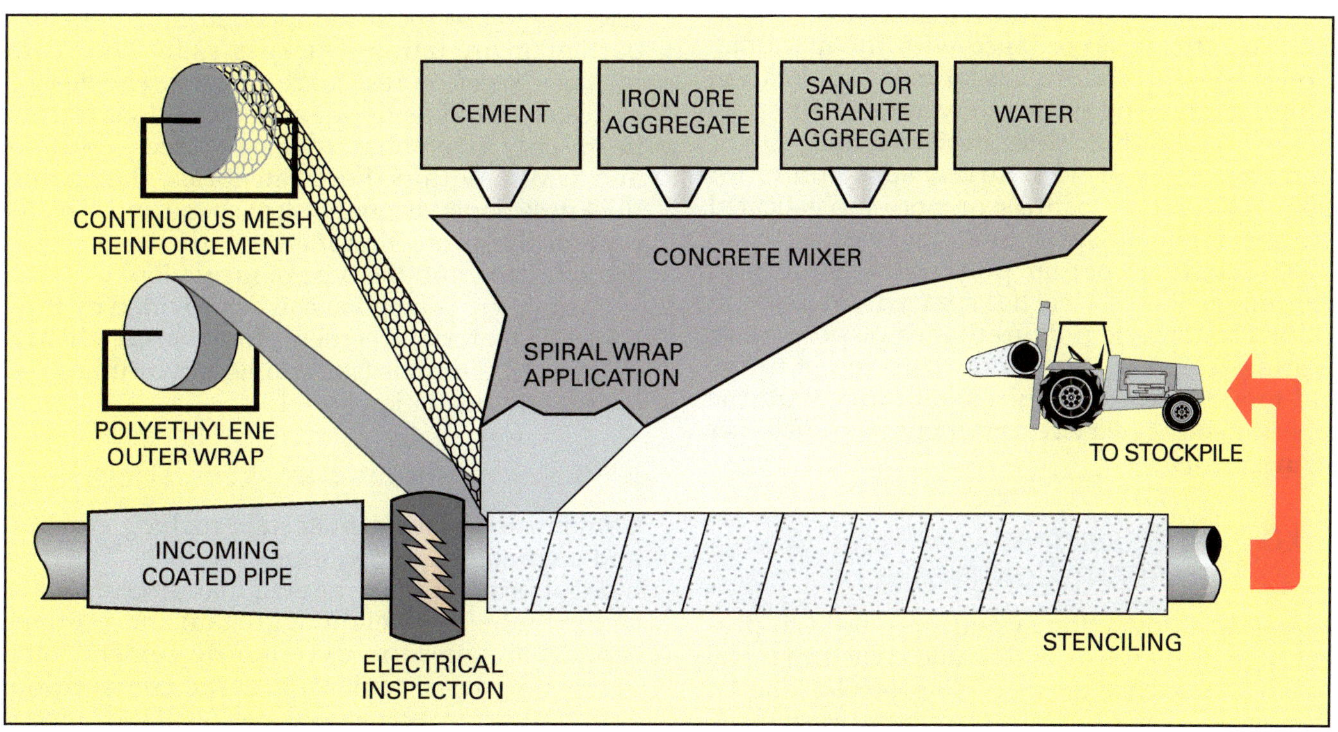

206F20.EPS

Figure 20 Extrusion/compression weight coating process.

prior to compression. The wire mesh is tensioned to position the mesh in the center of the concrete. The polyolefin outer wrap is tensioned to compress the concrete. Because the concrete is not applied at a high velocity, a mechanical barrier coat is not required over the pipe's protective coating, and very little concrete is wasted. If a nonslip epoxy coating is required, it is applied immediately prior to concrete application. The outer wrap allows the pipe to be handled immediately and assures complete curing of the concrete without any other intervention. The outer wrap is removed after curing is complete. For coating protection in rocky terrain, a thinner reinforced-concrete rock jacket that is bendable can be applied in the same manner as shown in *Figure 20*.

4.8.0 Liquid Pipeline Coatings

Liquid coatings on pipelines can be solvent-based coatings such as coal-tar cutbacks and vinyls, or chemically cured coatings, called multi-component liquid (MCL) coatings, such as inorganic zinc, coal-tar epoxies, high-solids epoxies, 100-percent-solids epoxy coal tars, coal-tar polyurethanes, and non-immersion 100-percent-solids elastomeric polyurethanes or polyureas. They are used as field-joint coatings and field-applied pipe rehabilitation, as well as coating on new pipe installations. These coatings have been covered in previous sections of this module.

Both solvent-based and chemically cured coatings must be applied over near-white metal blast surfaces in accordance with the manufacturer's specific recommendations. Application equipment may be simple conventional air-spray rigs for use with coal-tar cutbacks and vinyls; high-ratio (30:1 or 45:1) airless spray units for 100-percent-solids coatings; or more sophisticated plural-component spray units for two-part coatings, including coal-tar polyurethane and, for non-immersion service, hot-sprayed 100-percent-solids elastomeric polyurethanes or polyureas. Coating selection, number of coats, recoat times, and total thickness requirements vary with the generic material, service conditions, and customer requirements.

5.0.0 Fire-Resistant/Retardant Coatings

Passive fire-protection systems are used to protect high-risk and/or high-value structures. The materials used are designed to protect property, prevent structural damage or collapse of steel, allow critical fire-control systems to operate, and provide egress time for occupants and firefighters while a fire is brought under control. Protection time for high-risk and/or high-value structures at temperatures of 1,000°F (538°C) is set at three hours, as defined by *Underwriters Laboratories (UL) 1709, Standard for Rapid Rise Fire Test of Protection Materials for Structural Steel.* This is equivalent to four hours of protection, as defined by *ASTM E 119, Methods for Fire Tests of Building Construction and Materials.* The requirements can be exceeded for certain installations, such as reactors that handle large volumes of flammable material in congested areas.

Three-hour protection is applied to main structural support members. For low-risk or low-pressure areas (less than 500 psig or 232 kg/cm²), the fire protection coating is generally carried no higher than 20 feet (5.3 m) above ground, stopping at a structural intersection, platform, or other structural break. In areas where the fire could be carried higher due to fan drafts or air coolers, the structure should be fire-protected at least to the equipment support. This level is also used in high-pressure (500 psig or higher) or high-risk areas. Fires in high-pressure areas can cause damage to exposed structures 30 feet (8 m) above the escaping liquid or vapor. No set elevation limit exists, and fireproofing can be carried as high as necessary on structural supports based on specific situations. In areas where three-hour protection is not necessary, thinner coatings can be used. For formed concrete protection, the cost difference is negligible, but with other materials, it is more cost effective to use the recommended thickness for a particular fire rating. Thicknesses for hourly ratings are based on resistance designs. The designs are categorized as floor assemblies, floor or roof-beam-only assemblies, roof assemblies, columns, and walls. Within these categories, the coating thickness depends on the steel shape and size, the type of fire protection material, the thickness of the substrate, and the type of insulation.

Passive fire protection can be provided by three types of materials: cementitious based, ablative or non-cementitious (subliming and intumescent) based, and insulation based.

5.1.0 Cementitious-Based Materials

Cementitious-based materials include concrete and commercial cementitious products. The commercial products vary among manufacturers and can include lightweight aggregate and portland cement, magnesium oxychloride cement, inorganic magnesia cement, organic cementitious vermiculite, or vermiculite and portland cement. Various binders, inorganic fibers, and refractory fillers may also be in the material.

Concrete fire resistance is typically specified as *ASTM C 150, Type II*. This type of coating, along with commercial lightweight aggregate/portland cement mixtures, does not need top coating, except in severe freeze/thaw cycles. Other commercial lightweight cementitious coatings require top coating to prevent premature degradation. In contact with water, materials based on magnesium oxychloride can cause severe corrosion problems. Epoxy coatings are recommended for use on steel underneath cementitious fireproofing materials. Top coating, as well as water caps, are also required to prevent water absorption.

Cementitious-based materials are not necessarily destroyed after a fire. Concrete is a good insulator. Much of the remaining concrete after a fire is in good condition and can be patched or recoated as needed to its original thickness with standard concrete-repair methods. The same is true of commercial products where reapplication of the coating or special restoration coatings can be used to restore the product to its specified fire-rating thickness.

Application of cementitious-based materials to steel usually requires abrasive blasting to remove rolling compounds, mill scale, and other contaminants. Priming the steel with epoxy, inorganic zinc, or an epoxy/zinc combination is normally required. The application methods used include poured-in-place materials, shotcreted (slurry) spray, trowel, and precast panel systems. Shotcreted material is cost effective for low-risk, low-value areas in which aesthetics are not important. Curing of cementitious-based materials may require the application of a curing compound, wrapping with burlap or plastic sheets, or continuous spraying with a fine mist so that proper hydration occurs. Except for magnesium oxychloride materials, top coating is not usually recommended.

5.2.0 Non-Cementitious Materials

Non-cementitious materials are ablative coatings that consist of subliming or intumescent material. During a fire, ablative coatings gradually deteriorate over a specified amount of time. Both coatings require replacement after a fire.

Subliming materials absorb massive amounts of thermal energy from a fire to cause the chemical transformation of a solid material directly into a vapor without a liquid stage, a process that is called sublimation. The temperature of the underlying substrate is maintained at the temperature of the sublimation during the reaction, and the counterflow of noncombustible gases that is created blocks the heat input of the fire. A porous char layer also forms on the surface of the coating, providing cellular insulation and a radiating heat shield that delays heat penetration to the substrate. The subliming coating thickness mostly determines the fire rating. The coatings are lightweight, relatively maintenance free, and are top-coated. Subliming coatings are epoxy-based coatings and, in addition to fire protection, provide corrosion protection and excellent chemical resistance.

These coatings are applied to steel that has been abrasive-cleaned to a near-white metal blast surface with a specified profile. When the steel is inspected and found to be clean and dry, it is coated to a specified DFT with the primer recommended by the coating manufacturer. After the primer has dried and been inspected, subliming coatings are typically spray-applied; however, they may also be trowel-applied. After half the coating has been applied, fiberglass, carbon fiber, or wire-mesh reinforcement is placed. Following application of the next coat, fiberglass reinforcement is embedded using a solvent-moistened roller. If metal reinforcement is used, it can be held in place by staples that are removed when the next coat is applied, or it can be held with pins that were welded to the substrate. Coating application continues until the specified fire-rating thickness is achieved. A compatible top coat should be applied to the subliming coating within the time recommended by the manufacturer, or the subliming coating must be solvent-activated before top coating.

Intumescent materials become plastic when heated and produce nonflammable gases, such as carbon dioxide and ammonia. The gases convert the coating film into foam that is about fifty times as thick as the original coating. After this occurs, the foam solidifies and chars into a hard, thick, highly insulating layer of carbon that shields the substrate and reduces heat gain. Intumescent coatings are available as epoxy-based material, 100-percent-solids epoxies, or vinyl-based mastics. Epoxy primers are recommended for use with these coatings, but vinyl-based mastics require an epoxy top coat. Intumescent coatings resist solvents, acids, alkalies, salts, and abrasion. Thin coatings of 16 to 130 mils (406 to 3,302 microns) are used in facilities in which aesthetics are important because they can be top-coated with a decorative finish.

These coatings are applied to steel that has been cleaned to a commercial blast surface finish with a specified profile. When the steel has been inspected and is found to be clean and dry, it is coated with the primer recommended by the coating manufacturer to a specified thickness. Common primers that can be used are a thermally-sprayed aluminum or aluminum alloy with an epoxy tie coat, a polyamide epoxy, a zinc-rich epoxy, or a zinc-rich epoxy with an epoxy tie coat. When the primer is inspected and is found to be clean and dry, the intumescent coatings are normally applied using heated airless spray equipment at high pressure. Because the coating material can be highly viscous, the heat and pressure reduce the viscosity and enhance liquid flow. The recommended substrate temperature range for application is 70°F to 90°F (21°C to 32°C). The coatings are applied to a specified fire-rating thickness. As with subliming coatings, fiberglass or carbon fiber mesh is embedded when half the coating thickness has been applied. Final thickness is calculated as either the minimum allowed at any measuring point or as an average of multiple readings, with no thickness less than 90 percent of the specified thickness. The final coating surface may have a matte or orange-peel appearance that can be smoothed, if desired, by rolling, using a solvent recommended by the manufacturer. Decorative top-coating of the surface is usually desired, especially in moist environments. Top coats can be polyurethane, epoxy, or acrylic coatings.

5.3.0 Insulation-Based Materials

A number of commercial insulation-based fire-protection materials are available. They usually consist of incombustible mineral fiberboard products that are made of lime, silica, and asbestos-free reinforcement fibers. These products are used to enclose structural steel and provide protection for columns, walls, shafts, and ductwork, as well as protection for firewalls. They may require weather jacketing for exterior use. Manufacturers normally recommend an appropriate top coat and primer for their product; however, epoxy primer is typical. Mineral fiberboard is installed dry and exhibits very good stability, thermal shock resistance, low thermal conductivity/heat storage, and good compression resistance. It is resistant to chemical attack, except for strong alkalies and hydrofluoric or phosphoric acids.

6.0.0 Powder Coatings

Powder coatings are non-liquid and do not contain a solvent. They can be applied manually with propane flame-spray guns or electrostatic spray guns. They are also applied in stationary plant systems that use fluidized beds and cloud chambers, as well as spray guns. If applied electrostatically, the powders are charged electrically and sprayed onto a surface that is electrically grounded. The difference in electrical charge makes powder particles stick to the surface. This allows the thickness of a powder coating to be much greater than many conventional liquid coatings. Most often, the surface is preheated before the coating is applied. Usually just before the surface is coated, it is heated to a high temperature so the coating melts, creating a durable finish when it cools.

The most common uses for powder coatings are pipelines, reinforcing bar (rebar), household appliances, and automotive exteriors. In addition, some powder coatings are chemically resistant and may be used on pump impellers, agitation blades, or other devices that are used with chemicals.

Powder coatings are similar to liquid coatings in their makeup. Powders generally contain some of the following elements:

- Pigment
- Curing agents (partially reacted solid resins)
- Wetting agents
- Flow-control agents
- Fillers and extenders
- Foam breakers and other additives

All powder coatings are dry particles that look like a fine powder—like flour or talc—except that they may be colored with pigments. Powders fall into two broad categories: thermoplastic and thermosetting. Thermoplastic materials soften whenever they are heated and harden when cooled. Thermosetting materials harden when heated and stay hard after they have cooled.

Thermoplastic coating materials include the following:

- Polyvinyl chloride (PVC)
- Polypropylene
- Kynar®
- Halar®
- Polyethylene
- Teflon® (polytetrafluoroethylene or PTFE)

Thermosetting coating materials include the following:

- Epoxy
- Urethane
- Polyester
- Acrylic
- Alkyd

A number of these coatings, along with their application sequence, were described earlier in this module for pipeline coatings. Powder coating application to other steel objects follows the same general sequence given for pipeline coatings. When handling or applying powder coatings, always wear proper respiratory protection to avoid inhaling very fine powders.

7.0.0 Pipeline and Specialty Coatings Standards

A number of standards exist for pipeline and specialty coatings. In addition to the standards referenced in previous sections, NACE International publishes standards for coatings and linings. The following is a list of standards directly related to coatings and linings:

> **NOTE**
> When these standards are revised, recommended practices (RP) will be reidentified as standard practices (SP) using the same base number.

- *NACE No. 10/SSPC-PA 6, Fiberglass-Reinforced Plastic (FRP) Linings Applied to Bottoms of Carbon Steel Aboveground Storage Tanks*
- *NACE No. 11/SSPC-PA 8, Thin-Film Organic Linings Applied in New Carbon Steel Process Vessels*
- *NACE No. 12/AWS C2.23M/SSPC-CS 23.00, Specification for the Application of Thermal Spray Coatings (Metallizing) of Aluminum, Zinc, and Their Alloys and Composites for the Corrosion Protection of Steel*
- *RP0105-2005, Liquid-Epoxy Coatings for External Repair, Rehabilitation, and Weld Joints on Buried Steel Pipelines*
- *RP0191-2002, The Application of Internal Plastic Coatings for Oilfield Tubular Goods and Accessories*
- *RP0198-2004, The Control of Corrosion Under Thermal Insulation and Fireproofing Materials—A Systems Approach*
- *RP0303-2003, Field-Applied Heat-Shrinkable Sleeves for Pipelines: Application, Performance, and Quality Control*
- *RP0375-2006, Field-Applied Underground Wax Coating Systems for Underground Pipelines: Application, Performance, and Quality Control*
- *RP0394-2002, Application, Performance, and Quality Control of Plant-Applied, Fusion-Bonded Epoxy External Pipe Coating*
- *RP0399-2004, Plant-Applied, External Coal-Tar Enamel Pipe Coating Systems: Application, Performance, and Quality Control*
- *RP0402-2002, Field-Applied Fusion-Bonded Epoxy (FBE) Pipe Coating Systems for Girth Weld Joints: Application, Performance, and Quality Control*
- *RP0602-2002, Field-Applied Coal-Tar Enamel Pipe Coating Systems: Application, Performance, and Quality Control*
- *SP0108-2008, Corrosion Control of Offshore Structures by Protective Coatings*
- *SP0176-2007 (formerly RP0176), Corrosion Control of Submerged Areas of Permanently Installed Steel Offshore Structures Associated with Petroleum Production*
- *SP0181-2006 (formerly RP0181), Liquid-Applied Internal Protective Coatings for Oilfield Production Equipment*
- *SP0185-2007 (formerly RP0185), Extruded Polyolefin Resin Coating Systems with Soft Adhesives for Underground or Submerged Pipe*
- *SP0298-2007 (formerly RP0298), Sheet Rubber Linings for Abrasion and Corrosion Service*
- *SP0302-2007 (formerly RP0302), Selection and Application of a Coating System to Interior Surfaces of New and Used Rail Tank Cars in Molten Sulfur Service*
- *SP0386-2007, Application of a Coating System to Interior Surfaces of Covered Steel Hopper Railcars in Plastic, Food, and Chemical Industries*
- *SP0487-2007 (formerly RP0487), Considerations in the Selection and Evaluation of Rust Preventives and Vapor Corrosion Inhibitors for Interim (Temporary) Corrosion Protection*
- *SP0892-2007 (formerly RP0892), Coatings and Linings over Concrete for Chemical Immersion and Containment Service*

Summary

Placing a protective coating or lining between a corrosive environment and a material that is the substrate to be protected is a fundamental method of corrosion control. It is the most widely used method of protecting metals and other substrates. This module provided additional details about protective industrial coatings and linings. It also provided detailed coverage for pipeline coatings, fire-resistant/retardant coatings, and powder coatings.

Review Questions

1. Linings are primarily used for the purpose of _____.
   a. decoration
   b. immersion
   c. fire protection
   d. insulation

2. Coatings that provide galvanic protection for steel are known as _____ coatings.
   a. immersion
   b. barrier
   c. inhibitive
   d. sacrificial

3. Permeability is a coating's ability to resist _____.
   a. moisture absorption/transmission
   b. color fading
   c. surface deterioration due to UV light
   d. surface refraction

4. Mineral spirits and VMP naphtha are common _____ solvents.
   a. aromatic hydrocarbon
   b. ketone
   c. aliphatic hydrocarbon
   d. glycol ether

5. A coating that uses chemical polymerization for curing is called a(n) _____.
   a. coreaction-cure coating
   b. condensation-cure coating
   c. oxidation-cure coating
   d. hydration-cure coating

6. One type of hydration-cure coating is a _____ coating.
   a. phenolic
   b. vinyl ester
   c. concrete
   d. furan

7. One type of evaporative coating is a(n) _____ coating.
   a. vinyl acetate
   b. urethane
   c. polyester
   d. epoxy

8. Material normally used to seal concrete is known as a(n) _____ primer.
   a. thin-film
   b. penetrating
   c. thick-film
   d. elastomeric

9. A mainline coating is usually applied in a plant.
   a. True
   b. False

10. One advantage of pipeline FBE coatings is that they provide the best _____.
    a. bonding to steel
    b. moisture resistance
    c. damage resistance
    d. protection for short design life pipelines

11. One advantage of petrolatum pipeline tapes is their _____.
    a. exceptional resistance to cathodic disbondment
    b. excellent seal at the spiral seams
    c. minimal surface preparation requirements
    d. high resistance to soil-stress disbondment

12. Shrink-fit sleeves used on the field joints of non-mastic pipe coatings in soil where significant pipe movement or soil stress is a problem should use a(n) _____ adhesive.
    a. acrylic
    b. mastic
    c. hot-melt
    d. polyolefin

13. A fire-resistant/retardant coating that may be repairable after a fire is a(n) _____-based material.
    a. cementitious
    b. non-cementitious
    c. insulation
    d. ablative

14. A fireproofing material that absorbs large amounts of heat and transforms directly into a vapor is called a(n) _____ material.
    a. cementitious
    b. intumescent
    c. subliming
    d. insular

15. Powder coatings may be applied with flame-spray guns or _____ spray guns.
    a. conventional
    b. airless
    c. electrostatic
    d. plural-component

Trade Terms Introduced in This Module

Barrier coating: A coating that keeps moisture and oxygen away from a substrate.

Condensation-cure coating: A coating in which two or more organic molecules react, forming a new bond and giving off a molecule of water.

Coreaction-cure coating: A coating that requires the addition of an activator or catalyst to the base material. Also called a plural or multi-component coating.

Hydration-cure coating: A coating that cures by a process called hydration, in which water combines chemically with the coating and its constituents.

Inhibitive coating: A coating that contains pigments that react with absorbed moisture within the coating and then react with the steel surface, passivating the metal and interfering with the corrosion process.

Opacity: A coating's ability to cover or hide a substrate.

Oxidation-cure coating: A coating containing vegetable oil that cures by a process called oxidation, in which coating ingredients react with the oxygen in the air.

Surfactants: Additives used in oil-and-water emulsions that break down the surface tension of the liquid.

Additional Resources

This module presents thorough resources for task training. The following resource material is suggested for further study.

NACE International for corrosion control resources and information. www.nace.org.

The Protective Coating User's Handbook, 2004. Dr. Louis D. Vincent. Houston, Texas: NACE International.

Figure Credits

Praxair Surface Technologies, Module opener

Graco Inc., 206F01

Topaz Publications, Inc., 206F02

Rogers & Associates Consulting Services, 206F03

CRC-Evans Pipeline International, Inc., 206F15

NACE International, 206T01, 206T02

Shaw Pipe, a ShawCor Company, 206F09–206F14, 206F16A, 206F16B, 206F17A, 206F17B, 206SA01, 206SA02, 206F20

NCCER CURRICULA — USER UPDATE

NCCER makes every effort to keep its textbooks up-to-date and free of technical errors. We appreciate your help in this process. If you find an error, a typographical mistake, or an inaccuracy in NCCER's curricula, please fill out this form (or a photocopy), or complete the online form at **www.nccer.org/olf**. Be sure to include the exact module ID number, page number, a detailed description, and your recommended correction. Your input will be brought to the attention of the Authoring Team. Thank you for your assistance.

Instructors – If you have an idea for improving this textbook, or have found that additional materials were necessary to teach this module effectively, please let us know so that we may present your suggestions to the Authoring Team.

NCCER Product Development and Revision

13614 Progress Blvd., Alachua, FL 32615

Email: curriculum@nccer.org
Online: www.nccer.org/olf

❑ Trainee Guide ❑ Lesson Plans ❑ Exam ❑ PowerPoints Other ____________________

Craft / Level: Copyright Date:

Module ID Number / Title:

Section Number(s):

Description:

Recommended Correction:

Your Name:

Address:

Email: Phone:

Coating Applications Two

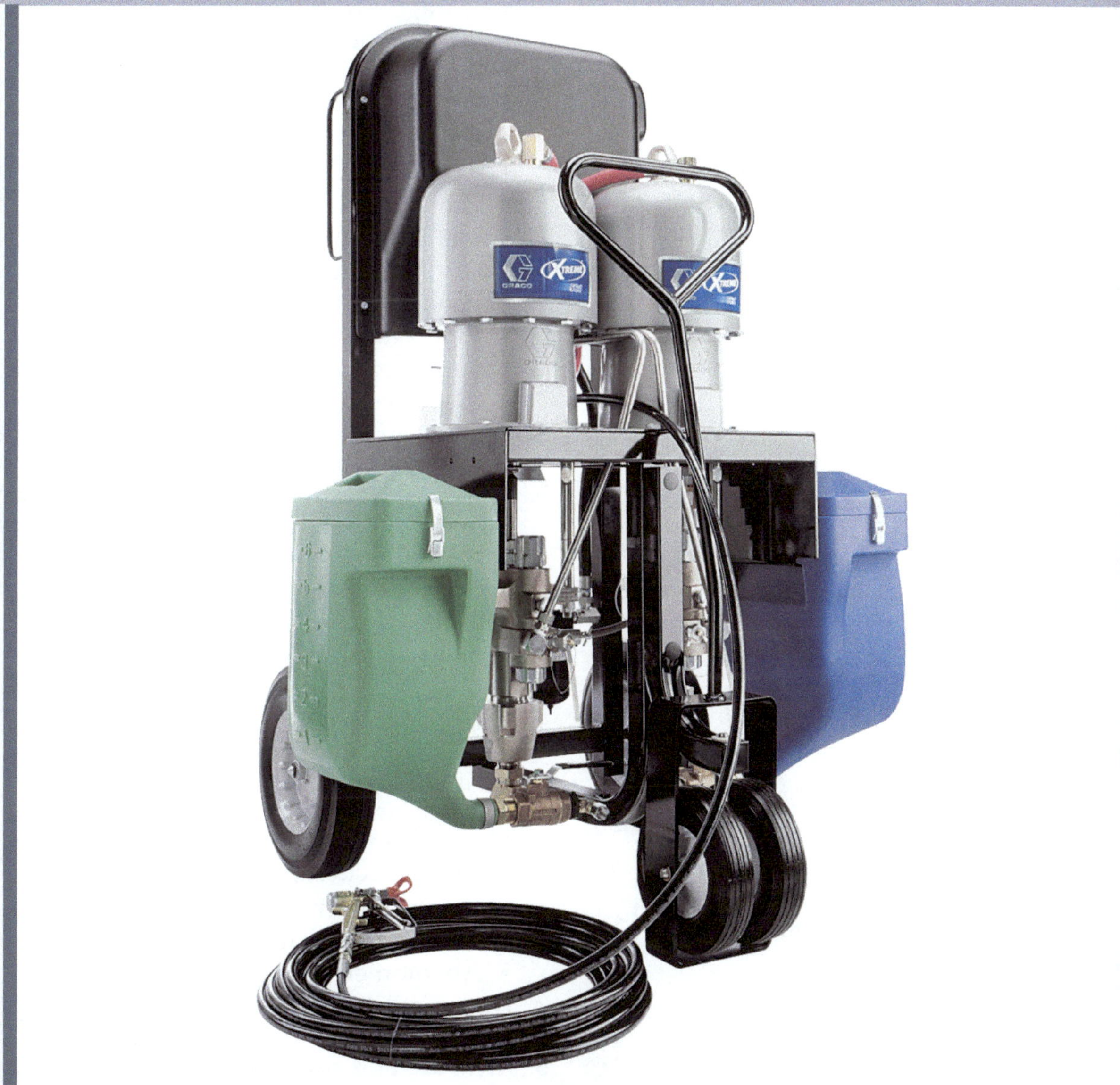

69207-10

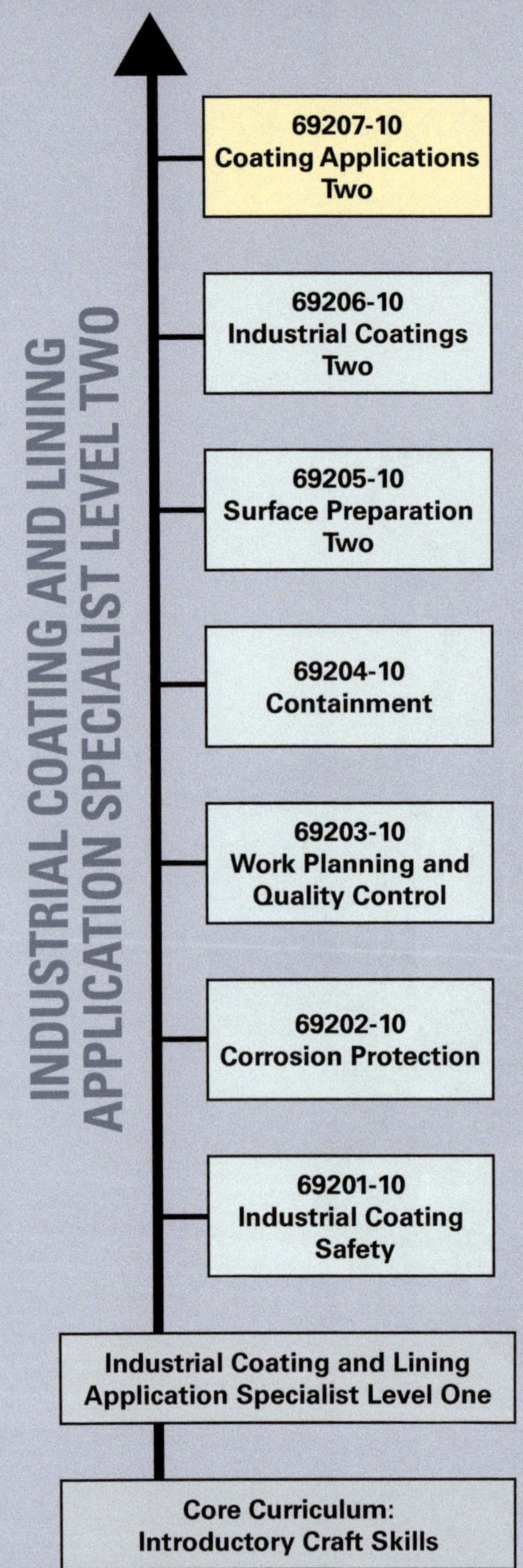

This course map shows all of the modules in *Industrial Coating and Lining Application Specialist Level Two.* The suggested training order begins at the bottom and proceeds up. Skill levels increase as you advance on the course map. The local Training Program Sponsor may adjust the training order.

Objectives

When you have completed this module, you will be able to do the following:

1. Demonstrate the use of a PDS to extract the following information:
   - Storage requirements
   - Mixing and thinning procedures
   - Induction time requirements
   - Pot life
   - Viscosity control requirements
   - Application equipment requirements
   - Recoat/curing times
   - Wet Film Thickness (WFT)
   - Dry Film Thickness (DFT)
   - Ambient condition requirements
   - VOC data
   - Repair procedures
   - Cleanup procedures
2. Employ procedures for visual verification of striping processes.
3. Select, demonstrate, and document setup, operation, adjustment, troubleshooting, maintenance, and cleanup of basic spray equipment on simple and complex structures, including:
   - Conventional air-spray
   - Airless spray
   - Air-assisted airless spray
   - HVLP
4. Analyze the need for auxiliary heating, humidification/dehumidification, and/or ventilation to aid coating drying/curing.
5. Select appropriate repair procedures for application defects.
6. Measure and document, prior to coating application, appropriate inspection procedures utilizing the following:
   - Psychrometer
   - Thermometers (air and coating temperatures)
   - Anemometer
7. Document batch/lot numbers and identify defective material.
8. Measure and document coating film quality, during and/or after coating application, using the following:
   - Wet film thickness gauge
   - Dry film thickness gauge
   - Low-voltage and high-voltage holiday detectors
   - Hardness test instrument
   - Solvent sensitivity test
9. Demonstrate the appropriate procedures for overcoating existing coating systems.
10. Identify common field and laboratory methods used for evaluating coatings and conducting coating failure investigations.
11. Demonstrate proper mixing and thinning methods.
12. Demonstrate coating coverage calculations using a calculator.

Trade Terms

Atomizing
Blast and coat cycle
Horn
Plasma
Short-term exposure limit (STEL)
Threshold limit value (TLV)
Time-weighted average (TWA)
Upper explosive limit (UEL)

Prerequisites

Before you begin this module, it is recommended that you successfully complete *Industrial Coating and Lining Application Specialist Level One* and *Industrial Coating and Lining Application Specialist Level Two,* Modules 69201-10 through 69206-10.

Contents

Topics to be presented in this module include:

Contents *(continued)*

Figures and Tables

Figures and Tables *(continued)*

1.0.0 Introduction

Coating application depends on the knowledge and skill of the coating and lining application specialist. The quality of the coating depends on the selection of equipment, the skill used in the application itself, the quality of the coating materials, surface preparation, and conditions under which the coatings are applied. Problems that arise during the process must be resolved correctly and efficiently in order to achieve a dependable coating.

2.0.0 Coating Safety

Working with coatings can expose technicians to a number of safety hazards. Information about specific hazards for a coating can be found on the product Material Safety Data Sheet (MSDS). A sample MSDS for a plural-component coating is provided in *Appendix A*. Safety issues related to coatings generally fall under two categories:

- *Fire hazards* – The flash point of a flammable liquid is the lowest temperature at which it can ignite when exposed to an ignition source. Sources of ignition include open flames, static electricity, heat from welding or grinding operations, lit cigarettes, sparks from tools, and chemical ignition sources. For example, the use of MEK (methyl ethyl ketone) and its vapors in the presence of activated charcoal filters, such as those commonly used in HVAC systems, creates a fire and explosion hazard.
- *Health hazards* – All coatings and solvents should be handled with care. *Section 8, Exposure Controls/Personal Protection* of any MSDS provides critical safety information for ventilation and the PPE recommended by the manufacturer for handling or applying a particular coating material. The National Institute of Occupational Safety and Health (NIOSH) approves PPE for use when working with industrial coatings. Typical PPE is shown in *Figures 1* through *5*.

In certain cases, hard hats, high-top steel-toed shoes, hearing protection, body covering closed at the wrist and neck, or solvent-resistant skin cream may be required. In addition, radiation protection clothing and personal radiation exposure devices may be required in certain environments. In confined areas, OSHA requires monitoring of flammable and/or hazardous solvent levels to prevent explosion and fire hazards or excessive displacement of oxygen. Additionally, OSHA requires that all respirators be periodically cleaned, sanitized, inspected, and tested.

Specific terminology is used when describing the safety concerns of working with or around coatings and solvents. An understanding of the following terms and definitions will help maintain safe working conditions:

- The explosive limit of a gas or a vapor is the limiting concentration that is needed for the gas to ignite and explode. There are two explosive limits for any gas or vapor, the lower explosive limit (LEL) and the **upper explosive limit (UEL)**. At concentrations in air below the LEL there is not enough fuel to burn; at concentrations above the UEL the vapor has displaced so much air that there is not enough oxygen to permit burning. Between the LEL and UEL there is danger of an explosion.
- **Threshold limit value (TLV)** is the amount and concentration of an airborne vapor that can be safely inhaled by workers during a specified period.
- **Short-term exposure limit (STEL)** is the highest concentration to which workers can be exposed over a stated period of time.
- **Time-weighted average (TWA)** is the threshold limit to which workers can be exposed for a specific work period, usually given in hours allowed to work per day or week.
- Volatile organic compound (VOC) is an organic chemical compound that vaporizes under normal conditions and enters the atmosphere as pollution. The term is often used in a regulatory context, such as those found in the Environmental Protection Agency (EPA) guidelines.

3.0.0 Mixing and Thinning

Before mixing any coatings, inspect the coating product to see if it appears to have been damaged by age or improper storage and handling.

207F01.EPS

Figure 1 Chemically safe goggles.

Products in cans that appear to have been opened should be examined carefully. The pigment in coatings can settle over time, so coatings that are prone to settling may require repeated mixing. Coatings that have separated or gelled should also be carefully examined. If in doubt, the coating should not be used. Set it aside and call the manufacturer for information. Mix the coatings in accordance with manufacturer's specifications. The specific results of failing to mix thoroughly vary between different types of coatings, but generally a poorly mixed coating will not provide the desired protection. Failure to mix components in the proper proportions can cause inadequate curing or the separation of components after application.

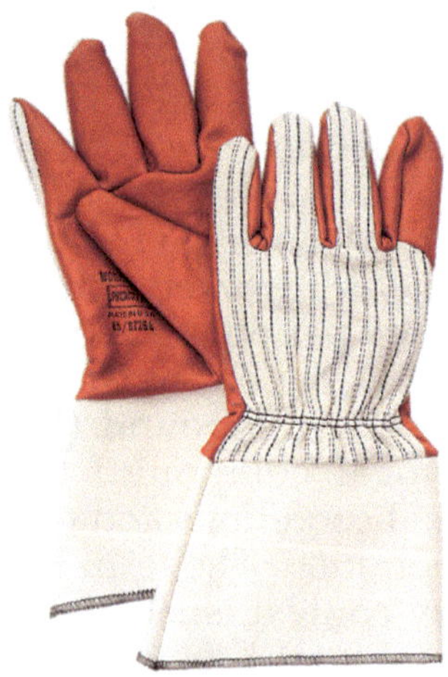

HEAT-RESISTANT GLOVES

CHEMICAL-RESISTANT GLOVES

207F02.EPS

Figure 2 Heat- and chemical-resistant gloves.

REUSABLE HALF-MASK RESPIRATOR

REUSABLE FULL FACEPIECE RESPIRATOR

207F03.EPS

Figure 3 Reusable half-mask and full-face air-purifying respirator.

207F04.EPS

Figure 4 Powered air-purifying respirator.

SELF-CONTAINED BREATHING APPARATUS

SUPPLIED-AIR RESPIRATOR

207F05.EPS

Figure 5 Self-contained breathing apparatus with full chemical-/fire-resistant body suit and supplied-air respirator.

WARNING

Two additives used with esters and polyesters are methyl ethyl ketone peroxide (MEKP) and cumene hydroperoxide, which is a catalyst to start the polymerization process. MEKP and cumene hydroperoxide must be stored and handled carefully because they are extremely reactive and irritating to the skin. Always refer to the MSDS for the use and PPE requirements. When mixing MEKP or cumene hydroperoxide into resin, start mixing the resin and then carefully pour the MEKP or cumene hydroperoxide into the vortex. Continue mixing, being certain to scrape material from the sides of the container, until the components are thoroughly mixed.

3.1.0 Mixing Equipment

A variety of tools can be used to properly mix coating materials, depending on the coating type and the amount of material. For larger projects or other applications involving large quantities of a coating, power mixing is commonly used. *Figure 6* shows examples of mixing tools that can be used with portable air motors or explosion-proof electric motors for hand mixing of 1- to 5-gallon (4 to 19 liter) containers. In other cases, portable or semi-portable mechanical shakers for use with metal or plastic containers are available, as shown in *Figure 7*. These allow mixing without opening the containers. The electrically powered one-gallon unit is useful for maintenance departments and smaller projects. Similarly sized units that are powered by an air motor are also available. The 1- to 5-gallon (4 to 19 liter) unit is electrically powered and can be floor-mounted or mounted on a trailer, heavy cart, or van. Both units are equipped with an adjustable timer switch to set the mixing duration.

207F06.EPS

Figure 6 Mixing tools.

FIVE-GALLON UNIT

ONE-GALLON UNIT

207F07.EPS

Figure 7 Mechanical shakers.

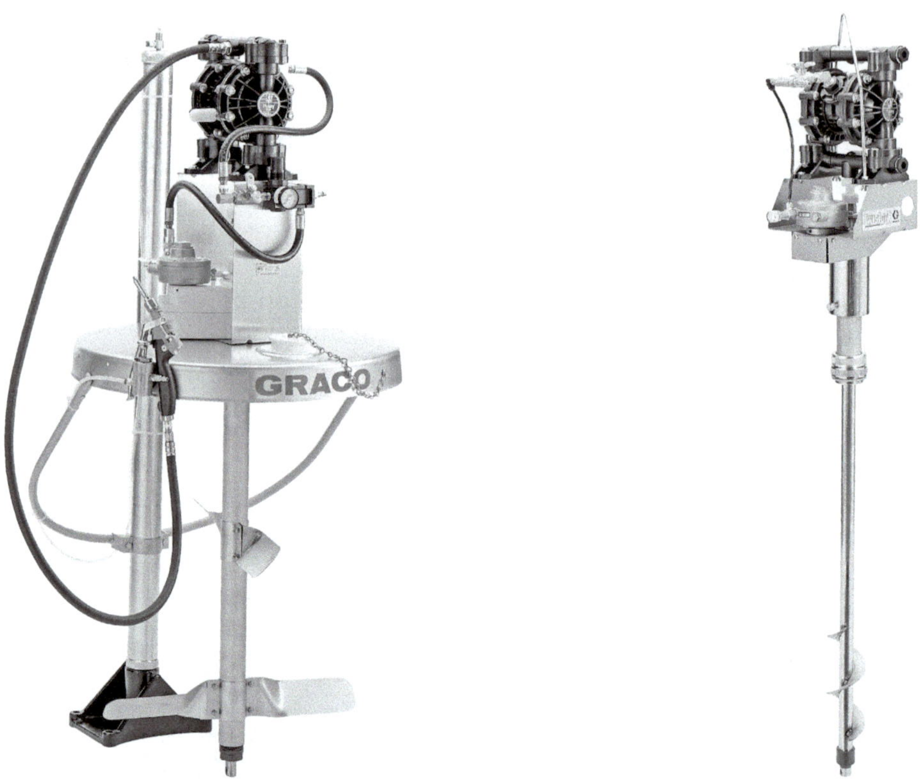

STAND-MOUNTED OPEN DRUM
PADDLE MIXER WITH AIR-POWERED
DRUM COVER LIFT AND PUMP

DRUM-MOUNTED SPIRAL MIXER
FOR CLOSED DRUMS WITH AIR-POWERED
MOTOR FOR SPIRAL AND PUMP

207F08.EPS

Figure 8 Air-motor powered drum mixers with pumps.

WARNING

Always use non-sparking mixer motors to prevent ignition of any flammable or explosive solvent vapors.

Figure 8 shows examples of drum mixers with stirring paddles or spirals powered by variable-speed air motors, along with air-powered coating material pumps. Drum mixers are not used for initial mixing of coatings, which must be accomplished by other means. They are used for 55-gallon (208-liter) drums of coating material. To prevent separation, they continuously mix coating material that is pumped from the drums and applied during coating operations.

NOTE

Always use the manufacturer's recommended mixing blade style, size, and speed.

All mixing tools and equipment must be promptly cleaned after usage with the proper solvents specified in the coating manufacturer's product data sheet (PDS) or on the container label. An example of a PDS is provided in *Appendix B*.

3.2.0 Mixing Guidelines

This section covers guidelines for mixing and thinning various types of coating materials, including single-component coatings, multi-component liquid coatings, and liquid coatings with a powder component.

3.2.1 Single-Component Liquid Coatings

For single-component liquid coatings, any one or more of the tools or equipment previously discussed can be used to mix the coating. Follow the manufacturer's instructions for mixing. Any air bubbles should be allowed to disappear before using the coating. For pigmented single-component coatings, the amount of mixing required depends on the amount of pigment settling (separation from the liquid resin and solvent) that has occurred. The degree of pigment settling depends on the coating type, formula, storage time, and storage temperature. Always use freshly stocked coatings or coating components that have been properly stored. Specifications may prohibit the use of partially used containers if skin has formed on top of the liquid.

When using a hand power mixer, mix at a slow rate to help prevent air entrainment and foaming of the coating. Some coatings may foam after being shaken, becoming unusable. The proper mixing rate is when a small whirlpool forms around the shaft of the mixer. After mixing, the coating should appear to be smooth with a uniform color. Make sure that there is no discernable settled pigment remaining at the bottom of the container. Allow the coating to sit for a short time so that any air bubbles can rise to the surface.

NOTE

When hand power mixing any coating, always insert the mixing tool into the coating container before starting the mixing motor. When mixing is completed, stop the mixing motor before withdrawing the mixing tool. The mixing tool can be cleaned by operating it in a can of the coating solvent for a short time.

The color of a pigmented coating can vary slightly from one container to the next, especially if the containers are from different production lots. It is normally a good practice to intermix the coating material if more than one container of a coating will be used. This is called boxing, and is accomplished after initially mixing two or more containers of a coating. To box a coating after initially mixing all containers, pour about half of the first container into a clean container. Then, pour about half of the second container into the first container and remix the first container. This process is repeated as necessary for any subsequent containers. Finally, fill the last container with the coating that was poured off the first container and remix the last container.

The coating should be strained after mixing to remove any lumps or other contaminants. If the coating will be sprayed, it should be thinned, if required. Thinning and straining are covered in a later section.

Single-component coatings that are not in their original container after they are mixed need to be labeled. An employer will most likely have a policy for this, so be sure to follow it. When in doubt, label the coating with all of the relevant information so that anyone looking at the label knows exactly what the coating is and how it was mixed. Include whatever information is required by federal, state, and local requirements, such as the following:

- Name of coating
- Manufacturer's name
- Batch number
- Date mixed
- Proportion and names of any thinners or additives
- Name of person doing the labeling
- Identity code for tracking sample, if required

3.2.2 Multi-Component Liquid Coatings

Multi-component liquid coatings have two or more components that must be mixed together before application. The coatings begin to cure over a certain amount of time when one of the components, called a curing agent, converter, hardener, or activator, is mixed into the resin. These agents cause a chemical reaction that starts an irreversible curing process, and the time available to thin (if required) and apply the coating is limited. The limited amount of time available for application of the coating is called pot life. Mix only the amount of coating that can be thinned and/or applied within the coating's pot life. When the end of the pot life is reached, discard the mixed coating and clean all the tools and equipment.

After mixing, some multi-component coatings also require an induction time, often called sweat-in time, to allow the curing agent, converter, hardener, or activator in the resin to undergo a chemical reaction and reach a turning point before thinning and/or application. Induction time is part of the pot life for the coating. Induction time, as well as pot life, can be affected by the ambient temperature at the time of mixing, and to a lesser extent, by humidity. High temperatures shorten induction time and cold temperatures extend it. The manufacturer's PDS will specify induction time (if required) and pot life at various temperatures. Failure to observe the pot life can result in defective coating application, wasted material, and damaged application equipment.

For large applications in the field or in manufacturing facilities, many of the multi-component coatings, especially those with a very short pot life, are kept in separate containers and mixed in a manifold just before spraying, or they are mixed in the plural-component gun itself. This eliminates wasted material and allows rapid flushing of the mixing manifold, hoses, and/or spray gun. If the components are required to be hot-sprayed, the component containers and all supply lines must be heated. *Figure 9* shows a field application with two 55 gallon (208 liter) barrels with barrel heaters containing the separate components of a plural-component coating. Each barrel is equipped with a barrel pump/agitator. The twin high-pressure pumps and a manifold mixer are shown on the left. Note the heated hoses for the spray guns.

Smaller quantities of multi-component coatings can be hand-mixed provided that the total quantity of each component container is mixed. To prevent waste, one or more sets of the smaller-sized containers of the coating components should be procured for the desired coating coverage. Depending on the coating pot life, only one set, or possibly two sets of containers should be mixed at the same time.

Always follow the manufacturer's label instructions or PDS for the proper mixing time, application data, pot life, and, if applicable, thinning and induction time. Pay careful attention to the component mixing sequence because different manufacturers specify and label the components differently. Always use a separate clean metal container for mixing, if not the original resin container. Do not use plastic, and do not use the curing agent, converter, hardener, or activator containers as mixing containers. Individual components are usually mixed first by shaking or power-mixing them separately. Then, the components are power-mixed together to start a chemical curing reaction. They are usually mixed only long enough to provide uniform color and consistency. This is because any induction time, as well as the overall pot life, has already started diminishing. It is good practice not to mix large quantities of a multi-component coating in a big container. Chemical reactions between the components may generate large amounts of heat, and if more of the coating is mixed, more heat is generated. This can result in temperatures that may constitute a safety hazard.

3.2.3 Liquid Coatings with a Powder Component

Liquid coatings with a powder component are multi-component coatings where one component is a metallic dust (powder), such as zinc. Normally, the dust is slowly added through a sifter or sieve to the liquid component and slowly power-mixed into the liquid. When all of the dust has been added and mixed for a short time, the coating should be strained to remove any lumps. When handling or mixing the dust component into the liquid, appropriate respirators must be worn to avoid inhaling the dust, which is regarded as hazardous. The manufacturer's PDS (or label instructions) and the MSDS should be carefully followed for the mixing of the components and any thinning that is required for spraying. Usually, zinc-rich coatings must be constantly and slowly mixed during the application process to prevent settlement.

WARNING

Never mechanically shake a metallic powder-filled coating. Shaking can cause the container to explode.

207F09.EPS

Figure 9 Field application using plural-component spray equipment.

3.3.0 Thinning Coatings

Coatings are normally supplied with the proper consistency for the most common methods of application. However, thinner or reducing solvents may be required to reduce the viscosity of the coating for spray application. Viscosity is a measure of the resistance of a material to flow. The viscosity of coatings can be manually measured by the use of various viscometers, such as the Shell cup viscometer (*Figure 10*). The cup of the viscometer is dipped into the coating and held for a short period until the cup reaches the temperature of the coating. It is then completely withdrawn and the time it takes for the coating to drain out of a calibrated hole at the bottom of the cup is clocked in seconds from the time the cup breaks the surface until the stream first stops. The elapsed time along with temperature graphs determine the viscosity of the coating in Shell cup seconds. Conversion charts can be used to convert the Shell cup viscosity to Zahn cup seconds, centipoises, and other viscosity units as desired. Shell cup viscometers are similar to Zahn cup viscometers, but are more accurate because they don't dribble when the last of the coating drains from the cup. This makes measuring the time duration of the flow more precise. Cups with different-sized holes are available to cover a wide range of viscosities.

Because spray equipment used for coating application may require specific coating viscosities to achieve proper spraying characteristics, the coating viscosity may have to be checked. If the coating is not thin enough to be sprayed, the viscosity must be reduced by heating or thinning. Some multi-component coatings must be heated to reduce viscosity. Other multi-component coatings and single-component coatings may be thinned, if necessary. A viscometer is used to check the viscosity in either case. The manufacturer's PDS must be consulted to determine the methods recommended for controlling viscosity. If thinning is permitted, the PDS will specify the maximum amount of thinner or reducer that can be added as a percentage by volume. The maximum amount should never be exceeded. When thinning a coating, it must be remixed and retested each time until the proper viscosity is obtained. For coatings with a pot life, this must be done quickly.

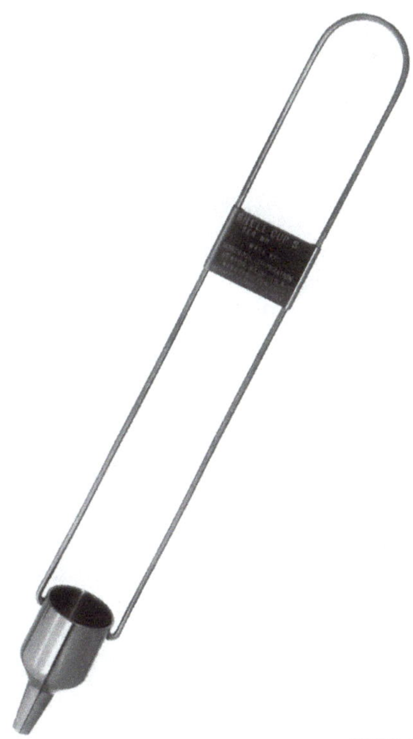

207F10.EPS

Figure 10 Shell cup viscometer.

Adding thinner or reducer for spraying will increase the wet film thickness (WFT) required for a specific dry film thickness (DFT). For instance, if the unreduced solids volume of a coating is a certain percentage, the DFT resulting from thinning/reducing can be calculated as shown in Equation 1. If the specific DFT desired is known, then the WFT needed for a reduced coating can be calculated as shown in Equation 2. In any case, minimal thinning should be employed because thinning can affect the quality of the coating. In addition, thinning with solvents must be done in compliance with VOC regulations.

Equation 1

WFT to DFT thickness:

(No thinner added)

$$\text{WFT} \times (1 + \%\ \text{thinner by volume})$$

(Thinner added)

$$\frac{\text{WFT 3 (\% solids by volume)}}{1 + \%\ \text{thinner by volume}} = \text{DFT}$$

Equation 2

DFT to WFT thickness:

(No thinner added)

$$\frac{\text{DFT}}{\%\ \text{solids by volume}} = \text{WFT}$$

(Thinner added)

$$\frac{\text{DFT} \times (1 + \%\ \text{thinner by volume})}{\%\ \text{solids by volume}} = \text{WFT}$$

4.0.0 Calculating Coating Coverage

To calculate the amount of coating needed for a job, the relationship between WFT and DFT must be understood. DFT is the usual specification criterion. WFT measurements help to determine how much of a coating must be applied to reach the specified DFT. WFT measurements are made at the time each coat is applied so that errors in thickness can be corrected early. However, WFT is only a guide. DFT is the defining measurement.

The dry-film to wet-film ratio is based on the percentage of solids by volume in the coating being used. Manufacturers' data sheets sometimes list solids by weight as well as solids by volume, but volume measurement must be used to calculate WFT. A handheld calculator, such as the one shown in *Figure 11*, with at least the functions for addition, subtraction, multiplication, division, percentage, and memory storage, can be used in the examples that follow. Some advanced-function calculators can also be programmed by the user for any desired formulas or equations that are often used. The WFT formula is:

$$\frac{\text{DFT} = \text{WFT}}{\%\ \text{solids by volume}}$$

NOTE

In the following calculations, the answers are rounded to the nearest whole number. This is because fractional results are insignificant in actual practice.

For example, the coating specification calls for each application to have a DFT of 4 to 6 mils (102 to 152 microns). The coating has 30-percent solids volume. First, convert 30 percent to a decimal as follows:

$$30\ \text{percent} = \frac{30}{100} = 0.30$$

Next, put the information into the formula:

$$\text{WFT} = \frac{4\ \text{mil}\ (102\ \mu\text{m})}{0.30\ \text{solids by vol.}} = 13\ \text{mils}\ (330\ \mu\text{m})$$

$$\text{WFT} = \frac{6\ \text{mil}\ (152\ \mu\text{m})}{0.30\ \text{solids by vol.}} = 20\ \text{mils}\ (508\ \mu\text{m})$$

207F11.EPS

Figure 11 Typical handheld advanced-function calculator.

One gallon of the example coating will cover 1,604 square feet (149 square meters) to 1 mil (25.4 microns) thickness when it is wet. To determine how much coating is needed, divide 1,604 by the desired WFT. For example, using the above WFT of 20 mils, calculate how much area one gallon of the example coating covers, as follows:

$$\frac{1{,}604}{20} = 80 \text{ ft}^2 \text{ per gallon } (7.45 \text{ m}^2)$$

or

$$\frac{149 \text{ m}^2}{20} = 7 \text{ m}^2 \text{ per gallon}$$

To determine how many gallons of the coating are needed to coat a 1,200 square foot (111 m^2) surface, use the following formula:

$$\frac{1{,}200 \text{ ft}^2}{80 \text{ ft}^2 \text{ per gallon}} = 15 \text{ gallons (57 liters)}$$

A simpler, faster method of calculating the coverage requirement is given by the following equations:

Equation 1

Number of U.S. gallons =

$$\frac{\text{Area (ft}^2) \times \text{DFT (mils)}}{1{,}604 \times \text{\% solids by volume (in decimal form)}}$$

Equation 2

Number of liters =

$$\frac{\text{Area (m}^2) \times \text{DFT (microns)}}{1{,}000 \times \text{\% solids by volume (in decimal form)}}$$

Substituting the previous example of 1,200 ft^2 (111 m^2) with a DFT of 6 mils (152 microns) and a coating with 30 percent solids by volume in the equations yields a total of about 15 gallons or 56 liters of coating necessary for the coverage. The slight difference in the number of liters required is due to accumulated rounding differences in the previous example.

No matter how carefully a calculation is made, there is always a small amount of waste. The amount of waste is determined by the application method and configuration of the structure.

If the coating is thinned, this must also be taken into account. Thinner increases the total volume of the coating without increasing the amount of solids. This introduces a new variable called percentage of thinner by volume (percent TBV). The formula used when thinner has been added is:

$$\frac{\text{DFT (1 + percent TBV)}}{\text{\% solids by volume}} = \text{WFT}$$

If a coating must be thinned by a percentage of volume, put the percentage into the formula. For example, using the WFT of 4 mils with a solid of 30 percent thinned 25 percent (0.25), the WFT is calculated as follows:

$$\text{WFT} = \frac{4 \text{ mil } (1 + 0.25)}{0.30 \text{ solids by vol.}} = 17 \text{ mils } (432\ \mu\text{m})$$

If the coating must be thinned by adding a specific volume of thinner to each gallon of coating, calculate the percentage of thinner before calculating the WFT. For example, if 1 pint (0.5 liter) of thinner must be added to each gallon of coating, calculate the percentage as follows:

$$1 \text{ gallon} = 8 \text{ pints, so } \frac{1 \text{ pint}}{8 \text{ pints}} = \frac{1}{8} = 0.125\%$$

5.0.0 Selection of Application Equipment

Coating application is one of the factors that determines the performance of any coating system. The quality and physical properties of a coating material are determined by the manufacturer, but optimum performance can only be acheived if the coating is properly applied. The coating film must be dense, resistant to the passage of moisture and other potentially damaging or corrosive materials, and must dry or cure to its solid state. There are many methods of applying protective coatings, including brush, mitt, roller, and spray. Of these, spray application is probably the most widely used for industrial protective coatings. *Figure 12* shows types of cart-mounted airless spray systems and equipment used for spray applications.

One or more of the following factors may influence the choice of the application method used:

- *Size and type of job* – To be competitive, equipment must be adjusted to the scale of the work, as well as to the type of application. Bigger jobs are more likely to use more equipment, and that equipment is likely to be more sophisticated. The type of job will determine what application method is required or most suitable.
- *Access to the surfaces to be protected* – Many times the project location limits the type of equipment that can be used for practical reasons. For example, an elevated water tank may be difficult to paint with spray equipment due to the height of the job.
- *Configuration of the surface* – More complex surfaces may be difficult to cover with spray

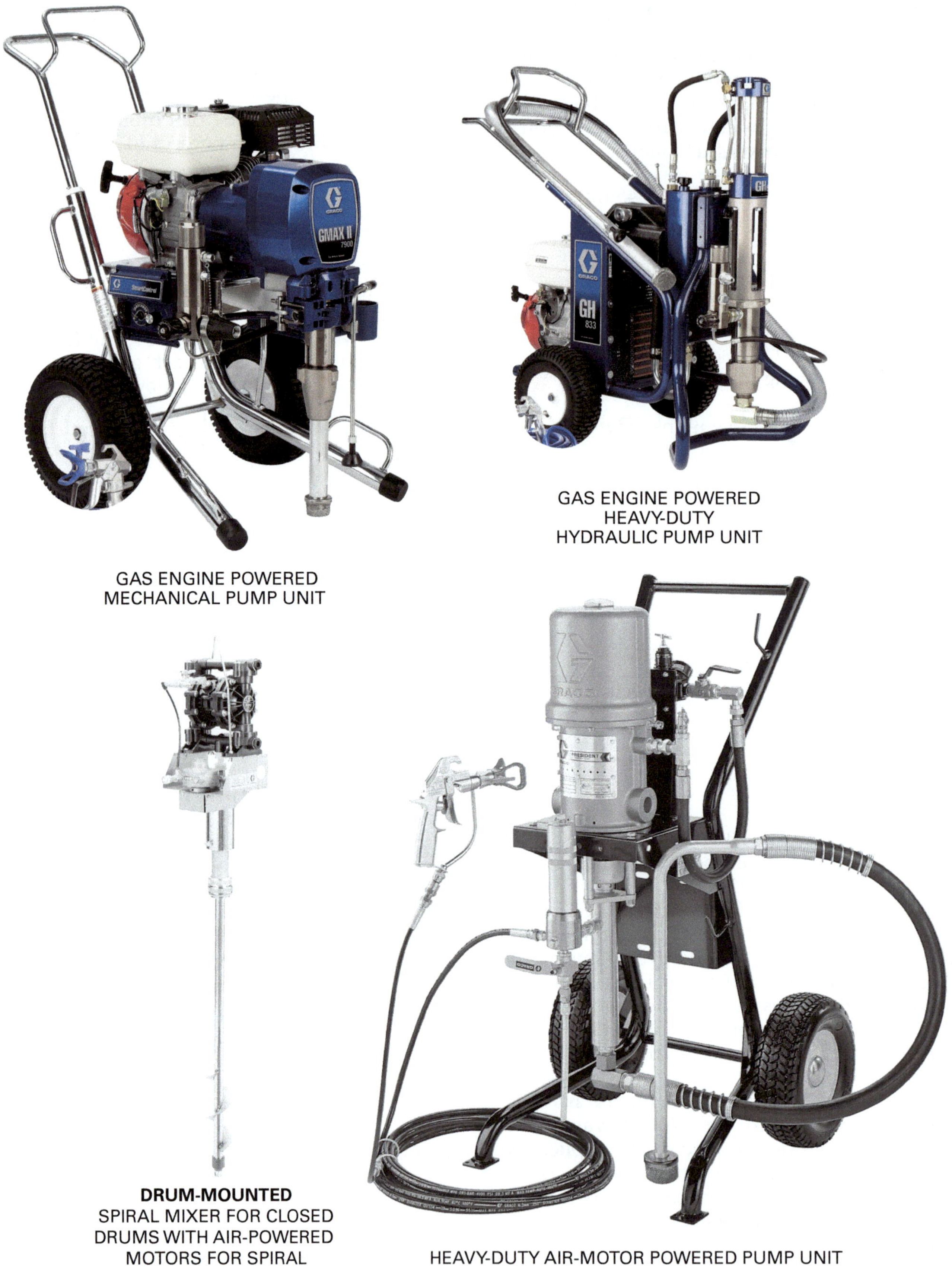

Figure 12 Cart-mounted airless spray systems and equipment used for spray applications.

equipment. Spray applications are better suited for large, flat areas.

- *Presence of critical areas or surrounding environment that can be damaged by overspray from spray equipment* – If the job area cannot be fully contained to prevent overspray from drifting off the job site, spray equipment may not be the best option. Many times the use of brushes and rollers provides a more economical solution than going to the expense of fully containing the project.
- *Type of coating* – Some coatings are designed for application only by spray techniques. Brush or roller application may not be recommended and should only be used when the PDS specifies that these methods are acceptable.
- *Availability of skilled laborers* – More sophisticated coatings and equipment require significant expertise by the individual applicators. In many places, the necessary number of skilled workers may not be available.
- *Budget constraints* – If money is not available to pay for the more expensive coatings, application equipment, or necessary skilled workers, then simpler, less expensive application methods must be used.

The following sections describe four major factors in determining the type of coating application equipment to use, including:

- Manufacturer's recommendations and PDS specifications
- Environmental/facility restraints
- Size and configuration of surfaces
- Production considerations

5.1.0 Product Data Sheets

Coating manufacturers supply a PDS for each of their products that must be read by all applicators. This sheet contains specific information, including a product description, product characteristics (spreading rates, drying time, WFT, DFT, and so on), technical data, and recommended uses. The PDS also normally lists the application methods and equipment recommended for a coating. Some may also provide application tips and guidelines. The sheets also contain safety references or precautions, surface preparation and coloring information, along with application criteria, such as temperature and humidity requirements. The layout of a PDS varies from manufacturer to manufacturer, but the information provided is generally similar. *Appendix B* is a typical PDS from one manufacturer. This PDS can be used as a reference when a PDS is mentioned in other sections of this module. Applicators should always cross-check the latest PDS data from the coating and solvent manufacturer against the requirements in the project coating specification. If any conflicts are noted, they should be resolved before the project starts.

> **NOTE**
> Some coating manufacturers may also supply a separate PDS for each component and primer used for a coating.

Figure 13 is a section from the PDS application bulletin in *Appendix B* that pertains to the application equipment. Note that recommendations are given for conventional air spray and airless spray.

When the technician is using conventional air spray, this PDS lists the type of spray gun that should be used, along with the type of fluid tip, the air cap, the recommended sizes for the air hose and the material hose, the atomizing pressure, and the pot pressure. Similar information is given for airless spray applications along with notes pertaining to specific conditions such as low-temperature applications, applications requiring longer hoses, and varying weather conditions.

APPLICATION EQUIPMENT

The following is a guide. Changes in pressures and tip sizes may be needed for proper spray characteristics. Always purge spray equipment before use with listed reducer. Any reduction must be compatible with the existing environmental and application conditions.

Reducer/Clean Up .... Reducer #132, R7K132

Airless Spray
Pressure .... 2400 - 3000 psi
Hose .... 3/8" ID
Tip .... .013" - .017"
Filter .... 60 mesh
Reduction .... As needed up to 10% by volume

Conventional Spray
Gun .... Binks 95
Cap 63P
Tip 66
Atomization Pressure ... 50 - 60 psi
Fluid Pressure .... 20 - 30 psi
Reduction .... As needed up to 10% by volume

Brush
Brush .... Natural Bristle
Reduction .... Not recommended

Roller
Cover .... 1/4" woven with phenolic core
Reduction .... Not recommended

If specific application equipment is not listed above, equivalent equipment may be substituted.

207F13.EPS

Figure 13 Sample PDS application recommendations.

The PDS provides information regarding specific applications along with recommended primers. Note that this particular product can also be applied using a brush or roller, and the PDS gives specific instructions and specific types of devices that can be used.

The following list describes the location of information that pertains to applicators from the PDS in *Appendix B*:

- *Product description* – Provides an overview of the product characteristics.
- *Recommended uses* – Covers the types of surfaces and applications for which the product was designed.
- *Product characteristics* – Covers the important points needed for product application including volume of solids, VOC emissions, mix ratio, recommended spreading rates, drying schedule, pot life, sweat-in time, storage information, fire flash point, and reducer/clean up recommendation.
- *Performance characteristics* – Covers the standard tests that have been conducted on the product and their results.
- *Recommended systems* – Covers the primer and product combinations for various substrates.
- *Surface preparation* – Briefly covers surface preparation specifications for various substrates. In this example, detailed preparation is referenced to the application bulletin.
- *Tinting* – For this product, provides mixing instructions for coloring the Part A base before adding the hardener.
- *Application conditions* – Covers the minimum and/or maximum temperatures and humidity for application of the product. For this product, detailed application information is referenced to the application bulletin.
- *Ordering information* – Provides packaging sizes available for this product.
- *Safety information* – For this product, reference the MSDS shown in *Appendix A*.

The application bulletin sheets (*Appendix B*) provide the following information:

- *Surface preparation* – Provides detailed surface preparation guidelines for various substrates including references to applicable standards.
- *Application conditions* – For this product, a repeat of the information in the product information sheets.
- *Application equipment* – Lists various types of recommended application equipment and their required characteristics.
- *Application procedures* – Provides detailed mixing procedures and, for this product, repeats recommended spreading rates, drying schedule, pot life, and sweat-in time from the product information sheets.
- *Performance tips* – For this product, covers important application information when coating surfaces along with additive information.
- *Cleanup instructions* – Provides applicable cleanup instructions with recommended solvents.
- *Safety precautions* – For this product, a reference to the MSDS(s).

5.2.0 Environmental/Facility Restraints and Monitoring

Before applying an industrial coating to any surface, the surface must be clean, dry, and otherwise properly prepared. For guidelines on preparing different kinds of surfaces, refer to the coating manufacturer's recommendations for the type of coating being used and the type of surface to which it will be applied. If a long time has elapsed between the time of surface preparation and coating application, the surface should be checked again for signs of dirt, contamination, flash rusting (on steel), or ultraviolet damage. For container-mixed coatings with short pot lives, only the surface area that can be coated within the pot life is normally prepared just prior to the coating application.

Before applying the coating, it is important to check the current environmental conditions, including the air and substrate surface temperature, humidity, dew point (the temperature at which the water vapor in the air condenses), and wind speed. Sun, rain, fog, and wind can result in a variety of problems for outdoor coating application. Wind can blow dust and dirt onto the wet surface; even a light wind of 3 to 5 mph (5 to 8 kmph) as measured by an anemometer (*Figure 14C*) can blow the overspray around and cause damage. Sunlight can raise substrate temperatures too high unless shaded. Rain and fog can wet the substrate or raise the dew point too high for coating operations.

For example, the PDS for a particular coating may state that the surface temperature should be 65°F to 85°F (18°C to 29.5°C), the ambient temperature 60°F to 85°F (15.5°C to 29.5°C), and the relative humidity (RH) 35 to 85 percent. Applying coatings outside the recommended ranges may result in coating system failure. Coating steel normally requires a surface temperature at least 5°F (15°C) above the dew point. Condensation or surface moisture can interfere with adhesion or proper curing of the coating.

Surface temperature of a substrate can be measured with a thermometer (*Figure 14A*) in direct contact with the substrate. Air conditions can be checked using a sling psychrometer (*Figure 14B*) which contains two thermometers, one to measure the dry air temperature (dry bulb) and the other to measure the wet air temperature (wet bulb). The measured wet-bulb and dry-bulb temperatures can be used to find the percent of RH and dew point of the measured air. This is done by using a built-in chart on the psychrometer or a separate psychrometric chart or table. A digital psychrometer is shown in *Figure 14D*. It provides the temperature, dew point, and humidity readings directly. An electric hygrometer (*Figure 14E*) can also be used to measure the air temperatures and give a direct reading of the RH. Some are also capable of giving a direct reading of the dew point.

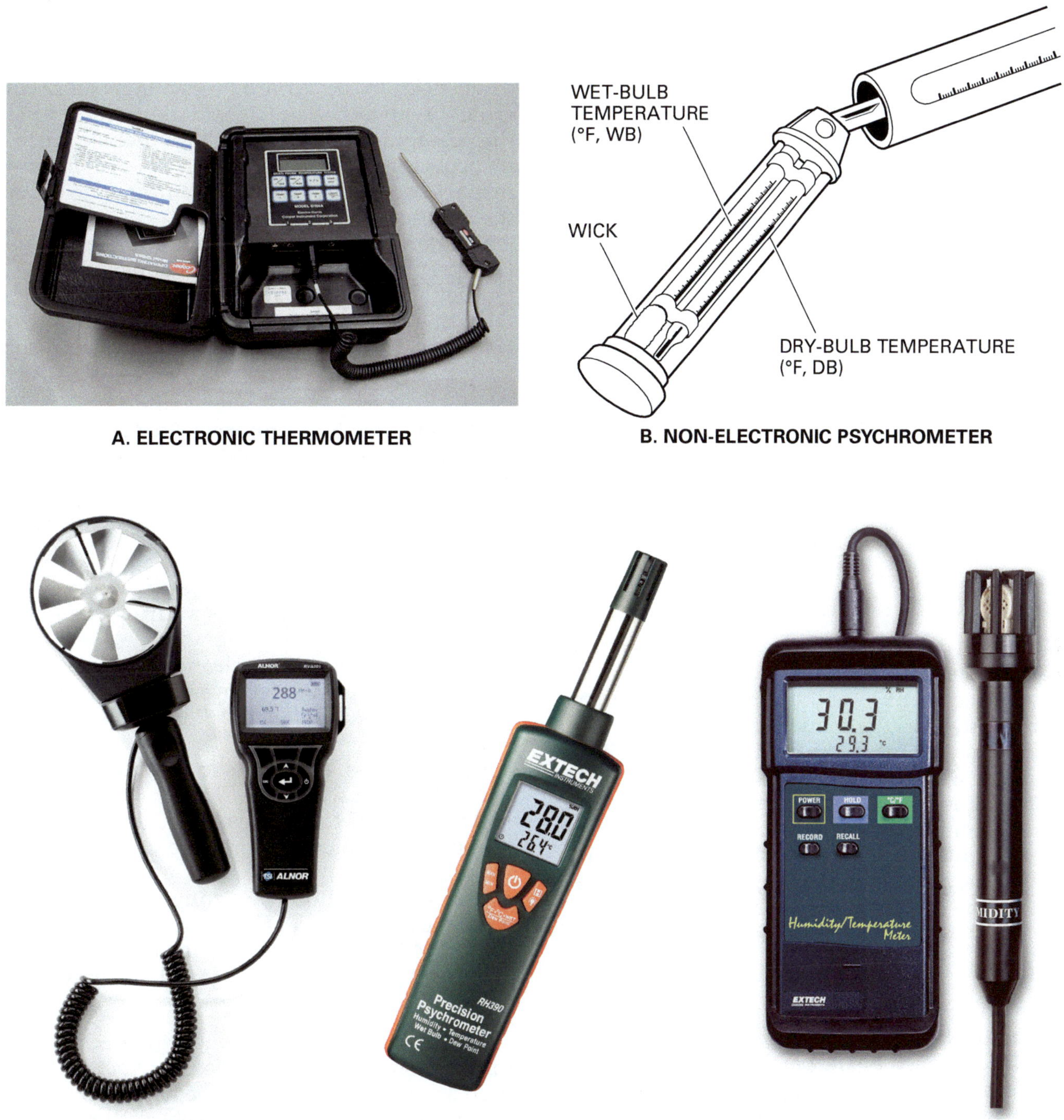

A. ELECTRONIC THERMOMETER

B. NON-ELECTRONIC PSYCHROMETER

C. ANEMOMETER

D. DIGITAL PSYCHROMETER

E. ELECTRONIC HYGROMETER

207F14.EPS

Figure 14 Temperature, humidity, and wind measurement devices.

5.3.0 Size and Configuration of Surfaces

The size and configuration of the surface to which the coating will be applied is a major factor in determining the type of application equipment to use. Brushes can be used for stripe coating in difficult areas including edges, around bolts, corners, and in crevices. They can also be used to repair holidays and to coat small, scattered areas. Use brushes in critical areas that cannot be effectively masked with sheeting or where spray might cause damage, such as in coating-sensitive machinery, motors, instruments, and gauges. Brush work is slow, however, and the surface quality may not be as aesthetically pleasing as rolling or spraying. Mitts can used for small projects, including hand rails, piping, and odd-shaped objects that are not practical to coat by roller or spray. Rollers are used to cover large flat, or nearly flat, areas including tank shells, roofs, floors, ceilings, and walls because they hold much more coating material than a brush, and application is two to three times faster. They are ideally suited for coating chain link fencing. Squeegees and trowels are used to spread coatings such as 100 percent solids on floors. Again, these are used where spraying is not available, allowed, or practical. Generally, spray application is the best method to use for rapid application of coatings to a large surface area and to complex structures, such as bridges. With appropriate equipment, a highly uniform application of coating is possible in nearly all situations.

5.4.0 Production Considerations

Production considerations must also be taken into account when determining the type of equipment to use. If the coating application is for a small job where waste must be minimized, a brush, roller, or mitts may be more practical than a spray system. If the application is in a large environment with multiple jobs, a faster, more efficient method should be employed. Always choose the application equipment to optimize transfer and build efficiencies, as long as the performance requirements of the coating system are maintained.

6.0.0 PRE-APPLICATION PREPARATIONS

The type of pre-application perparations will vary from job to job. The specifics of the preparations are governed by the particular requirements of the coating job. Some companies and clients require inspections performed by members of their own staff, while others require that outside firms specializing in inspections perform the examination. General items included in the pre-application examination include surface inspections, coating materials preparation, application equipment inspections, and documentation requirements.

6.1.0 Surface Preparation

The quality of the surface preparation has a direct bearing on how well the coating performs and the resilience of the coating. The majority of coating failures are due to either faulty surface preparation, or faulty coating application, or both. Therefore, it is imperative to properly examine the surface to determine how well it has been prepared to receive the coating, and to correct any deficiencies in the surface preparation before covering it with a coating. General items included in the pre-application examination include surface inspections for fabrication defects, pre-cleaning inspections, and surface preparations to accept the coating.

Standards and examination techniques for surface preparation have been established to help protect against premature coating failures. The following is a brief overview of some generalized standards for surface preparation of various substrates.

6.1.1 Masonry and Concrete Surfaces (SSPC-SP13/NACE 6)

Some generalized standards for surface preparation of masonry and concrete surfaces are as follows:

- Concrete must cure for a specified amount of time before coating unless a curing compound or accelerant is used.
- The surface should be cleaned with water, detergent, or steam to remove oil, grease, and dirt (*ASTM D4258, Standard Practice for Cleaning Concrete*).
- Cracks, holes, and voids in the surface should be filled and sealed.
- Any type of form-release agents and curing compounds must be completely removed from the surface.
- The surface should be roughened, if necessary, by mechanical abrading, abrasive blasting (*ASTM D4259*), or acid etching (*ASTM D4260*).
- The surface should be dry and either swept or vacuumed before coating.

Visual inspection is enough to check conformity to some of these standards. Other methods, such as magnification, must be used for others. A litmus test (*ASTM D4262*) will check the pH of the surface to determine whether it has been etched with acid. A moisture meter, calcium chloride test (*ASTM F1869*), or the plastic sheet method (*ASTM D4263*) can be used to check for moisture.

6.1.2 Steel Surfaces

The proper preparation of steel surfaces for grit blasting includes the following:

- Removal of slag, flux, weld splatter, and rough edges by specified methods
- Removal of grease, dirt, oil, debris, and other loose contaminants by specified methods

> **NOTE**
> To test for the presence of oil or grease, water can be applied to the surface. If it beads, it indicates the presence of oil or some other contaminant that prevents the water from wetting the surface.

The two primary areas of concern regarding the surface preparation of steel are cleanliness and profile. SSPC and NACE have developed joint standards for evaluating five degrees of abrasive blast cleaning. Starting at the lowest degree, they are:

- Brush-off blast cleaning (*SSPC-SP7/NACE 4*)
- Industrial blast cleaning (*SSPC-SP14/NACE 8*)
- Commercial blast cleaning (*SSPC-SP6/NACE 3*)
- Near-white blast cleaning (*SSPC-SP10/NACE 2*)
- White metal blast cleaning (*SSPC-SP5/NACE 1*)

As a guide for comparing surface cleanliness to these standards, SSPC has developed reference photographs to illustrate each grade (*SSPC-VIS1*). These visual standards are for steel that has not been previously painted.

6.1.3 Existing Coatings

To evaluate existing coatings, check the quality of the coating adhesion, the film thickness, and the condition of the substrate. Adhesion can be evaluated visually or by using adhesion testing equipment, the knife test (*ASTM D6677*), portable adhesion testers (*ASTM D4541*), or a tape test (*ASTM D3359*). The condition of the substrate can be checked by removing a little bit of coating and visually inspecting it for evidence of mill scale or rust on steel, and efflorescence on concrete. A variety of gauges are available to measure film thickness, which indicates the quantity of the coating and how well it was applied.

6.2.0 Coating Materials Preparation

Coatings should be mixed as directed in the mixing instructions section of the coating manufacturer's PDS or application bulletin to ensure uniform color and texture of the coating material(s). If plural-component coatings are being mixed, make sure to follow the manufacturer's instructions for mixing the components in the correct ratios and which component must be added to the other. Thinning or heating (preferably) a coating is sometimes necessary to obtain the required WFT or viscosity for the application equipment being used. Thinning must always be done according to the manufacturer's PDS and within the current EPA volatile organic compound (VOC) restrictions.

6.2.1 Coating Mixtures

When mixing coatings, adhere to the following general guidelines:

- Wear the proper PPE and follow any fire/explosion prevention specified in the coating manufacturer's PDS or MSDS.
- For multi-component coatings, always stir the components separately first and then mix them together in the correct proportions. Allow the mixed coating to stand for any required induction time before applying. Make sure the coating is applied within the recommended pot life.
- When using a coating that consists of two or three packages, the entire package, or kit, should be mixed at the same time. Do not mix together partial kits.
- Follow the manufacturer's PDS very closely. Some mixtures must be continually agitated slowly after mixing. With short pot-life coatings, the individual components are usually slowly agitated prior to being mixed at the spray equipment or gun.
- Ensure the even dispersal of pigment during the mixing process.

Once mixed, some types of coating mixtures have a short pot life. Plan ahead so that they are not mixed too soon and begin curing too early. Mixing sequence and pot life vary from product to product. Mixing of multi-component coatings with long pot lives can be done in large containers provided the chemical reaction temperatures do not reach dangerous levels. Heated multi-component coatings with very short induction and/or pot lives must be mixed in special spray equipment or spray guns as they are applied.

Epoxies have some of the most complex coating preparation requirements, including induction or sweat-in times and procedures. The induction or sweat-in time is the time that must be allowed from the beginning of the component mixing process, and after the mixing process is completed, until the coating can be applied. This time period is required to allow the components to begin the chemical reaction that will result in cured, dry film for coatings after their application. Most induction times depend on the temperature of the coating when it is mixed. Some are also affected by the RH.

6.2.2 Coating Viscosity

Coating material viscosity can be affected by two main factors: the amount of solvent the material contains, and the temperature of the material. Coating material kept at a higher temperature is less viscous. Coatings stored in the cold must be allowed to warm up before testing and use. This may be done with hot water baths or by electrical means.

To reduce viscosity, heat or thin the material. Heating is the preferred method for many applications. A common method for checking viscosity of a properly mixed coating is by timing a known amount of material while it flows through an orifice in a viscosity cup (*Figure 15*). Common viscosity cups are the Shell cup, Zahn cup, DIN cup, and the Ford cup. The flow time through the cup orifice is normally measured with a stopwatch, while the temperature of the material is measured with a thermometer. Using the measured flow rate and temperature, the viscosity of the material can be determined. For example, at a temperature of 75°F (24°C), a 15-second material refers to a measured quantity of paint that flowed through the viscosity cup for 15 seconds at 75°F (24°C). The testing and determination of viscosity should be done according to the instructions supplied with the viscosity cup being used.

If thinning is required, follow these guidelines:

- Assess the need for viscosity control. Document the need in the job report or log.
- Only use thinners specified by the coating manufacturer. Never use substitutes.
- Never exceed the amount of thinner specified by the coating manufacturer.
- Add thinners to the coating very slowly, with constant stirring. Retest the viscosity often to prevent excessive thinning of the material.

6.3.0 Preparation of Application Equipment

Quality inspectors must be familiar with the equipment and methods required for use on a specific job. The coating manufacturer's product sheets and application bulletins should be consulted, and the manufacturer's recommendations must be followed when selecting application equipment. The equipment types commonly recommended are brush, roller, and spray. Not all coatings can be successfully applied with all three. In the case of spraying equipment, the manufacturer's literature will recommend the type of equipment, such as conventional or airless, tip sizes, and pressure.

6.3.1 Brushes

The quality of a brush can range from poor to excellent based on its materials and construction. Use only quality brushes, because they hold more paint, leave a better finish, and do not shed their bristles. They also last longer and generally make painting easier and faster. Price and feel are both factors in selecting a quality brush. Normally, the higher the price, the better the quality of the brush. To inspect a brush, follow these guidelines:

- Pull gently on the bristles. A good brush will only lose one or two when pulled on. If several bristles come out, the brush should not be used.
- Bounce the bristles in the palm of your hand. A good brush has bristles that feel soft and springy, and bounce back into shape quickly.
- Make sure the bristles are thick and plentiful.
- Ensure that the brush has a substantial handle that is balanced, comfortable, and in good shape.
- Ensure that the metal barrel that holds the bristles is free of rust and fastened securely to the handle. Make sure that it does not move when the bristles are wiggled back and forth.

207F15.EPS

Figure 15 Viscosity cups.

6.3.2 Rollers

A large variety of rollers are used for coating applications. As with brushes, the quality of a roller is normally directly related to its price. Follow these guidelines to ensure that a roller is in good condition and suitable for the coating application:

- Make sure the roller cover is free of loose nap, dust, dirt, and any other particles that may show up on the coated surface.
- Ensure that the roller spins freely and does not squeak or vibrate, indicating poor quality.
- Always rub down new roller covers by wringing the cover tightly with bare hands to remove all loose nap.
- Before use, ensure that the roller tray or roller grid is clean and does not contain any dirt, sand, dirty solvent, or previously used coating residue.

6.3.3 Spray Equipment

Spray equipment and supply lines must be checked before use. Always follow the manufacturer's operation manuals for specific items to inspect. OSHA requires that an operator's instruction manual be at the job site whenever spray equipment is used. The following general items are guidelines for checking spray application equipment:

- Check that hoses are either ground or conductive type hoses.
- Always shut off and release any fluid and/or air pressure before checking the gun. If present, apply the trigger lock. Be very aware of where the gun nozzle is pointed.
- Ensure that there is adequate ventilation before starting. Mechanical ventilation must be provided when spraying in enclosed areas. Ventilation equipment must be located so that it will not circulate contaminated air back to the work area or other areas inside of the building.
- Know the specific hazards of the coating being used. Ensure that the sprayer is compatible with the chemicals contained in the coating.
- Make sure that appropriate warning signs are posted in all hazardous areas.
- Make sure that the spray area is free of open flames, sparks, and other ignition sources.
- Make sure that the spray area is free of combustible materials and residue.
- Check the equipment carefully before use. Remove damaged equipment from service.
- Check the spray equipment's safety shutoff valve and pressure gauges to be sure that they are in good working condition prior to each use.
- Close the safety shutoff valve and relieve pressure when spraying operations are not being done.
- Make sure that the system has been properly flushed according to the manufacturer's instructions for the coating that was used.
- Check filter and oil traps.
- Drain both oil and water traps.

6.3.4 Conventional Spray Systems

Conventional spray systems require air supply checks as well as equipment checks. Items that should be checked before using conventional spray systems include the following:

- Make sure that the lines are properly connected and that the supply of fluid is adequate.
- Ensure that the compressed air supply is clean, dry, and free of oil (use a blotter test). Most spray systems use an extractor of some sort to remove water and oil from the air supply. The extractor is usually between the compressor and the gun, because the compressor may introduce moisture or oil to the air. Make sure the filters are clean and clear.
- Check the cleanliness and working order of spray tanks and other components before use. If dirt is clogging the paint equipment, coating material may be lost due to elapsed pot life or the presence of contamination.
- Inspect the spray lines, tips, and nozzles for the presence of moisture. Moisture contamination can cause such problems as spray gun spitting, spatter, bubbles, pinholes, and other film defects.
- Inspect the spray lines, tips and nozzles for the presence of oil. Oil contamination can cause adhesion problems, drying problems, soft, tacky films, and other defects.
- Ensure that the proper nozzle combination has been selected based on the type of spray gun being used, the expected flow rate, and the type of material being sprayed. This information is usually available in the manufacturer's instructions.

6.3.5 Airless Sprayers

Airless spray systems must also be thoroughly checked before use. Items that should be checked before using airless sprayers include the following:

- Check the condition of all hoses. Ensure that all hose connections and fittings are tight and leak-free, and hose whips are attached. Use only equipment that is rated for airless spraying and is in adequate condition to withstand the high pressure of airless spraying.
- Examine the gun filter to be certain that it is not clogged or compromised.
- Make sure the pump, spray gun, and object being sprayed are all properly grounded to prevent static sparks and possible explosions.

- Ensure that the correct voltage is applied to the sprayer and the electrical supply circuit is properly grounded (when using an electric airless sprayer).
- Check the hydraulic oil level, if applicable.
- Check the engine oil level, if applicable.
- Inspect the return line filter to ensure that it is not clogged.

6.4.0 Documentation Requirements

A work plan specifically details how the contract is to be performed, and is most likely required before beginning any part of the coating work. The work plan should also explain the documentation requirements associated with the job. Documentation, starting with the work plan, provides a clear paper trail for the completion of the job.

Documentation requirements vary from job to job. Documentation refers to all of the required recordkeeping and reports of the inspection process. The coating inspector is almost always required to provide the client with some documentation of the inspection. Even if documentation is not specifically required, good practice dictates that accurate detailed records be kept.

Detailed and accurate records of each step of the whole process of preparation and application should also be recorded. These records should show all items, activities, and materials involved in surface cleaning, pre-treatment, and coating application. These records should also include reasons for work stoppages. Good documentation can provide valuable information on the durability of coatings and the economical protection they provide. The documentation requirements should be listed in the job specification.

Make sure to document and record the lot numbers of the coating containers. The lot numbers on the shipping manifest and the containers should match. The lot number can be decoded from the identification number on the container or shipping manifest by contacting the coating manufacturer. Typically the identification number is decoded as follows: the first character is the month, followed by the year, the plant, then the lot number. Make sure that all containers for the same coating have the same lot number.

7.0.0 Application Procedures

Before coating an interior or exterior area or surface, always make sure that all adjacent surfaces have been properly masked against unwanted coating or splatters. The coatings being used should be those recommended by the specification and/or manufacturer for use on the surfaces to be coated. If required, make sure that the lot numbers for all containers of a coating match. If applying multiple coats such as a primer, intermediate industrial coating, and a final finish to a surface, all coatings must be compatible.

7.1.0 Spray Gun Application Techniques

The general sequence for spray gun coating of surfaces is to coat the edges, outside corners, and inside corners first (*Figure 16*). This is called striping or, in some cases, banding. The primary purpose of striping is to ensure coverage on complicated geometries such as weld areas, corners, crevices, and areas difficult to spray. Striping reduces overspray dusting when the remaining flat surfaces are coated, especially at the edges, because the spray gun pattern can be stopped before it passes beyond the edges. Any weld seams, irregular surfaces, and protruding shapes, such as bolt heads, should be cut in using a brush to prevent corrosion and ensure coating coverage.

Some specifications require that all striping be done using a brush in order to break surface tension. They may also require a color difference for striping. Striping should be carefully inspected visually before further application of coatings. Note that the coating may pull away at the edges and that the inside of a corner may pull out of the corner. Therefore, it is important to examine all such areas for problems, especially because the areas where brush striping is required are usually the areas where corrosion cells are most likely to form. If the striping is the same color as the finish coat, no recoat drying time is required.

After the edges and corners are coated, begin coating the flat, vertical surfaces at the top and work down, using horizontal strokes (passes). See *Figures 16* and *17*. For narrow, vertical surfaces use vertical strokes. Coat long and level panels or surfaces using horizontal strokes starting at the closest edge and moving toward the furthest edge (*Figure 18*). Some slight gun tilt is required so that overspray falls on the uncoated part of the surface. For long panels or surfaces, try to work from one natural breakpoint to the next. If a long panel or surface is vertical, work from the top down. Where there are no natural breaks, work across the width of the surface, coating only as much length as will allow blending the coating before it dries. *Figure 19* shows the techniques for coating cylinders and slender surfaces.

Ideally, indoor or outdoor work areas should be protected by containment and with the use of temperature and moisture-control equipment. In lieu of specific requirements in the job specification or manufacturer's recommendations, the

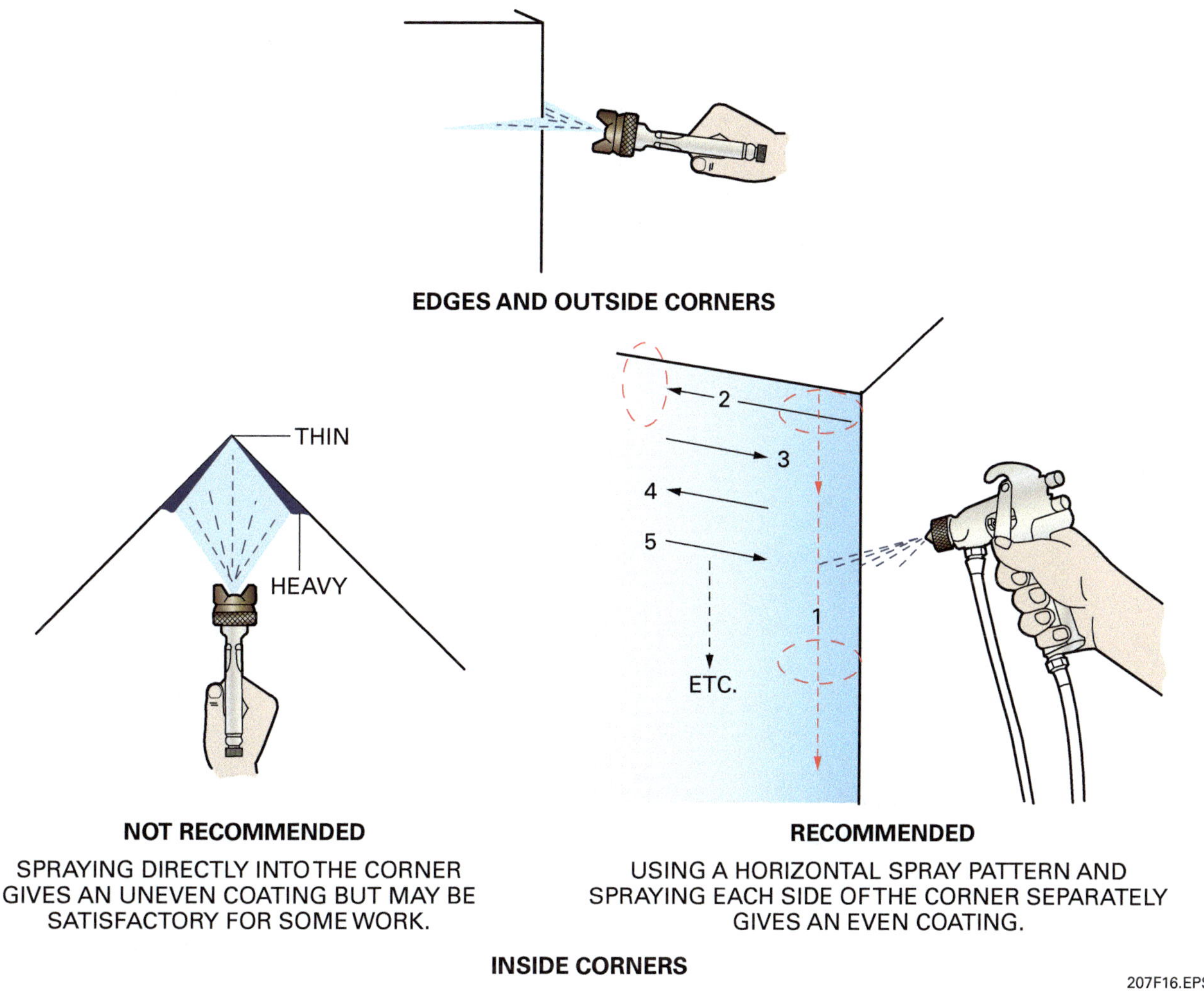

Figure 16 Spraying edges and corners.

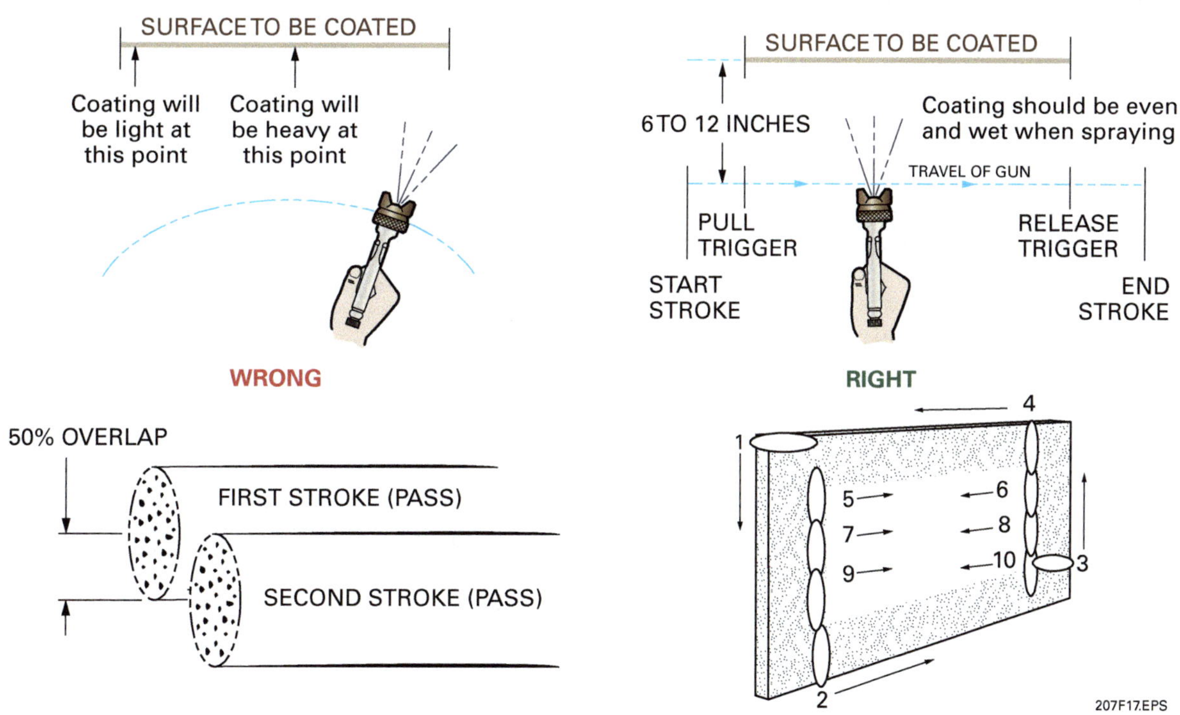

Figure 17 Spraying flat vertical surfaces.

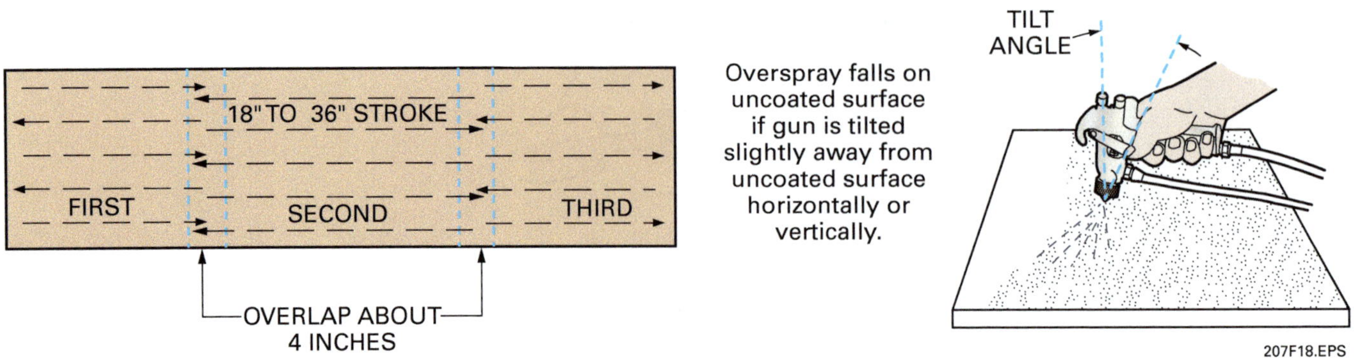

Figure 18 Spraying long and level panels or surfaces.

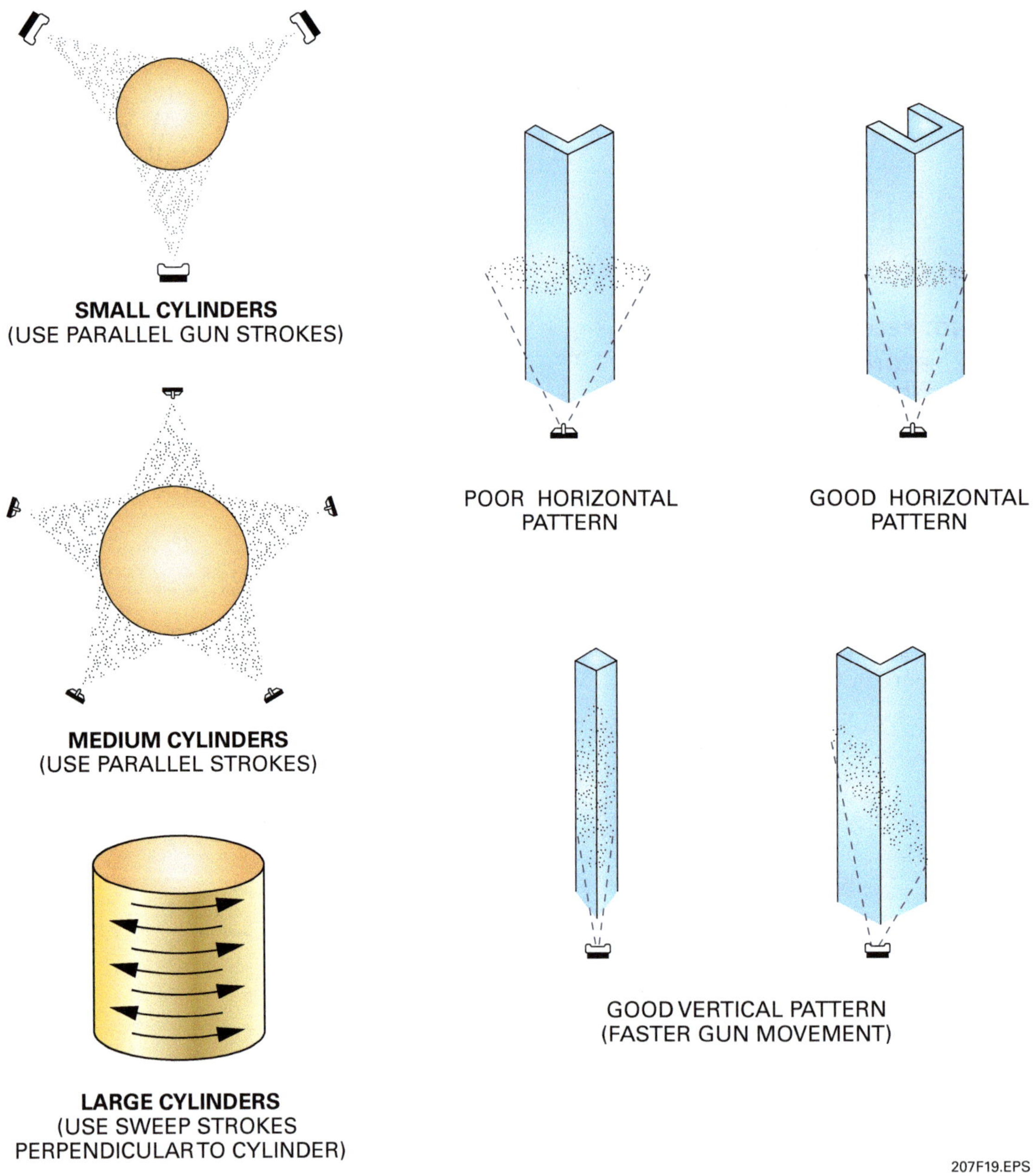

Figure 19 Spraying cylinders and slender surfaces.

following are general guidelines for spray gun coating applications:

- Be cautious of sudden changes in temperature. Apply coatings only when the weather moderates to a fairly even temperature.
- Allow any damp or wet surfaces to dry and reach proper temperatures for coating application.
- Avoid working in windy conditions because fresh coatings can collect leaves, dirt, dust particles, and insects. In addition, drift can cause coatings to damage adjacent surfaces.
- Do not work in the direct rays of a hot summer sun because coatings may not spread correctly on a hot surface and may dry too fast.

7.2.0 General Spray Equipment Setup and Operation

Proper equipment setup is a combination of inspection, troubleshooting, and maintenance. The first step is to examine the spray system equipment and connections. Any obvious conditions, such as damaged hoses or tubing, or visible damage to the gun, are the first indication of a problem to be corrected. If there are no visible flaws, the system should be assembled and tested in a safe place to determine whether it is working correctly. It is also good practice to flush the system and gun with an appropriate solvent before use.

It is very important to use the right hoses for each system. Normally, fluid hoses are larger than air hoses, so it would be very difficult to hook up the hoses to the wrong inlet, but different systems use different hoses, with different pressure ratings. Make sure that the right ones are used for each connection and purpose. Fluid hoses are usually either 5⁄16 inch or 3⁄8 inch (7.9 to 9.5 mm), while air hoses are usually 1⁄4 inch to 5⁄16 inch (6.35 to 7.9 mm).

WARNING

Many spray guns are made of aluminum, or have aluminum-wetted parts. Do not use these guns with halogenated solvents. Common halogenated solvents such as carbon tetrachloride, dichloroethane, chlorobenzene, ethylbromide, or ethylchloride can react with aluminum and cause explosions under certain circumstances. Check the MSDS and manufacturer's instructions very carefully.

NOTE

All spray guns must be assembled by hand and hand-tightened first to prevent cross threading and damaged threads. Spray guns are delicate, precise tools, and must never be forced.

7.3.0 Conventional Spray System Setup, Operation, and Shutdown

Conventional spray systems for industrial applications normally use pressure-feed spray guns in conjunction with stainless steel pressure tanks (pots) that meet ASME standards as the fluid containers for air-powered low-pressure transfer pumps positioned in or near an unpressurized fluid container. The pots used for conventional spraying must be rated for high-pressure use. Pumps or pots may also be equipped with optional air-powered agitators. Normally, pressure-feed conventional spray guns use atomization and a fluid pressure in the range of 20 to 70 psi (0.1 to 0.5 megapascal [MPa]) and most equipment is rated for use up to 100 psi (0.69 MPa). These are two-stage (lead and lag) guns, because when the trigger is fully depressed or released, the fan air flow occurs before and after the fluid flow to prevent an improper pattern when starting and stopping spraying operations.

WARNING

Make sure that all equipment and hoses to be used are rated for the coating material atomization and fluid pressure that will be required as determined from the coating material manufacturer's PDS.

7.3.1 Initial Setup

Two methods are generally used for pot systems. One method uses air pressure regulation and metering for the gun where the air line comes from an oil and water filter (extractor) as shown in *Figure 20*. The other method uses a regulator mounted at the pot for the gun as shown in *Figure 21*. A setup for an air-powered low-pressure transfer pump system is shown in *Figure 22*. In this particular application, a container-mounted pump is used. Like the pressure pot systems, optional regulators at the transfer pump can be used for gun air along with an optional agitator.

Step 1 Shut off the air from the compressed air source.

WARNING

Make sure that the pot or pressure pump is grounded and that a conductive fluid line and/or air line is used from the pot or pump to the spray gun. This prevents static electricity from building up and causing ignition of volatile vapors in the area.

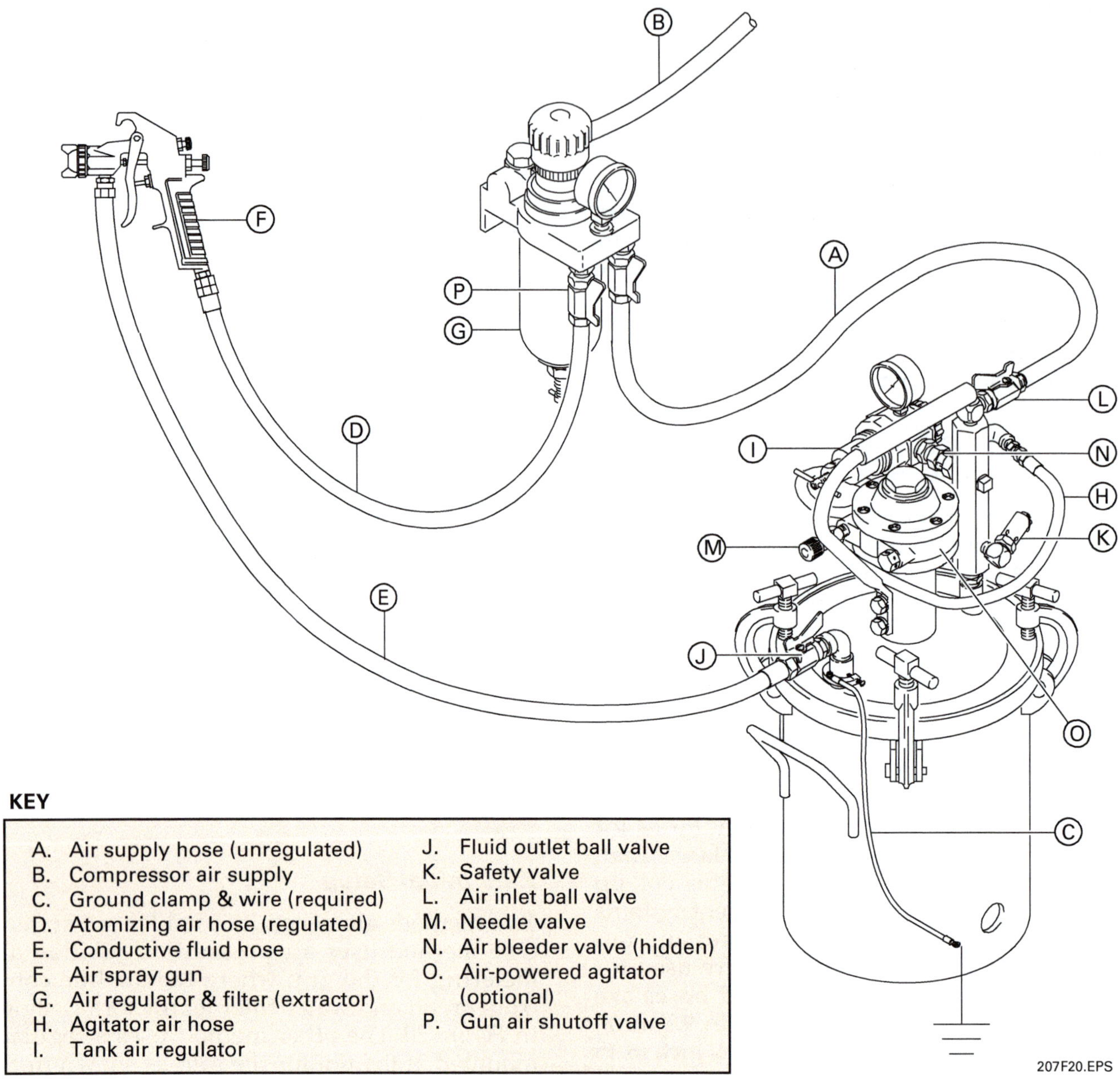

Figure 20 System setup for regulated gun air from an extractor.

> **NOTE**
> Make sure that the compressed air source will supply the required air flow in cubic feet per minute (CFM) for the spray system(s) being used. Check the equipment manufacturer's technical data for the CFM requirements of all equipment including the required CFM for the gun air cap and fluid nozzle to be used.

Step 2 Connect the air and conductive fluid lines to the equipment as shown for the appropriate system being used.

Step 3 Connect the ground wire from the pot or pressure pump to an earth ground (metal cold-water pipe or ground rod).

Step 4 Close the spray gun and pot or pump regulators by turning the regulator knobs counterclockwise until loose. If equipped with an agitator, turn the needle valve on the air motor clockwise until it seats.

Step 5 Open the air bleeder valve at the pump or pot, close the fluid outlet valve, and allow any pressure to bleed off. For pots, remove the cover after any pressure has bled off.

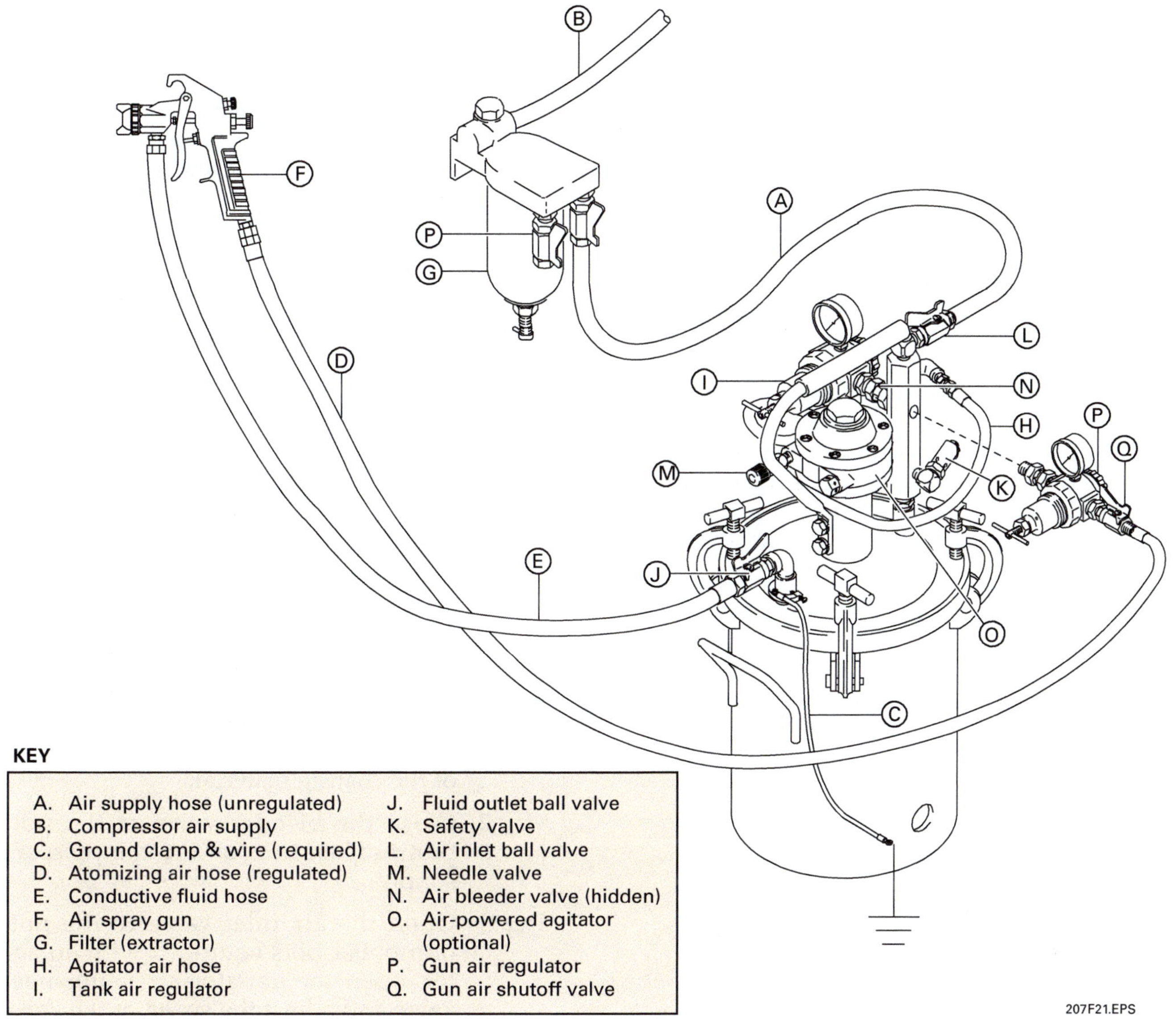

Figure 21 System setup for regulated gun air from a pot.

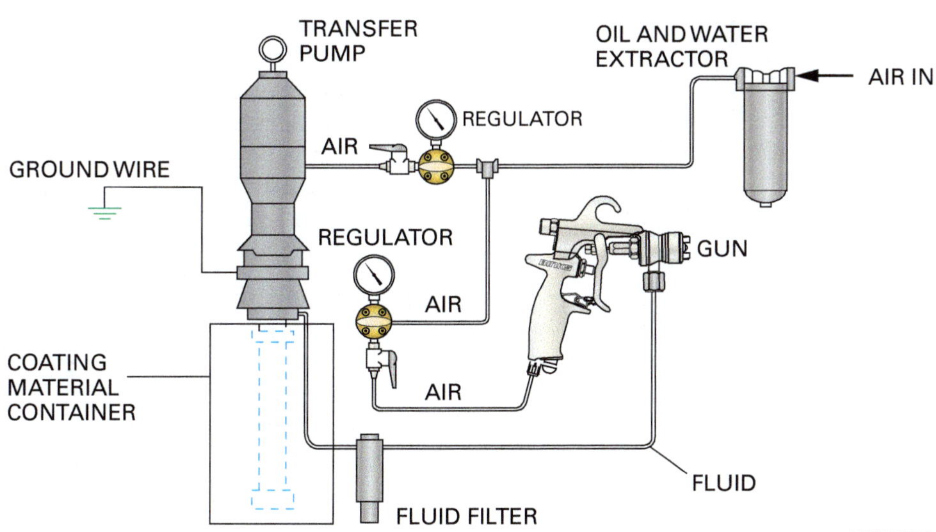

Figure 22 System setup for an air-powered low-pressure transfer pump.

Step 6 Check that the spray gun (*Figure 23*) is equipped with the coating PDS or equipment manufacturer's recommended air cap, fluid nozzle, and fluid needle. *Appendix C* shows an illustrated breakdown of a conventional air spray gun. Set the air cap for the initially desired fan pattern (vertical or horizontal) position.

> **NOTE**
> If the pattern orientation must be changed during operation, loosen the air cap locking ring, rotate the cap 90 degrees, and tighten the locking ring.

Step 7 For a full fan pattern, turn the pattern-control valve on the gun fully counterclockwise. For a round pattern, turn the pattern-control knob fully clockwise.

Step 8 Establish the correct fluid flow for the installed fluid nozzle and needle in the gun by turning the fluid-control knob counterclockwise until no restriction of the trigger movement is felt, then turn an additional half turn.

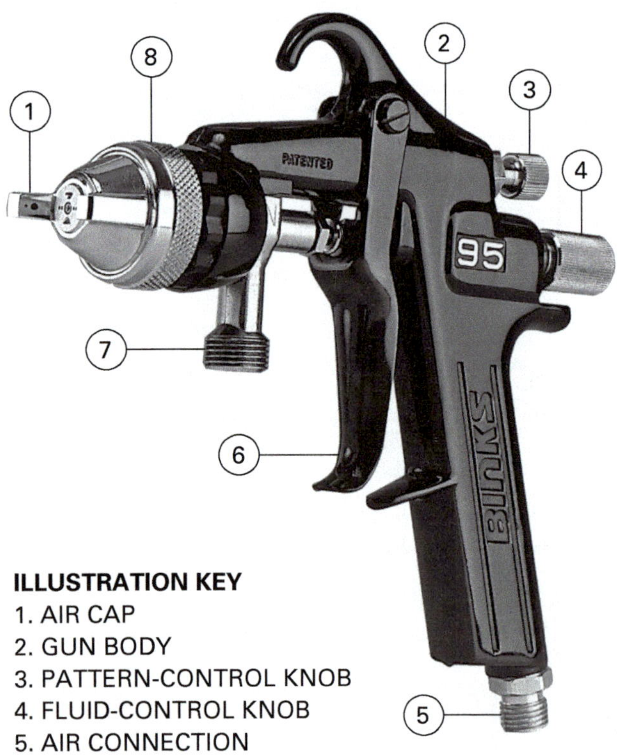

ILLUSTRATION KEY
1. AIR CAP
2. GUN BODY
3. PATTERN-CONTROL KNOB
4. FLUID-CONTROL KNOB
5. AIR CONNECTION
6. TRIGGER
7. FLUID INLET
8. FLUID NOZZLE AND NEEDLE

207F23.EPS

Figure 23 Typical conventional air spray gun.

7.3.2 Startup and Operation

> **WARNING**
> Check the coating manufacturer's MSDS and make sure to use the recommended PPE when performing the following procedures.

> **CAUTION**
> If a premixed plural-component coating will be sprayed, make sure that it is mixed as specified in the coating manufacturer's PDS. Also make sure that all spray system startup and operation procedures are completed and equipment shutdown and cleaning is initiated before the pot life of the coating expires.

Step 1 For pots, insert a compatible anti-static polyethylene bag in the pot. As necessary, mix, strain, and box the coating material as appropriate. Then fill the pot. For pressure pumps, remove the lid of the coating container(s). As necessary, mix, strain, and box the coating material as appropriate. Then insert the pump in a container of the coating material.

Step 2 Close the bleeder valve at the pot or pump. Then turn on the compressor air supply.

Step 3 Open the air inlet valve to the pot or pump. For pots equipped with an agitator, open the agitator air motor needle valve and adjust the speed of the agitator to about 40 to 60 rpm if the coating must be constantly mixed. Set the regulator for the coating or equipment manufacturer's recommended pot pressure. For pumps, set the fluid regulator for the coating or equipment manufacturer's recommended pressure. If the pump is equipped with an air-powered agitator, adjust the agitator air motor needle valve for a speed of 40 to 60 rpm if the coating must be constantly mixed.

Step 4 Open the atomizing air shutoff valve at the regulator. With the gun trigger depressed, adjust the gun air regulator to the coating material or equipment manufacturer's recommended atomizing pressure, then release the trigger.

Step 5 Open the fluid outlet valve at the pot or pump.

Step 6 Trigger the gun and spray test patterns at a distance of 6 inches (15 cm) to 12 inches (30 cm) depending on the coating material and atomizing pressure. Adjust the pattern (sometimes called fan-air) control knob as desired. If the spray is too fine and excessive dusting occurs, slightly reduce the gun air pressure at the regulator; if too coarse, increase the pressure slightly. If necessary, the fluid flow can be reduced slightly by turning the fluid-control knob clockwise a small amount; however, if the correct fluid flow cannot be achieved by minor adjustments of the atomization pressure and fluid flow, a different size fluid nozzle and needle may be necessary.

Step 7 After a correct test pattern is achieved, begin spraying the desired surfaces.

NOTE

Provided the coating being sprayed does not have a relatively short pot life, some short breaks in spraying can be accommodated by placing the fluid nozzle of the spray gun in a container of appropriate coating solvent to prevent drying of the coating in the nozzle.

7.3.3 Shutdown and Cleaning

Step 1 Turn off the air supply to the pot or pump and open the pot or pump bleeder valve.

Step 2 Loosen the air cap on the spray gun and press a rag over the orifices on the air cap. Trigger the gun for a time to force the fluid in the fluid hose back into the pot or out of the pump.

NOTE

The following cleaning should be done when changing colors, changing from a water-based to solvent-based coating, changing from a solvent-based to water-based coating, and before storage.

Step 3 For pots, remove the cover. Then remove the antistatic liner containing any remaining coating material and pour a suitable solvent into the pot. Replace the pot cover and secure. Properly dispose of the liner and remaining coating material.

Step 4 For pumps, remove the pump from the coating container and install in a grounded metal container of suitable solvent. Properly dispose of remaining coating material and its container.

Step 5 Turn off the bleeder valves and turn on the air supply to the pot or pump.

WARNING

Due to VOC restrictions and/or explosion hazards associated with certain solvents, the following steps may be required to be performed in an enclosed area with a grounded container.

Step 6 Hold the metal part of the gun against a grounded container, if required, and trigger the gun until clean solvent comes from the gun.

Step 7 Turn off the air supply to the pot or pump and open the bleeder valve. Then repeat Step 2 above to flush the solvent from the fluid hose and pump.

Step 8 Remove the pump from the solvent container or remove the pot cover. Properly dispose of the solvents.

Step 9 Disassemble and clean the air cap and fluid nozzle of the gun with a rag and solvent. Clean the pump or pot and pot cover with a rag and solvent. If present, clean the agitator shaft and paddles with a rag and solvent.

Step 10 Trigger the gun several times and reassemble the fluid nozzle and air cap.

Step 11 Turn off the main air supply to the gun and the pot or pump.

Step 12 Disassemble the system setup as necessary.

Step 13 Lubricate the gun and air-powered pump and/or agitator as specified by the manufacturer.

Step 14 For pots, replace the pot cover, then properly store any equipment disassembled from the system setup.

7.4.0 Airless Spray System Setup, Operation, and Shutdown

The distinguishing characteristic of airless systems is the delivery pressure of the fluid. The fluid is atomized by high pressure as it exits the tip of the spray gun. Fluid is delivered to the gun at a pressure typically about 1,000 to 5,000 psi (7 to 34 MPa) or higher in the airless system. The fluid supply hose is a special conductive high-pressure type, to handle the fluid pressures used and to prevent static electricity buildup at the gun. The tip guard on the gun is designed to keep the spray away from flesh, and to prevent accidental injections.

WARNING

Do not place gloved or bare hands or any other part of the body in the fluid stream from the tip of an airless spray gun. Do not aim the gun at another person. Do not attempt to check for fluid hose leaks with a gloved or bare hand. Hose leaks can be nearly invisible. If a hose leak is suspected or detected, immediately shut down and depressurize the sprayer system following the manufacturer's recommended pressure-relief procedures. If the fluid stream from a gun tip or a hose leak comes into contact with skin, even through a glove, see a physician immediately. Even if the skin does not appear to be broken or injured, an injection injury may have occurred. Injection can take place at pressures as low as 100 psi (0.7 MPa). If fluid is injected into any part of the body, severe injuries can result.

An example of initial setup, startup, operation, shutdown, and cleaning is provided in the following sections for one model of a cart-mounted gasoline-engine mechanical airless sprayer (*Figure 24*) with a typical industrial airless spray gun (*Figure 25*).

WARNING

Never operate a gasoline engine in a closed building unless the exhaust is piped outside. The exhaust contains carbon monoxide, a poisonous, odorless, and invisible gas that can cause serious illness and possible death.

WARNING

Gasoline-engine mechanical airless sprayers are never used in nuclear or petrochemical facilities.

7.4.1 Initial Setup

Step 1 Prepare the gasoline engine for operation. Do not start the engine.

WARNING

Make sure that the sprayer is grounded and that a conductive fluid line, rated for the pressure of the sprayer, is used from the sprayer to the spray gun. This is to prevent static electricity from building up and igniting any volatile vapors in the area.

CAUTION

To prevent damaging the sprayer control unit, use only nylon-reinforced fluid hoses. Never use wire braided hoses because they are too rigid to act as pulsation dampers.

Step 2 Connect at least a 50 foot (15.2 m) conductive fluid hose to the fluid outlet of the sprayer and to the fluid inlet of the airless spray gun.

Step 3 Connect the sprayer grounding clamp and wire to an earth ground (metal cold-water pipe or ground rod).

Step 4 Fill the sprayer pump packing nut/wet cup with piston lubricant, commonly called throat seal liquid (TSL).

Step 5 Lock the spray gun trigger with the safety lock.

Step 6 Loosen the tip guard retaining nut and remove the tip guard and spray tip. Make sure that the tip orifice size matches the coating PDS or equipment manufacturer's recommendations. *Appendix D* shows an illustrated breakdown of a typical airless spray gun.

WARNING

If the pattern orientation or spray tip is changed during operation, make sure to lock the gun trigger and relieve spray pressure as covered in the following startup and operation procedures for installing the tip and tip guard before attempting to change the tip and/or tip guard orientation.

207F24.EPS

Figure 24 Cart-mounted gasoline-engine mechanical airless sprayer.

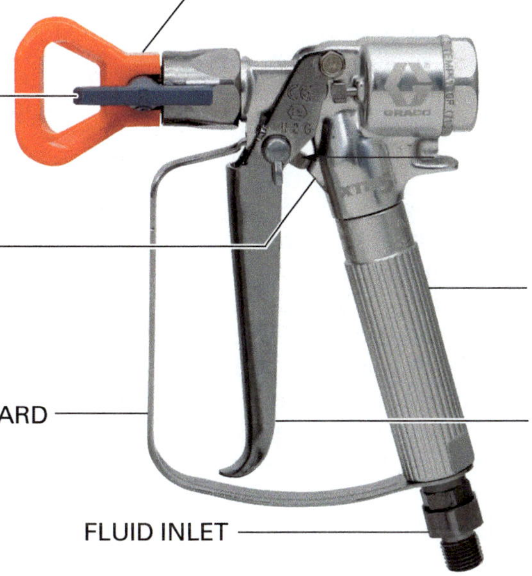

Figure 25 Typical industrial airless spray gun components.

7.4.2 Startup and Operation

WARNING

Check the coating manufacturer's MSDS and make sure to use the recommended PPE when performing the following procedures.

CAUTION

If spraying a premixed plural-component coating, make sure that it is mixed as specified in the coating manufacturer's PDS. Also, make sure that all spray system startup and operation procedures are completed and equipment shutdown and cleaning is initiated before the pot life of the coating expires.

Step 1 Set the spray pump controls to the following positions:

- Pressure-control switch to Off.
- Pressure-adjusting knob to the lowest pressure setting.
- Pressure-drain valve to Open.

Step 2 Place the sprayer suction tube (or immersion tube) into a grounded metal container filled with coating material that has been properly mixed in accordance with the coating manufacturer's PDS.

Step 3 Place an empty container under the drain valve.

Step 4 Set the engine switch to On and start the engine.

Step 5 Prime the pump by turning the pressure-control switch to On and turning the pressure-adjusting knob about ¼ turn from the minimum position. Run the pump until coating material flows smoothly from the drain valve; then close the pressure-drain valve.

WARNING

If spraying flammable solvents or coatings, use an earth-grounded container and hold the gun against the side of the container during the following step.

Step 6 Disengage the safety lock and trigger the gun into a container until the coating flows from the gun.

Step 7 Release the gun trigger and engage the trigger safety lock.

Step 8 Set the pressure-control switch to Off, open the drain valve to relieve pressure, then close the drain valve.

Step 9 Install the tip guard with an appropriate spray tip and loosely tighten the tip guard retaining nut. Orient the spray tip and guard for the desired vertical or horizontal spray pattern, then hand-tighten the tip guard retaining nut.

Step 10 Set the pressure-control switch to On.

NOTE

The tip guard and spray tip for the gun shown in *Figure 25* is oriented for a vertical fan.

> **CAUTION**
> To increase sprayer life, always use the lowest required fluid pressure and the lowest engine speed. Higher settings cause excessive clutch cycling, premature tip wear, and premature pump wear.

Step 11 Set the pressure-adjusting knob to the pressure recommended by the coating manufacturer as a starting point. Disengage the gun trigger safety lock and spray a test pattern at a distance of 12" to 14" (30 to 36 cm) to check pattern and spray atomization. Turn the pressure-adjusting knob until a good spray pattern without bands or tails at the ends of the pattern is obtained. After a good pattern is obtained, lower the engine speed as far as possible without changing the spray pattern.

Step 12 Spray the surface to be coated.

The spraying techniques described earlier for conventional spray also apply to airless spray with the following exceptions:

- Because of the higher pressures involved, maintain a distance of about 12" to 14" (30 to 36 cm) from the spray gun tip to the surface. If held too close, runs and sags may occur; if too far, a thin coat and more overspray will occur.
- An airless spray gun operates in only two modes, on and off. Always quickly trigger the gun fully on and quickly release the trigger to fully off to avoid gun spitting.
- Gun travel speed can only be governed by the tip orifice size and the spraying pressure. If the gun must be moved too slowly to obtain proper coating coverage, raise the spray pressure slightly, if possible, or use a tip with a larger orifice size. If the gun must be moved too fast for proper coating coverage, lower the pressure slightly, if possible, or use a smaller orifice size.
- Unlike conventional spray guns, feathering is not an option. Parallel strokes may not need to overlap as much as with conventional guns because of the heavier edges.
- Provided the coating being sprayed does not have a relatively short pot life, some short breaks in spraying can be accommodated by shutting off the pump and placing the spray tip of the spray gun in a container of appropriate coating solvent to prevent drying or setting of the coating in the spray tip.

7.4.3 System Shutdown and Cleaning

Step 1 Engage the spray gun trigger lock.

Step 2 Set the engine and pressure-control switches to Off.

> **WARNING**
> Due to VOC restrictions and/or explosion hazards associated with certain coatings or solvents, you may be required to be perform the following cleaning steps in an enclosed area with a grounded container.

Step 3 Disengage the gun trigger lock and trigger the gun into a metal container to relieve pressure. Then place a container under the drain valve and open the drain valve to relieve any residual pressure.

> **NOTE**
> The following cleaning procedure should be done when changing colors, changing from a water-based to solvent-based coating, changing from a solvent-based to water-based coating, and before storage.

Step 4 Remove the filter bowl and screen from the fluid filter. If equipped, remove the filter from the gun handle. Reinstall the screen support and bowl, and as necessary, reassemble the gun handle. Clean the screen and/or filter separately.

Step 5 Place the suction tube (or immersion tube) in a grounded container filled with the appropriate solvent. Remove and properly dispose of the coating container.

Step 6 Engage the gun trigger lock and remove the spray tip and tip guard from the gun. Clean the spray tip and tip guard separately.

Step 7 Start the system as previously described and operate it at the lowest pressure.

Step 8 Open the sprayer drain valve until the solvent flows, then close the valve.

Step 9 Trigger the gun into the container, grounding the gun against a grounded container as necessary, until the solvent runs clear.

> **CAUTION**
> Do not let the pump run dry for more than 30 seconds during the following step.

Step 10 Remove the suction tube (or immersion tube) from the solvent and trigger the gun into the container to remove the solvent from the suction hose and pump.

Step 11 Shut down the system as previously described and engage the gun trigger lock. Open the drain valve to relieve any pressure and then close the valve.

CAUTION

If the coating solvent used was water, do not let any water remain in the pump. Flush the sprayer again using an oil-based solvent (mineral spirits or paint thinner) to displace any water. Then leave the pump filled with the oil-based solvent.

Step 12 Remove the fluid hose from the gun and sprayer. Allow the hose to drain into the container under the filter-drain valve.

Step 13 Clean the exterior of the suction tube and hose (or immersion tube) and suction tube tip strainer.

Step 14 Reassemble the gun including any handle filter, the screen in the sprayer filter, the suction tube and hose, and install the tube tip on the suction tube or immersion tube.

Step 15 Lubricate the gun as specified by the manufacturer. Properly store the system and any equipment left disassembled from the setup.

7.5.0 Air-Assisted Airless Spray System Setup, Operation, and Shutdown

When compared to airless spray systems, the distinguishing characteristic of air-assisted airless spray systems is the lower delivery pressure of the fluid. The systems include added atomization and spray pattern adjustment that is provided by the spray gun from a fan air hose in conjunction with the fluid hose. Fluid pressures for air-assisted systems typically range from 400 to 1,200 psi (3 to 8 MPa). The fluid spray tip determines the maximum pattern (fan) width. The fluid is then additionally atomized and the fan width is reduced by introducing fan air supplied by the fan air hose through adjustment of the pattern-(fan) adjustment knob.

Both fan air hoses and fluid supply hoses are special conductive types to prevent static electricity buildup at the gun. Fluid supply hoses are high-pressure types used to handle the fluid pressures required. The tip guard on the gun is designed to keep the spray away from flesh and to prevent accidental injections.

WARNING

Do not place gloved or bare hands or any other part of the body in the fluid stream from the tip of an air-assisted airless spray gun. Do not aim the gun at another person. Do not attempt to check for fluid hose leaks with a gloved or bare hand. Fluid hose leaks can be nearly invisible. If a fluid hose leak is suspected or detected, immediately shut down and depressurize the sprayer system following the manufacturer's recommended pressure-relief procedures. If the fluid stream from a gun tip or a hose leak comes into contact with skin, even through a glove, see a physician immediately. Even if the skin does not appear to be broken or injured, an injection injury may have occurred. Injection can take place at pressures as low as 100 psi (0.7 MPa). If fluid is injected into any part of the body, severe injuries can result.

An example of initial setup, startup, operation, shutdown, and cleaning is provided in the following sections for one model of an industrial cart-mounted, air-motor powered, air-assisted airless sprayer (*Figure 26*) with a typical industrial air-assisted airless spray gun (*Figure 27*).

Air-assisted airless spray guns have the pattern– (fan) adjustment knob at the rear of the gun like a conventional air spray gun but, like an airless spray gun, no fluid-control knob. In this spray system example, the fan air pressure is set by a pressure regulator at the sprayer, but it could also be supplied from a regulator at a separate location. Unlike a conventional air spray gun, the fan-adjustment knob of an air-assisted airless spray gun decreases the width of the fan spray pattern by a maximum of 25 percent when the knob is turned counterclockwise (opened) and increases the fan width if it is turned clockwise (closed). Like conventional air spray guns, these are two-stage (lead and lag) guns because the fan air flow occurs before and after the fluid flow to prevent an improper pattern when starting and stopping spraying operations. In this example, an additional pressure regulator on the sprayer is used to adjust the speed of the sprayer air-motor pump to control fluid pressure supplied to the gun.

7.5.1 Initial Setup

Step 1 Connect an air hose from a filtered air source with a closed, master bleed-type air valve (compressor or shop air) to a closed bleed-type air valve on the sprayer.

KEY

A. Bleed-type master air valve required for pump	J. Gun air hose (fan air)
B. Air filter	K. Pump air regulator
C. Air supply line to sprayer	L. Fan air regulator
D. Grounded 5 gallon metal pail	M. Bleed-type air valve
E. Air line moisture trap	N. Pump ground wire
F. Air-motor pump	O. Fluid pressure gauge
G. Air-assisted airless spray gun	P. Filter drain valve
H. Gun fluid hose	Q. Fluid outlet
I. Tip guard	R. Fluid filter
	S. Suction tube and hose

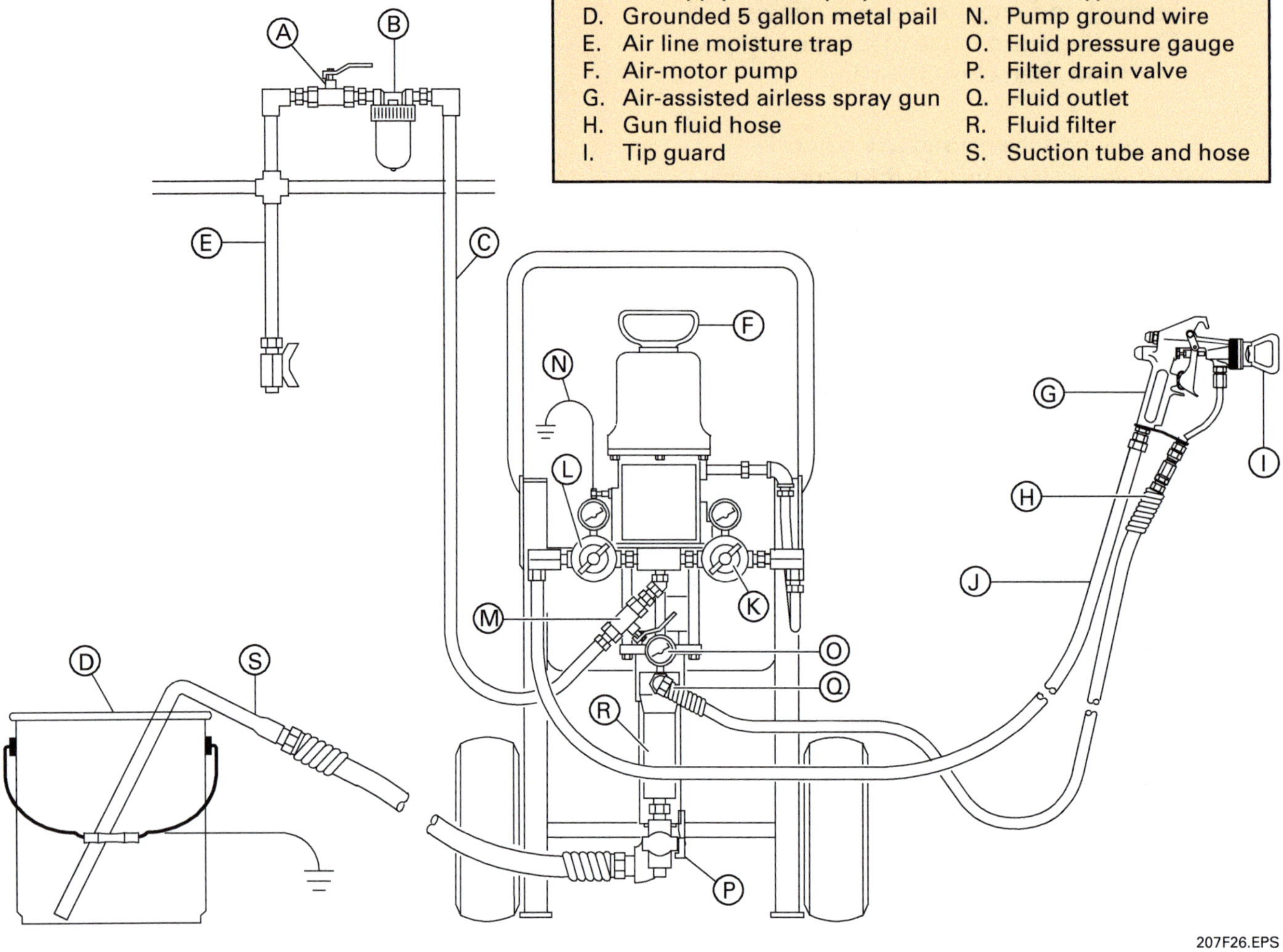

Figure 26 Industrial cart-mounted, air-motor powered, air-assisted airless sprayer components.

WARNING

If the pattern orientation or spray tip will be changed during operation, make sure to lock the gun trigger and relieve spray pressure as covered in the following startup and operation procedures for installing the tip and tip guard before attempting to change the tip and/or tip guard orientation.

CAUTION

Use only nylon-reinforced fluid hoses. Never use wire braided hoses because they are too rigid to act as pulsation dampers.

NOTE

Make sure that the compressed air source will supply the required air flow in CFM for the spray system(s) being used. Check the equipment manufacturer's technical data for CFM requirements of all equipment including the required CFM for the gun spray tip to be used.

Step 2 Connect at least a 50 foot (15.2 m) conductive fluid hose to the fluid outlet of the sprayer and to the fluid inlet of the air-assisted airless spray gun. Connect at least a 50 foot (15.2 m) conductive fan air hose from the output of the fan air regulator to the air inlet of the spray gun.

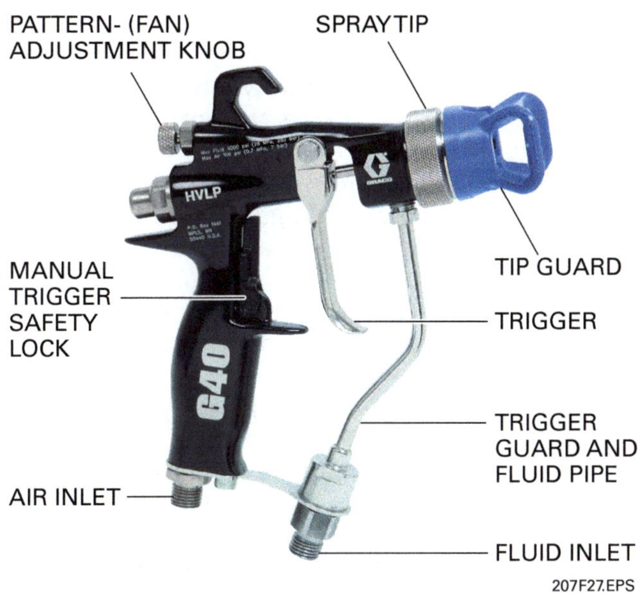

Figure 27 Typical industrial air-assisted airless spray gun components.

Step 3 Connect the sprayer grounding clamp and wire to an earth ground (metal cold-water pipe or ground rod).

Step 4 Fill the sprayer pump packing nut/wet cup with piston lubricant, commonly called throat seal liquid (TSL).

Step 5 Lock the spray gun trigger with the safety lock and turn the fan-adjustment knob fully clockwise (closed).

Step 6 Loosen the tip guard retaining nut and remove the tip guard and spray tip. Make sure that the tip orifice size matches the coating PDS or equipment manufacturer's recommendations. *Appendix E* shows an illustrated breakdown of a typical air-assisted airless spray gun.

7.5.2 Startup and Operation

WARNING

Check the coating manufacturer's MSDS and make sure to use the recommended PPE when performing the following procedures.

CAUTION

If spraying a premixed plural component coating, make sure that it is mixed as specified in the coating manufacturer's PDS. Also make sure that all spray system startup and operation procedures are completed and equipment shutdown and cleaning is initiated before the pot life of the coating expires.

Step 1 Close both regulators by loosening the adjustment stems and close the sprayer filter-drain valve. Then open the master and sprayer bleeder-type valves to supply air to the sprayer.

Step 2 Place the suction tube into a grounded metal container filled with coating material that has been properly mixed in accordance with the coating manufacturer's PDS.

Step 3 Place an empty container under the drain valve.

Step 4 Slowly open the air-motor pump regulator until the pump cycles slowly.

NOTE

The pump may cycle quickly until air is purged from the filter and fluid hose in the following steps.

Step 5 Prime the pump by running the pump until the coating material flows smoothly from the filter drain valve. Then close the valve.

WARNING

If spraying flammable solvents or coatings, use an earth-grounded metal container and hold the gun against the side of the container during the following step.

Step 6 Disengage the safety lock and trigger the gun into a container until the coating flows from the gun.

Step 7 Release the gun trigger and engage the trigger safety lock.

Step 8 Close the sprayer bleeder-type valve and open the filter drain valve to relieve pressure. Then close the valve.

Step 9 Install the tip guard with an appropriate spray tip and loosely tighten the tip guard retaining nut. Orient the spray tip and guard for the desired vertical or horizontal spray pattern. Then hand-tighten the tip guard retaining nut.

NOTE

The tip guard and spray tip for the gun shown in *Figure 27* is oriented for a vertical fan.

Step 10 Open the sprayer bleeder-type valve.

> **CAUTION**
>
> To increase sprayer life, always use the lowest required fluid pressure. Higher settings cause premature spray tip and pump wear.

Step 11 Open the pump regulator and set the reading on the fluid pressure gauge to the pressure recommended by the coating manufacturer as a starting point. Disengage the gun trigger safety lock and fully trigger the gun to spray a test pattern at a distance of 10" to 12" (30 to 36 cm). Check for good spray atomization. Tails or banding at the ends of the pattern should occur at this point. If not, lower the fluid pressure until they do occur.

> **CAUTION**
>
> Do not exceed the air pressure or fluid pressure ratings of the sprayer or gun in the following step.

Step 12 Partially open the gun fan-adjustment knob (counterclockwise), and partially trigger the gun so only air is emitted. Set the fan air regulator to 20 psi (0.1 MPa). Fully trigger the gun and use the fan-air adjustment knob to add air to eliminate just the tails, bands, or any spots. If tails, bands, or spots cannot be eliminated, retest the pattern as necessary by raising the fluid pressure in 100 psi (0.7 MPa) increments and adjusting fan air until a full, clean pattern is obtained.

> **NOTE**
>
> Always use the lowest fluid and air pressure required for good atomization and spray pattern to achieve maximum fluid transfer efficiency. Fan air pressures above 30 psi (0.2 MPa) will cause turbulence in the fan pattern, dirty spray tips, and slower production capability.

Step 13 Spray the surface to be coated.

The spraying techniques described for conventional spray also apply to air-assisted airless spray with the following exceptions:

- The fluid flow of an air-assisted airless spray gun operates in only two modes, on and off. Always quickly trigger the gun fully on and quickly release the trigger to fully off to avoid gun spitting.
- Gun travel speed can be governed only by the tip orifice size and the spraying pressure. If the gun must be moved too slowly to obtain proper coating coverage, raise the spray pressure slightly, if possible, or use a tip with a larger orifice size. If the gun must be moved too fast for proper coating coverage, lower the pressure slightly, if possible, or use a smaller orifice size.
- Unlike conventional spray guns, feathering is not an option. Parallel strokes may not need to overlap as much as with conventional guns because of the heavier edges.
- Provided the coating being sprayed does not have a relatively short pot life, some short breaks in spraying can be accommodated by shutting off the pump and placing the spray tip of the spray gun in a container of appropriate coating solvent to prevent drying of the coating in the spray tip.

7.5.3 System Shutdown and Cleaning

Step 1 Engage the spray gun trigger lock.

Step 2 Close the sprayer bleeder-type valve and allow the sprayer air pressure to bleed off.

Step 3 Close the fan air regulator and open the filter drain valve to relieve any residual pressure after making sure that a container is under the drain valve.

Step 4 Remove the filter bowl and screen from the filter. If equipped, remove the filter from the gun handle. Reinstall the screen support and bowl, and as necessary, reassemble the gun handle. Clean the screen and/or filter separately.

Step 5 Place the suction tube in a grounded container of appropriate solvent. Remove and properly dispose of the coating container.

> **WARNING**
>
> Due to VOC restrictions and/or explosion hazards associated with certain coatings or solvents, you may be required to be perform the following cleaning steps in an enclosed area with a grounded container.

> **NOTE**
>
> The following cleaning procedure should be done when changing colors, changing from a water-based to solvent-based coating, changing from a solvent-based to water-based coating, and before storage.

Step 6 Engage the gun trigger lock and remove the spray tip and tip guard from the gun. Clean the spray tip and tip guard separately.

Step 7 Open the pump air regulator until the pump cycles slowly. Operate the pump until solvent runs from the filter drain valve. Then close the drain valve.

Step 8 Disengage the gun trigger lock and trigger the gun into a metal container to relieve pressure.

Step 9 Trigger the gun into the container, grounding the gun against a grounded container, as necessary, until the solvent runs clear.

CAUTION

Do not let the pump run dry for more than 30 seconds during the following step.

Step 10 Remove the suction tube from the solvent and trigger the gun in the container to remove the solvent from the suction hose and pump.

Step 11 Shut down the system as previously described and engage the gun trigger lock. Open the filter drain valve to relieve any pressure and then close the valve.

CAUTION

If the coating solvent used was water, do not let any water remain in the pump. Flush the sprayer again using an oil-based solvent (mineral spirits or paint thinner) to displace any water. Then leave the pump filled with the oil-based solvent.

Step 12 Remove the fluid hose and fan air hose from the gun and sprayer. Allow the fluid hose to drain into the container under the filter-drain valve.

Step 13 Clean the exterior of the suction tube and hose and the suction tube tip strainer.

Step 14 Reassemble the gun including any handle filter, the screen in the sprayer filter, the suction tube and hose, and install the tube tip on the suction tube.

Step 15 Lubricate the gun as specified by the manufacturer. Properly store the system and any equipment left disassembled from the setup.

7.6.0 HVLP Spray System Setup, Operation, and Shutdown

High-volume low-pressure (HVLP) spray systems have become more prevalent over the years, largely because of environmental regulations that require limitation of VOC release. The high-solids coatings used in HVLP spray systems limit the use of solvents or thinning agents. For high-volume or large area industrial applications, pressure-feed HVLP spray guns (*Figure 28*) are used in the same type of system configurations previously described for industrial conventional air spray systems. The difference is that the air pressure supplied to the gun air inlet ranges from 5 to 50 psi (0.03 to 0.3 MPa) depending on the size of the air cap and fluid nozzle/needle used. The sizes of the air cap and fluid nozzle/needle are dependent on the viscosity of the coating material. Fluid pressure and final spraying air pressure for all air cap and fluid nozzle sizes ranges from 1 to 10 psi (0.006 to 0.07 MPa) depending on the viscosity of the coating material. Generally, the fluid and air pressures are kept as low as possible while still producing a good pattern. In some locations, regulations prohibit air cap atomizing pressure in ex-

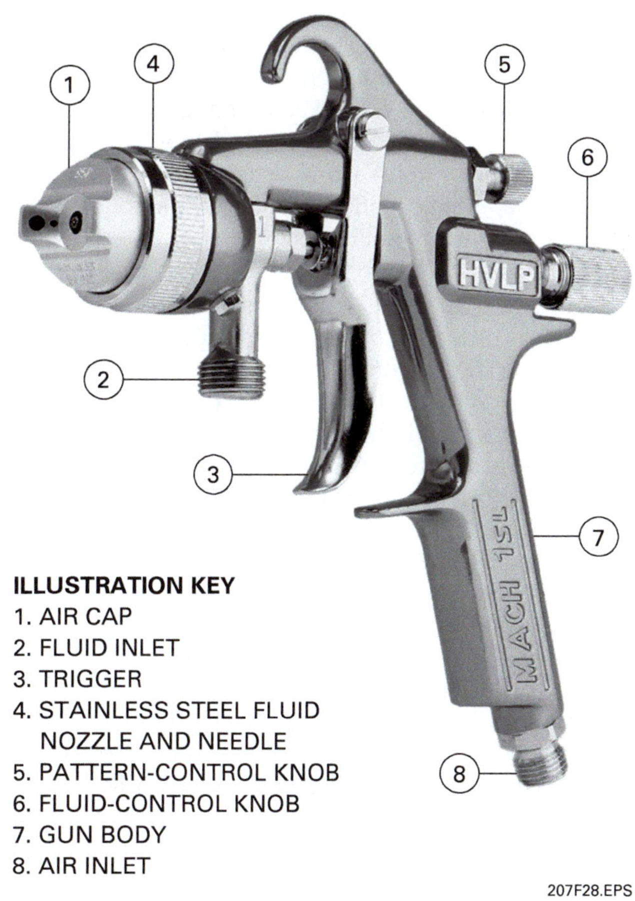

Figure 28 Typical HVLP spray gun.

cess of 10 psi (0.07 Mpa) or 50 psi (0.3 MPa) gun air inlet pressure. In these instances, a test gauge (*Figure 29*) should be used to confirm actual air cap operating pressure. Manufacturer's instructions will help in the selection of appropriate air cap and fluid nozzle/needle sizes. *Appendix F* shows an illustrated breakdown of a typical HVLP spray gun. HVLP spray guns are lead/lag like conventional air spray guns. For HVLP systems, use the same initial setup, startup, operation, shutdown, and cleaning procedures as previously provided for conventional air spray systems.

207F29.EPS

Figure 29 HVLP air cap test gauge.

7.7.0 Troubleshooting

> **WARNING**
>
> Do not place gloved or bare hands or any other part of the body in the fluid stream from the tip of an airless or air-assisted airless spray gun. Do not aim the gun at another person. For airless and air-assisted airless spray systems, do not attempt to check for fluid hose leaks with a gloved or bare hand. Hose leaks can be nearly invisible. If a hose leak is suspected or detected, immediately shut down and depressurize the sprayer system following the manufacturer's recommended pressure-relief procedures. If the fluid stream from a gun tip or a fluid hose leak comes into contact with skin, even through a glove, see a physician immediately. Even if the skin does not appear to be broken or injured, an injection injury may have occurred. Injection can take place at pressures as low as 100 psi (0.69 MPa). If fluid is injected into any part of the body, severe injuries can result.

If there are gun leaks, the pressure must be shut off before making any effort to repair the gun. If fluid is leaking from the back of the seal assembly, look at the condition of the seal and needle shaft. If there is a leak at the front of the gun, check the valve seat and the point of the needle. If fluid is coming from the air passages, check the spray tip seal and the fluid seal.

If material blocks or builds up on the tip, cut the pressure immediately. For airless or air-assisted airless guns, do not wipe the tip or the guard while pressure is on. If the gun is triggered, the spray will come through the wipe and cause an injection injury. If the tip doesn't come clean, place the tip of the gun in a container of approved solvent or thinner, and let it soak for a few minutes. Do not put the whole gun into the solvent, as some parts (such as the grips) may be soluble. If there is some obstruction, such as dried coating in the tip, shut down the pump and clean the tip with a non-metallic brush. It is a bad idea to use a metal object to try to clear blockage on any gun. The metal of the gun is often brass or aluminum, and the carbide tip can easily be scored, producing deflected spray patterns.

Conventional, HVLP, and airless guns can be diagnosed by examining the spray pattern delivered. It is preferable to troubleshoot the pattern on a spray sample at setup, on a surface other than the project itself. However, it is important to keep an eye on the pattern as the job progresses, to prevent unnecessary rework.

7.7.1 Troubleshooting Conventional and HVLP Spray Systems

Conventional spray systems are driven by air pressure, either pushing fluid through a line from a tank or can, or siphoning fluid from a can. The cans, used for small areas, may be siphon cans under the gun, or gravity feed cans on top of the gun. The fluid is atomized and mixed with air, and essentially blown onto a surface. The air cap has two protrusions, called wings or **horns**, which blow air and material in a spread pattern.

As shown in *Table 1*, the spray pattern and other symptoms can be used to diagnose problems. Suggested solutions for these spray problems are also provided in the table.

Both types of gun are nearly the same in terms of changing tips or removing the valve mechanisms. The knobs for the control of fluid and spray are held in place by a screw in the valve assembly or they are threaded into or onto the assembly. All of the assembly can be removed by unscrewing it. The tip can be removed in both cases by unscrewing the ring and the horns then unscrewing the fluid nozzle from the head insert.

Table 1 Conventional and HVLP Spray System Diagnostics

Problem	Cause	Solution
Pattern is shaped like a crescent.	One horn is blocked, probably by dried paint.	Wash ports with thinner and blow gun clean.
Pattern is split, or heavy on the ends and weak in the middle.	Air pressure may be too high, or too wide a pattern with thin coating.	Reduce air pressure, or open supply valve, decrease spray width.
Intermittent or spitting spray.	Swivel nut loose, packing on needle valve too dry, dirt in nozzle seat.	Tighten or replace swivel nut, add a few drops of oil to packing, clean nozzle seat and body.
Pattern heavy on one end, light on the other.	Blockage of nozzle tip.	Clean nozzle tip.
Pattern too heavy in the middle.	Too much fluid pressure, not enough air, fluid too thick.	Adjust air and fluid pressure, increase spray pattern width. Thin fluid mix.
Too much overspray.	Air pressure too high, improper gun handling (too far away, too fast).	Adjust air pressure, gun handling.
Sags and runs.	Material too thin, material flow rate too high, gun stroke too slow.	Adjust gun or fluid pressure, remix, correct gun handling.
Not spraying.	Air supply loss, wrong tip and air cap, fluid pressure too low, needle valve too tight, fluid too heavy (suction feed).	Check components, adjust pressure, thin material.
Orange peel.	Surface dirt or oil, gun too close, mix problems, low air pressure.	Clean surface, correct gun handling, mixing, air pressure.
Gun spitting.	Air getting into paint stream.	Check if fluid source is empty. Tighten fluid nozzle. Check and tighten needle packing nut. Check fluid nozzle for damage.
Gun fluid pressure is too high with gun triggered (cannot achieve desired flow rate).	Using needle/nozzle kit with too small orifice.	Use needle/nozzle kit with larger orifice.
Using a low fluid pressure setting, the fluid flow is too high, making it necessary to restrict needle travel to reduce fluid flow.	Using needle/nozzle kit with too large orifice.	Use needle/nozzle kit with smaller orifice.
Fluid system will not operate at low enough pressure (below 10 psi [70kPa, 0.7 bar]).	There is no fluid regulator, or air regulator is not sensitive enough at low pressure.	Add low pressure fluid regulator, or add more sensitive low pressure air regulator.
Fluid flow is fluttering while spraying.	Fluid filter is clogged. Fluid source is empty.	Check fluid filter. Refill.
Fluid flow fades while spraying high viscosity fluids.	Air hose size is too restricted for higher air flow being used.	Use 5⁄16 in (7.9 mm), inside diameter (ID) air hose if the hose is 25 ft (7.6 m) long. If longer hose is needed, use a 3⁄8 in (9.5 mm) ID hose.

207T01.EPS

7.7.2 Airless and Air-Assisted Airless Spray System Troubleshooting

In addition to the problems mentioned in the conventional troubleshooting section, there are additional troubleshooting diagnostics specific to airless and air-assisted, airless spray systems These are shown in *Table 2*.

The airless gun has similar tip, ring, nozzle removal, and assembly procedures as conventional and HVLP guns. Again, the control assemblies simply unscrew.

8.0.0 SPECIALTY COATINGS

The application of specialty coatings requires special training and certification. However, an introduction to the basics of such systems is appropriate at this level. The three most common specialty coatings are electrostatic spray, plural-component, and thermal spray.

8.1.0 Electrostatic Spray and Powder Coatings

Electrostatic sprayers most often use powder component coating methods involving fluidized

Table 2 Airless and Air-Assisted Airless Spray System Diagnostics (1 of 2)

PUMP TROUBLESHOOTING

Problem	Cause	Solution
Does not operate.	Valve closed or clogged.	Clear air line; increase air supply. Check that valves are open.
	Fluid hose or gun obstructed.	Clean hose or gun.*
	Dried fluid on displacement rod.	Clean rod; always stop pump at bottom of stroke; keep wet-cup filled with compatible solvent.
	Air motor parts dirty, worn, or damaged.	Clean or repair air motor.
	Runaway error on DataTrak tripped.	See DataTrak Operation-Runaway.
Output low on both strokes.	Air line restricted or air supply inadequate. Valves closed or clogged.	Clear air line; increase air supply. Check that valves are open.
	Fluid hose/gun obstructed; hose ID too small.	Clear hose or gun*; use hose with larger ID.
	Air motor icing.	Open de-ice control.
Ouput low on downstroke.	Open or worn intake valve.	Clear or service intake valve.
	High viscosity fluid.	Adjust intake spacers.
Output low on upstroke.	Open or worn piston valve or packings.	Clear piston valve; replace packings.
Erratic accelerated speed.	Fluid supply exhausted, clogged suction.	Refill supply and prime pump. Clean suction tube.
	High viscosity fluid.	Reduce viscosity; adjust intake spacers.
	Open or worn piston valve packings.	Clear piston valve; replace packings.
	Open or worn intake valve.	Clear or service intake valve.
Runs sluggishly.	Possible icing.	Stop pump. If equipped, open de-ice control.
Cycles or fails to hold pressure at stall.	Worn check valves or seals.	Refer to manufacturer's instructions for service.
Air bubbles in fluid.	Loose suction line.	Tighten. Use compatible liquid thread sealant or PTFE tape on connections.

*To determine if fluid hose or gun is obstructed, follow pressure-relief procedures. Disconnect fluid hose and place a container at pump fluid outlet to catch any fluid. Turn on air power just enough to start the pump. If the pump starts, the obstruction is in the fluid hose or gun.

207T02A.EPS

beds or powder spray guns with heated surfaces and bake ovens. Non-electrostatic thermal spray guns are also used for the powder coating of non-heated surfaces, however, electrostatic fluid spray systems are also used. *Figure 30* shows such a spray system driven by an air-motor diaphragm pump and equipped with an electrostatic fluid spray gun (*Figure 31*). This system is capable of spraying medium to high solids solvent-borne or water-borne primers, and single-component or premixed plural-component coatings that are designed for electrostatic spray application. The particular gun shown for this system is self-powered with an internal air-powered turbine and DC power supply that furnishes a high- or low-voltage electrostatic charge. As coating fluid exits the nozzle orifice, a positive electrostatic charge is given to the coating fluid via the electrode in the center of the orifice. The operating fluid and air pressures for this gun are similar to conventional air spray guns. Other types of electrostatic fluid spray guns operate at fluid and air pressures similar to HVLP and air-assisted airless guns. Guns with remote power supplies are also used.

Table 2 Airless and Air-Assisted Airless Spray System Diagnostics (2 of 2)

GUN TROUBLESHOOTING

Problem	Cause	Solution
Fluid leakage from back of fluid needle.	Worn packings or needle shaft.	Replace external needle packings or entire fluid needle cartridge.
Air leakage from front of gun.	Air valve not seating properly.	Clean/service air valve.
Fluid leakage from front of gun.	Fluid needle worn or damaged. Worn diffuser-seat housing.	Replace fluid needle cartridge. Replace the diffuser-seat and gasket. The gasket must be replaced whenever the diffuser-seat is removed.
Fluid in air passages.	Fluid tip seal leaking. Leaking around diffuser-seat housing. Fluid inlet fitting leaking.	Tighten or replace fluid tip. Replace the diffuser-seat gasket. The gasket must be replaced whenever the diffuser-seat is removed. Replace the fluid fitting gasket. The gasket must be replaced whenever the fluid fitting is removed.

SPRAY PATTERN TROUBLESHOOTING

Problem	Cause	Solution
Fluttering or spitting spray.	Insufficient fluid supply. Air in paint supply line.	Adjust fluid regulator or fill fluid supply tank. Check, tighten siphon hose connections, bleed air from paint line.
Poor finish or irregular spray pattern.	Incorrect fluid pressure at gun. Fluid is too thin or too thick. Dirty, worn, or damaged spray gun.	Read fluid manufacturer's recommendations. Adjust fluid viscosity; read fluid manufacturer's recommendations. Service spray gun.
Pattern pushed to one side, same side of air gets dirty.	Air horn holes partially or totally plugged.	Clean air horn holes with solvent and soft brush.

207T02B.EPS

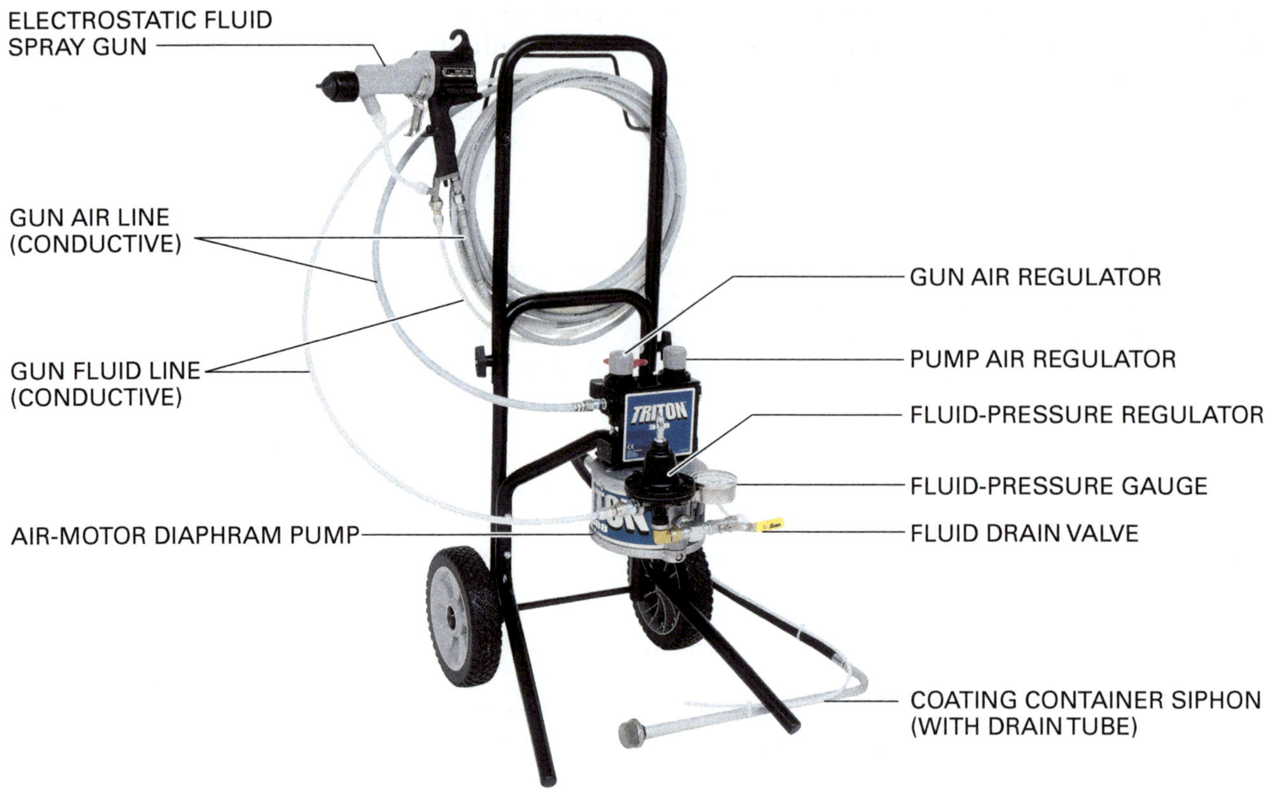

Figure 30 An airless spray system with an electrostatic gun.

In any electrostatic spraying system, the surfaces being sprayed must be a suitably conductive material, such as metal. Normally, only one layer of coating is applied because of the insulating characteristics of most coating materials. The particles of spray are electrically charged, and the object being sprayed is given the opposite charge. The particles are then drawn to the object, just as metal shavings are attracted to a magnet. Because of the charge, there is a risk of electrical arcs if any part of the operation can close a circuit that has not been properly grounded. Several precautions must be taken around this equipment, including the following:

- The operator, the floor of the booth, and every metal and potential conductor in the area must be grounded.
- Operators must remain in physical contact with the handle of the sprayer, and must be grounded. Often, the operator wears grounding leg straps, and insulating rubber-soled shoes are not permitted.
- Operators must not have any ungrounded metal objects on their bodies.

The risk of fire is high in this particular specialty, so there are a number of precautions associated with this hazard:

- Fire extinguishers must be present in the spray area, and the area must be kept clean.
- The voltage to the equipment must be turned off prior to any service or maintenance.
- The solvent used for flushing must have flash points as high as or higher than the coating itself.
- Special precautions also apply in the case of waterborne coatings, because of the additional risks involved. Water is usually a good conductor, so some hazard is involved in the splash from its use.

8.1.1 Electrostatic Spray Gun Coating

With an electrostatic spray gun, the fluid or powder is atomized and fed into a nozzle or bell that sprays it and gives it a positive charge at the same time. The target is grounded, and the charge causes the coating material to cover the surface. The particles don't attach to each other because of the like charge.

With many coatings, especially powder coatings, the target piece is heated afterwards, to melt the coating together. The result is a fairly hard coating.

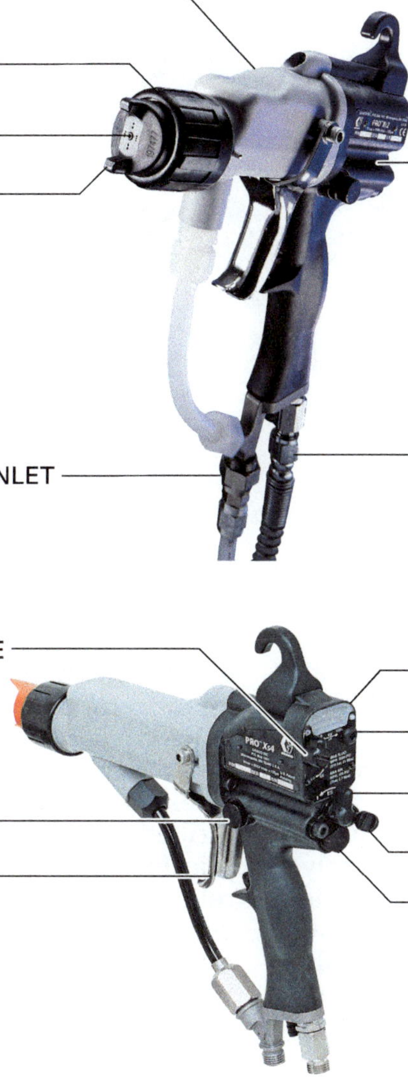

Figure 31 Electrostatic fluid spray guns.

8.1.2 Fluidized Bed Coating

The other procedure commonly used to form coatings from powder is the fluidized bed, in which a heated object is dipped into a vat of powder that has been made into a fluid. This produces a relatively thick single coat, and if the mass of the object does not retain enough heat to completely cure the coating, it can be removed to an oven for three or four minutes.

In the fluidized bed, the powder sits atop the air supply on a separating membrane. The membrane allows the air to blow the powder into a state that resembles a fluid. This allows the powder to coat the heated object completely. This technology is quite expensive in terms of equipment.

8.2.0 Plural-Component Spray Systems

Plural-component spray systems mix the components at the gun, the sprayer, or a remote manifold. If the components are mixed at the gun, the gun has multiple fluid inlet fittings. These types of airless guns are used to spray coatings with very short pot lives. The guns are available with a mechanical-purge, air-purge, or solvent-purge capability so that the tip and/or the mixing chamber can be cleaned when the gun is idle. The gun shown in *Figure 32* is a mechanical-purge type where a rod, passing through the mixing chamber when the trigger is released, clears the chamber up to the ceramic nozzle. During use, clean-off air is constantly expelled around the ceramic nozzle at

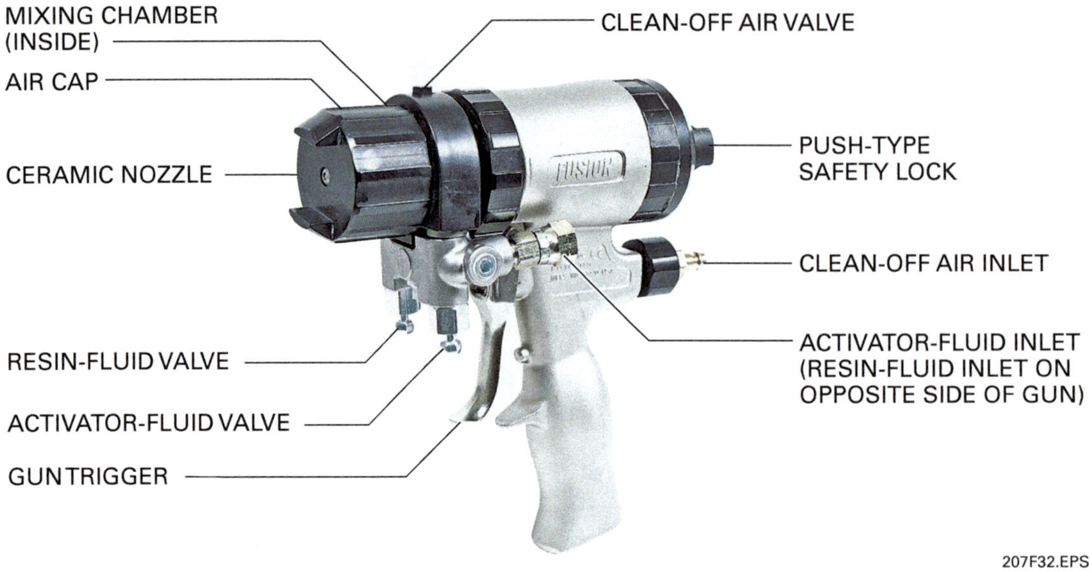

Figure 32 Mechanical-purge internal-mix plural-component airless spray gun.

the air cap to keep coating accumulation to a minimum. *Figure 33* shows a fast-set, plural-component, airless spray system used for the gun shown in *Figure 32*. The fluid components (hardener and resin) of the coating are fed from separate storage containers by transfer pumps to a hydraulically-powered proportioning pump unit, usually called a proportioner. The proportioner is set to supply the proper ratio of the fluid components for the coating. It has two electric preheaters to separately heat each fluid before they enter electrically heated fluid lines to the gun. An optional recirculation kit shown with this system allows fluids to be bypassed from the pumps back to their storage containers when the gun is idle.

Figure 34 shows another plural-component spray system that uses a standard airless spray gun for application of coatings that have more than a minute of set time. It is a mechanical proportioning system with a single air motor that operates both fluid pumps. Coating fluids, heated by band heaters on their storage containers, are pumped through heated hoses to a remote mixing manifold. The manifold allows proportioning adjustment and fluid shutoff and is shown feeding two serially connected static mixing tubes, a whip hose, and the spray gun. The static mixing tubes thoroughly mix the coating components and have replaceable mixing elements. This system is shown configured with a solvent flush capability for the remote mixing manifold, static mixing tubes, whip hose, and gun.

The preliminary procedure with the plural-component gun is to flush it with the appropriate solvent. The procedure is specific to the manufacturer of the equipment, but it is basically a matter of running the solvent through the gun and the mixing chamber. Having purged the gun, hook up the component lines from the mixer or the containers, test the pattern, and then start spraying.

If the operation is over for the night or for the weekend, prepare the gun and sprayer as described in the manufacturer's instructions. If applicable, leave the hoses hooked up and do not relieve the pressure from the containers. Whenever the spraying operation is complete, flush the gun and lines, empty the mixing chamber, and fill it with petrolatum jelly. The object is to keep the mixture from setting up hard inside the gun. Plural-component work frequently involves epoxies, and once the epoxy sets inside the gun, the gun is almost impossible to clear.

The adjustment procedure is much like that of an airless gun, but the pumps can introduce problems if one pump is working correctly and the other is not. It is also possible for the ratios to vary if one component is more viscous than the other. Read and closely follow the manufacturer's instructions for the equipment and components.

The solvent purge is often a source of VOCs. Be sure to use the solvent recommended by the manufacturer for the material being sprayed. Check the MSDS for the material and solvent, and use the correct PPE and procedures. Any negative effects from exposure to VOCs may be cumulative, taking place over a matter of years.

8.3.0 Thermal Spray

Thermal spray coating is the use of a high-heat gun to spray material in a molten state onto the target. This may be used to apply a metallic coat-

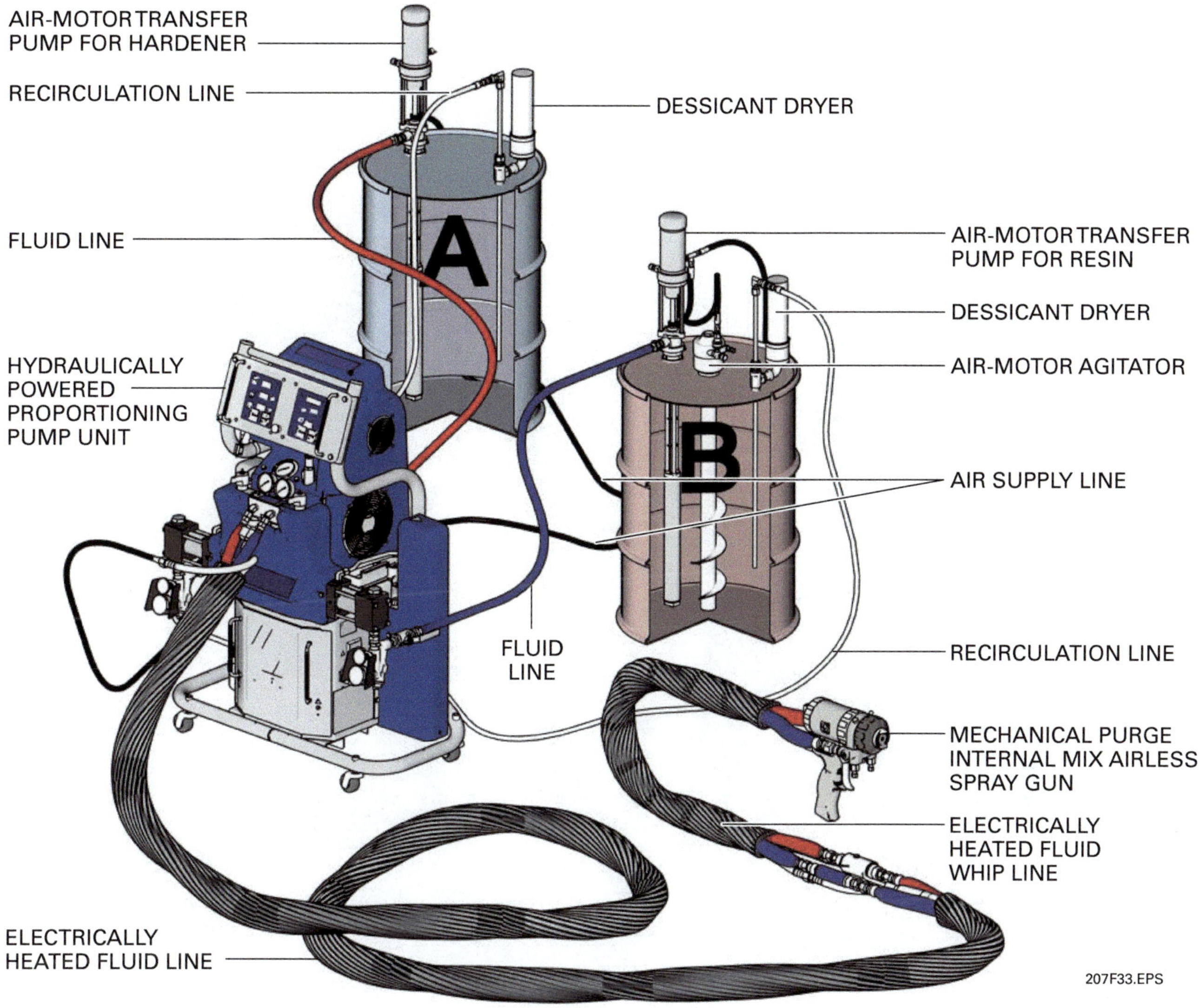

Figure 33 Fast-set plural-component airless spray system.

ing on metals and various kinds of ceramics. There are three basic types of thermal spray guns: **plasma**, combustion-flame, and two-wire electric arc. Feedstock may be powder, wire, or a rod. Plasma is the hottest and the most versatile thermal spray gun.

> **WARNING**
>
> The temperatures in a high-heat gun are hot enough to melt metals and powders of ceramic, or combinations of the two. Operation of a high-heat gun requires special training and certification.

9.0.0 Auxiliary Equipment

Environmental conditions, such as humidity and temperature, have a substantial impact on coating application operations. Sometimes, auxiliary equipment must be used inside a structure or containment to modify the ambient conditions and substrate surface temperature to provide more desirable conditions to apply industrial coatings. The substrate surface temperature, ambient air temperature, and RH affect the coating process and curing time. Air temperature and substrate surface temperature are often quite different. For solvent-based coatings, the rate of solvent evaporation and curing time can be greatly affected by the ambient conditions. When temperatures are too low, the coatings may not cure properly. When surface temperatures are too high, the coatings may not properly wet the surface and may cause problems with film formation. When the RH is too high, the solvents may take longer to evaporate and recoat time may be affected, resulting in solvent entrapment, blistering, delamination, and low adhesion. Many water-based coatings must be applied when the ambient air is above a specified minimum RH to cure properly.

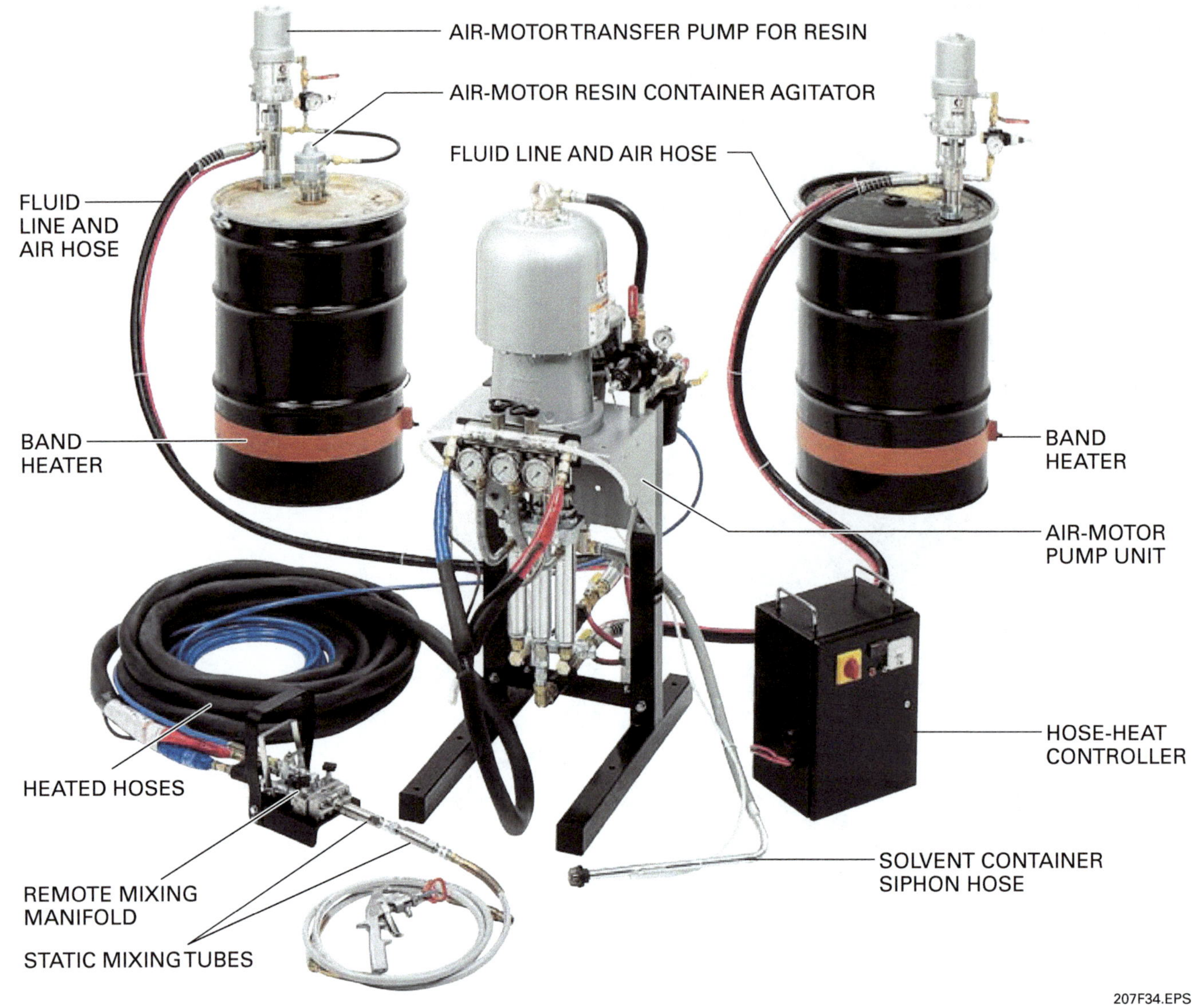

Figure 34 Quick-set plural-component airless spray system.

The PDS normally gives a range of substrate surface temperatures, ambient air temperatures, and RH that are acceptable for coating application. If any of the ranges cannot be met, auxiliary equipment must be brought in to improve conditions. This equipment includes HVAC units, humidifiers, and dehumidifiers.

9.1.0 Humidity and Ferrous Surfaces/ Substrates

In many industrial situations, the standard practice for ferrous surfaces/substrates, weather permitting, is to prepare and coat only the amount of area that can be done in one day. Contractor work crews clean and prepare an area that can then be painted before corrosion occurs (flash rusting or rust blooming), caused by high humidity or water vapor condensing on the surface. On days that the weather allows them to work, this requires contractors to change over from surface preparation operations to coating applications operations each day, with the associated cleanup involved with each operation. Since most industrial surface preparation involves abrasive blasting of some sort, this daily cycling is often referred to as a **blast and coat cycle**. The goal of this cycle is to get the coating onto the surface of the substrate before the rust blooming (flash rusting) becomes visible. The loss of surface sheen and the appearance of flash rusting are implied by the expressions "the blast has gone off" or "the blast is lost," which is sometimes used to describe this condition. Although some coatings can be applied over surfaces with rust bloom, most coatings, including metallic thermal-sprayed coatings, cannot tolerate flash rusting.

9.1.1 Humidity Control for Ferrous Surfaces/Substrates

To eliminate flash rusting, rust inhibitors must be applied to the ferrous surface or the humidity at the surface must be controlled so that the blasted surface remains clean. This is sometimes referred to as "holding the blast." Rust inhibitors can only delay flash rusting for a limited period. Many coatings are not compatible with rust inhibitors, including thermal-sprayed metallic coatings. The blast can be held for relatively long periods of time by controlling the humidity of the air in contact with the ferrous surface. This can be accomplished within confined areas by dehumidifying or conditioning the incoming air for surface preparation and coating purposes, and for worker safety and comfort. The same approach can be used for outdoor projects, but they require the use of some sort of sealed containment to isolate all, or incremental segments of, the ferrous surface from direct contact with the elements. Over an extended period of time, controlling the humidity allows completion of all the required surface preparation and cleanup first. Then continuous coating application can be accomplished, thus eliminating the daily blast and coat cycle. Some of the benefits of work area dehumidification for ferrous surfaces are as follows:

- The project can be conducted on a 24-hour around-the-clock basis, if desired. Crews can work at optimum production rates nearly any time, regardless of most weather conditions.
- Contractors can accurately predict when a project will be completed and will generally not be at the mercy of weather conditions.
- Environmental temperature and humidity control can extend the work season many months in some geographical areas. Work can start earlier in the spring and extend later in the fall, even when weather conditions fluctuate on a daily basis.
- Overlap problems from one coated surface to the next, which can occur with the daily blast and coat cycle, are eliminated.
- Dust and grit contamination of newly coated surfaces, which is a common consequence with the daily blast and coat cycle, are eliminated.
- If more than one coat is being applied, then all coats can be completed under ideal conditions. The typical two-day window between primer and final coat can be achieved under optimal conditions.

9.1.2 Determination and Selection of Dehumidification Equipment for Ferrous Surfaces/Substrates

The dehumidification solution to ferrous material corrosion problems caused by moisture is to reduce the moisture content of the air to a safe level, typically 50 percent RH or less, based on the temperature of the steel surface, not the temperature of the air flowing through the work area. The desired reduction can be determined by first measuring the temperature of the steel surface with a simple thermometer and then measuring the dry-bulb and wet-bulb temperature of the ambient air using a sling psychrometer to determine ambient air humidity. Humidity can also be measured directly with a humidity or dew point meter. A psychrometric chart can then be used along with these measurements to determine the dew point and the subsequent RH of the air in relation to the steel surface temperature.

Whatever the humidity, it is almost always possible to cool the air sufficiently to achieve saturation, and then condensation. Either mechanical refrigeration dehumidifiers, absorption-type desiccant dehumidifiers, or a combination of both are used to reduce the humidity of air for containments or confined areas.

The mechanical refrigeration process occurs in a closed, pressurized loop in which a refrigerant is circulated through a compressor, condenser, receiver, expansion valve, and evaporator. In this system, the refrigerant gas is compressed and fed through a water-cooled or air-cooled condenser. Heat is removed from the gas in the condenser, causing the gas to condense into a liquid. The liquid refrigerant passes through a receiver tank and is released through an orifice in the expansion valve to the low-pressure side of the valve where the liquid evaporates (expands) to a gas during its movement through the evaporator. Evaporation of the liquid refrigerant lowers the temperature of the refrigerant and evaporator to just above freezing, which causes heat to be extracted from the air passing through the evaporator. This causes significant cooling of the air and the condensation of moisture in the air onto the evaporator fins. The moisture, in the form of water, drains off the fins into a collection pan and then to an exterior drain port.

The absorption-type desiccant dehumidifiers operate on an entirely different principle. A water-absorbing desiccant is loaded into a honeycomb-type wheel that rotates at 8 to 10 revolutions per hour. The wheel rotates continuously between a dehumidifying sector and a regeneration sector.

In the dehumidifying sector, moisture is extracted from the incoming air and absorbed by the desiccant. The regeneration sector uses electric or fuel-fired heaters and a reverse, fan-forced flow of hot air to dry the desiccant as it rotates through the regeneration sector. In some units, the heaters can also heat the dry air coming out of the unit to raise the ambient air temperature in the working environment.

Selection of the right type of dehumidifier for a job is determined in large part by the low-temperature performance of the machine. Refrigeration dehumidifiers are used extensively, but they perform much less efficiently as temperature and humidity levels decrease during fall, winter, and spring. In some cases, the evaporator coils will freeze as water condenses on them, thus eliminating their effectiveness. For cold weather operation, these units usually requires pre-heater units to bring the incoming air to well above freezing and after-heaters to raise the air cooled by the evaporator to a level acceptable within the working area.

Absorption-type desiccant dehumidifiers have nearly the same performance during both winter and summer because the affinity of the desiccant for moisture is not greatly affected by temperature. In these units, moisture is removed in a vapor state so there is no water to potentially freeze. For fall, winter, and spring operations, the absorption-type desiccant dehumidifiers are the units used most often along with additional external heaters, if necessary, to condition the air supplied to the work area.

For summer use, the refrigeration-type dehumidifiers are used most often because they cool as well as dehumidify the air supplied to the work area. In extremely humid or fall and spring situations, refrigerant units may be supplemented with absorption-type desiccant dehumidifiers and external heaters. The power requirements for multiple large refrigerant units may require substantial on site power generation equipment.

9.2.0 Environmental Dehumidification and Conditioning Equipment

The following types and sizes of equipment are generally used for containment and confined-area dehumidification and conditioning. The selection and sizing of the equipment for a particular project in a specific geographical location is complicated and requires the services of an expert in the field of temporary workspace conditioning.

9.2.1 Desiccant Dehumidifiers

These all-temperature units *(Figure 35)* are available as fuel-fired or electric units (460 VAC, 3-phase power supply). They can be paralleled to provide dehumidification for any size project. These units typically have capacities ranging from 600 to 9,000 CFM (17 to 255 cubic meters/sec) each.

9.2.2 Refrigerant Dehumidifiers

These warm-weather units provide dehumidification as well as work-area cooling. They are available as 460 VAC units, ranging in cooling capacity from 3 to 20 tons (2.7 to 18 metric tons) each. Some units are stackable to obtain the total dehumidification/cooling required.

9.2.3 Chillers and Cooling Coils

These units are available for use individually or in conjunction with desiccant equipment. They can be paralleled together for any size project. Their use with desiccant equipment allows greater reduction of humidity and temperature within the work area. Chillers range in cooling capacity from 20 to 45 tons (70 to 158 KW). Cooling coils have capacities ranging from 1,125 to 9,000 CFM (32 to 255 cubic meters/sec).

9.2.4 Direct-Fired, Indirect-Fired, and Steam/Hot Water Heaters

Portable high-volume, direct-fired, indirect-fired, and steam/hot water heaters are available to provide heat to a work environment. Capacities range from 150,000 Btus to nearly 7 million (M) Btus (158 to 7,385 million joules [MJ]) depending on the type of unit and fuel. *Figure 36* shows a propane or natural gas direct-fired unit that is available in sizes up to 4.5 MBtus (4,748 MJ). This unit provides 100 percent makeup air for the work space. Makeup air is drawn from the outdoor air and exhausts air inside a building or containment. A disadvantage of direct-fired units is that they introduce combustion products into the air being supplied, which can cause problems for coatings. Problems for personnel can also occur if the air exchange volume is too low. *Figure 37* shows a propane or oil indirect-fired heater that is available in sizes up to 1 MBtu (1,055 MJ). It can provide any mix of makeup air as well as recirculated interior air because the combustion products are not mixed with the heated air and are vented outdoors. For very large projects, steam or hot water heaters and air handlers can be used. *Figure 38* shows a natural gas- or oil-fired steam boiler with an output of 6.9 MBtus (7,280

MJ). This unit can supply 6,902 lbs (3,131 kg) per hour of steam at up to 212°F (100°C) and pressures up to 135 psi (0.9 MPa) through insulated lines to one or more air handlers. The air handlers contain heat exchangers and blowers. The large one can heat up to 30,000 CFM of air. If the air intake is located outdoors, the air handlers can provide any mix of makeup air and recirculated air desired.

All of these units heat air for comfort but cause a lot of air turbulence that can stir up dust during coating operations. In addition, they may never bring surfaces or objects up to the proper temperatures for coating applications without making the surrounding air temperature uncomfortable. Heaters are required to have carbon monoxide monitoring and alarms.

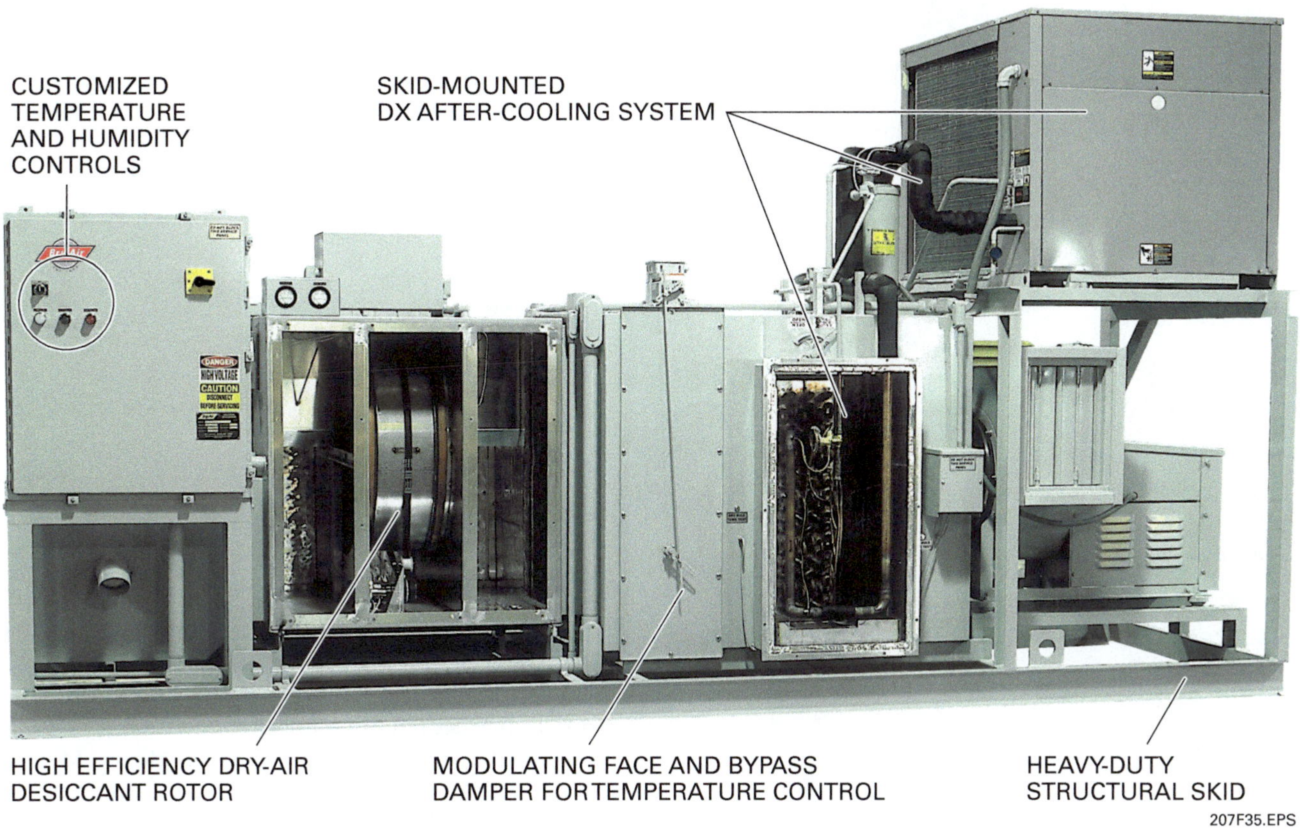

Figure 35 Desiccant dehumidifier.

Figure 36 Propane or natural gas direct-fired heater.

Figure 37 Propane or oil indirect-fired heater.

207F38.EPS

Figure 38 Natural gas- or oil-fired steam boiler.

9.2.5 Electric Heaters

Simple to operate, flame-free and compatible with desiccant dehumidifiers, electric heaters (*Figure 39*) are available with outputs of 15 to 150 kW (51,000 to 512,000 Btus). These units, like fired heaters, heat air for comfort but can cause air turbulence.

> **WARNING**
>
> Unless electric heaters are used for supplying 100 percent makeup air, flammable vapors may be ignited by the heat of electrical resistance elements. If electric heaters are used, a ground fault circuit interrupter (GFCI) must be used.

207F39.EPS

Figure 39 An electric heater.

9.2.6 Infrared Heaters

High-intensity infrared heaters emit radiant heat as electromagnetic energy in the near (bright), medium, and far (dark) infrared wavelengths. Near infrared is called bright because it is visible as a bright reddish light with a glare. Medium infrared may be visible as a faint, dull, dark red, and far (dark) infrared is generally not visible. None of the radiation directly heats air or causes air turbulence. The radiation is absorbed by objects and surfaces, causing them to heat up. The heated objects and surfaces then warm the air by convection. Near infrared is absorbed the most by steel or silica-based surfaces and is effective for bringing structural steel or concrete to the temperatures required for coating operations. The medium- and far-infrared wavelengths are good for drying applications, especially for water-based coatings.

For fixed industrial coating application purposes, fixed banks of electric or vented gas-fired tubes are usually used. For field construction operations, unvented gas-fired or electric portable units are normally operated in banks to obtain the desired level of heating.

WARNING

These heaters cannot be operated in areas where flammable vapors are or will be present. In addition, the unvented gas-fired types must be operated only in adequately ventilated areas. Prolonged exposure to these heaters can cause burns.

10.0.0 Coating Quality Control and Evaluation

The purpose of quality control inspection is to ensure that a quality coating application is obtained, as required by the coating specification. The contractor is responsible for all the quality control functions based on the coating specification. Proper quality control is the key to effective coating performance and should be the means to comply with the specification.

Because many critical coating operations are highly expensive and labor intensive, quality control is extremely important. This is the reason that inspection plays such an important role in a coating program. The quality of an applied coating is dependent on two very important items:

- The coating specification.
- Proper inspection during surface preparation. Also important are proper observation of the coating mixing, induction time, application time, recoat time, and ambient conditions for application and curing.

10.1.0 Inspection Gauges, Instruments, and Tests

A large number of destructive and nondestructive gauges, instruments, and tests are used for coating inspection. Some of these are described in the following sections.

10.1.1 WFT Gauge

Because an applicator must know the thickness of the coating being applied, WFT gauges, either metal or plastic, are used by applicators as well as inspectors. These gauges are normally calibrated on one side in mils with the other side in microns (µm). *Figure 40* shows how a WFT gauge, calibrated in microns, is used to check the WFT of a coating. The indentations in the wet coating left by a gauge must eventually be removed and overcoated.

NOTE

A WFT gauge cannot be used successfully on heavily pigmented or extremely fast drying coatings.

10.1.2 DFT Gauge

In addition to destructive DFT gauges, nondestructive DFT gauges are available to check dried/cured coating thickness on metal surfaces. As defined in *SSPC Paint Application Specification No.2 (PA-2)*, two types of gauges are used. Type 1 gauges, effective only for ferrous metal substrates, are pull-off gauges in which a permanent magnet is brought into contact with the coating surface. The force necessary to pull the magnet straight away from the surface is measured and interpreted by the gauge as the DFT value on a scale or display that, in most cases, is calibrated in both mils and microns. The pull-off force varies inversely with the thickness of the coating. Thin coatings require more force and thick coatings require less. Because the temperature of the coating and substrate affects the pull-off force, compensation charts or gauges that are specific for high, low, or normal temperature ranges must be used. *Figure 41* shows an example of a pocket-pencil version of a Type 1 pull-off gauge with a mil scale on one side and a micron scale on the other. It is accurate to ±10 percent plus 0.1 mil or 2.5 µm and is used primarily for spot checks or in narrow areas. This particular manufacturer provides gravity compensation in the pencil so that it can be used in any position. This manufacturer has gauges for normal temperature ranges and for extreme temperatures between –150°F (–100°C) to 450°F (230°C).

Figure 41 also shows a more accurate Type 1 magnet pull-off gauge that, when used on a flat surface, does not require physically lifting the gauge off of the surface. The lift-off is accomplished by a thumbwheel on the top of the gauge. This gauge is also gravity compensated and is accurate to ±1 percent. It is furnished with interchangeable gauge dials in mils or microns. Both this and the pencil gauge are typically verified for accuracy using an optional manufacturer-supplied coating thickness standard traceable to national standards (NIST). If necessary, when checking DFT on small test specimens or panels, the rear foot of the gauge must be shimmed to the same thickness as the specimen to obtain an accurate measurement.

Type 2 gauges are electronic gauges. All such gauges supplied in the U.S. are designed for ferrous metal substrates with non-magnetic coatings. With these types of gauges, the electromagnetic probe is held against the coating surface and electromagnetic induction principles are used by the gauge to detect the flux density of the magnetic field at the surface of the coating. Like pull-off gauges, flux density varies inversely with the coating thickness, and temperature compensa-

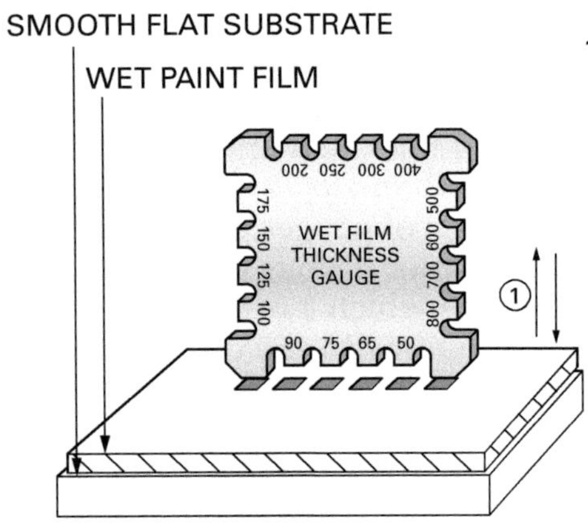

1. While holding perpendicular to the wet surface, gently press down onto the surface. The end notches should land on the substrate or previously cured primer or finish. Remove from surface without rocking the gauge. In this example, all of the notches (from 50-90μm) are covered with the coating. Clean off all residues of the coating on the gauge with a soft cloth.

VIEW OF A WFT GAUGE INSERTED INTO AND THEN REMOVED FROM A WET PAINT FILM

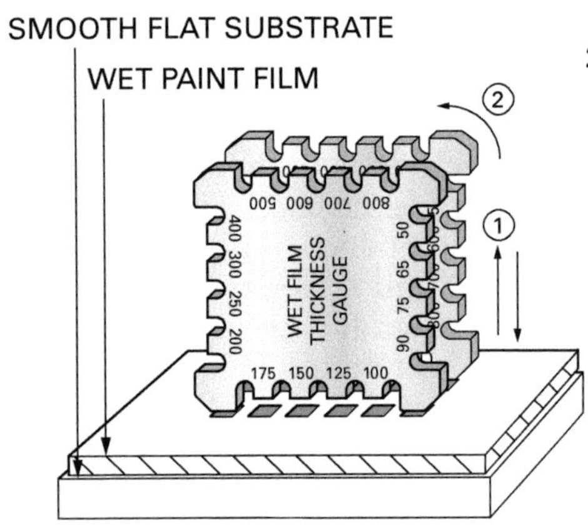

2. Rotate gauge around to the next higher set of readings. Repeat reading as per Step 1, taking the reading in close proximity to the first and second reading. Again in this example all of the notches (from 100 and 175μm) are covered with the coating. Clean off all residues of the coating with a soft cloth.

VIEW OF A WFT GAUGE INSERTED INTO AND THEN REMOVED FROM A WET PAINT FILM

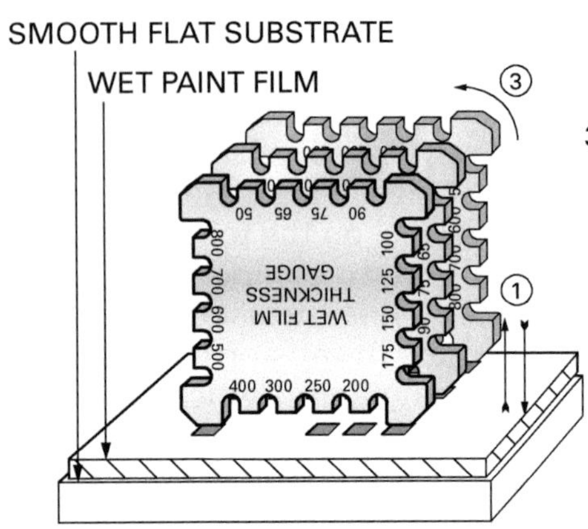

3. Rotate gauge around to the next higher set of readings. Repeat reading as per Step 1, taking the reading in close proximity to the first and second reading. In this example, the notches 200 and 250μm are covered with the coating as are the outside notches. The marks left on the coating surface show the same indents. This means that this reading is over 250μm but under 300μm. Clean off all residues of the coating with a soft cloth.

Repeat taking measurements in an area of around 1 sq/yd around the initial measurement. Average the result. This is the average WFT in this area. From this information, you should be able to determine whether the correct amount of paint is being applied to substrate.

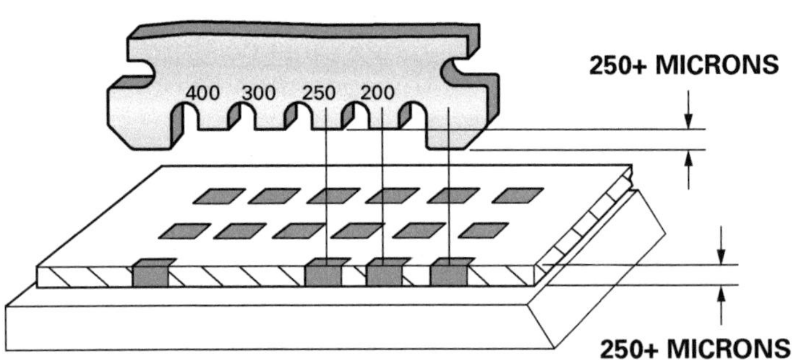

CROSS SECTIONAL VIEW OF A WFT GAUGE.
THE RESULT OF THIS MEASUREMENT IS BETWEEN 250-300 MICRONS.

207F40.EPS

Figure 40 Checking WFT with a gauge.

tion charts are usually required. *Figure 41* shows an example of a Type 2 programmable electronic gauge that can measure coating DFT for ferrous metal substrates. It is accurate to ±1 percent and covers a wide range of coating thicknesses. Ancillary software for recordkeeping and data analysis is also available. Precision plastic shims of known thicknesses are provided with the gauge. They are used over a bare metal substrate to verify the measurement range for the desired DFT checks.

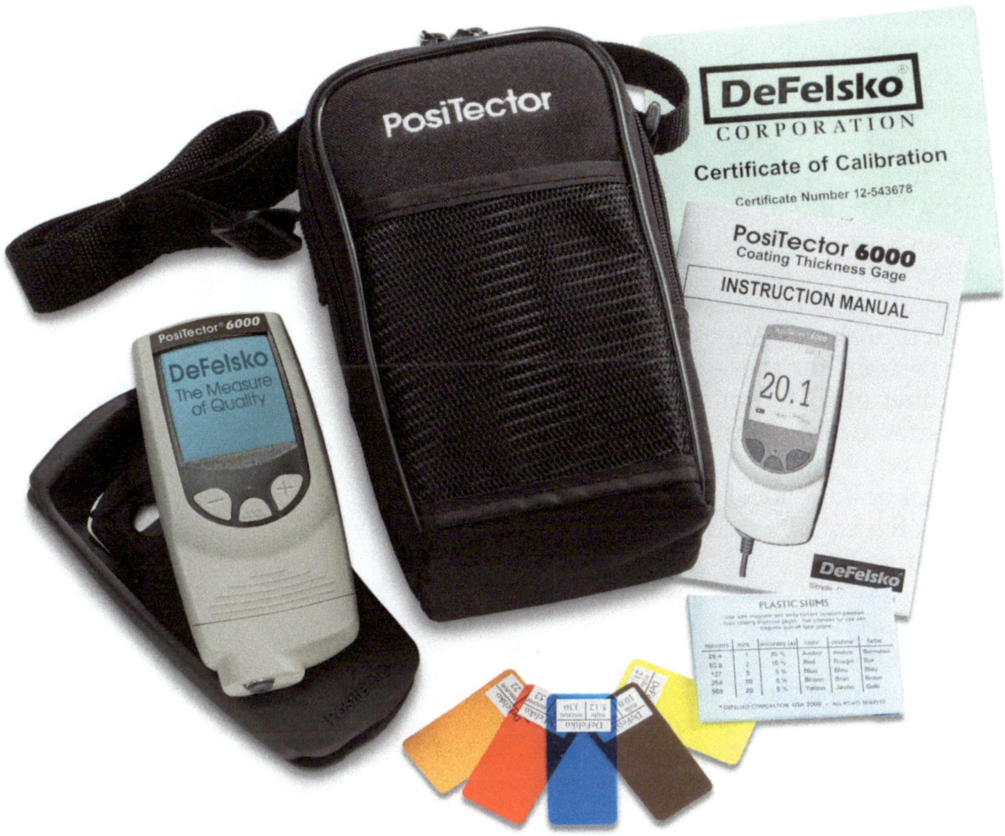

TYPE 2 ELECTRONIC GAUGE

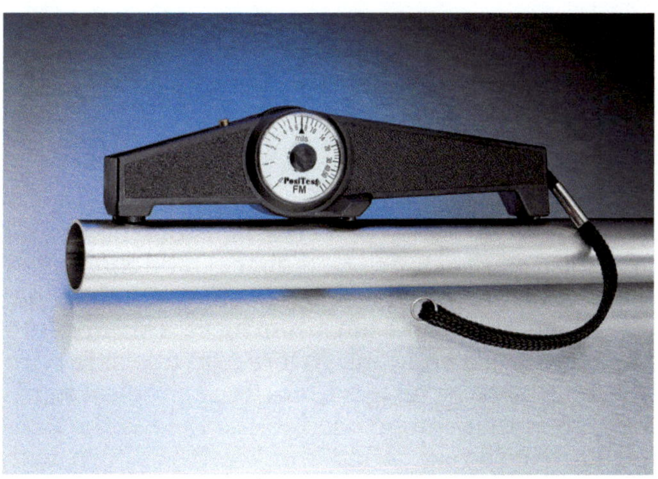

TYPE 1 MAGNETIC PULL-OFF GAUGE

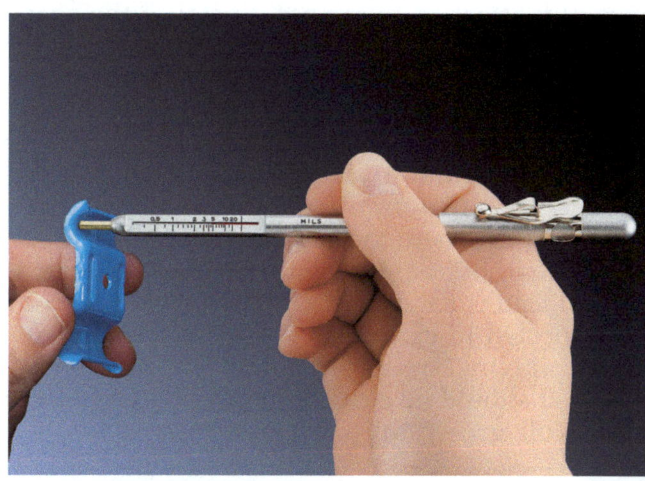

TYPE 1 MAGNETIC PENCIL PULL-OFF GAUGE

207F41.EPS

Figure 41 Type 1 and 2 DFT gauges.

10.1.3 Low- and High-Voltage Holiday/Pinhole Detectors

To achieve maximum protection of a coated metal substrate, any holidays/pinholes that are discontinuities in the coating film must be detected and eliminated. This is done to reduce the effects of environmental breakdown of the coating and/or corrosion of the substrate over time. During inspections, holidays/pinholes can be detected in a non-conductive cured film over a conductive substrate by means of low-voltage, wet-sponge detectors and high-voltage, wire brush detectors (*Figure 42*). Low-voltage detectors are normally used on coating systems with a DTF of 0 to 20 mils (0 to 508 μ). There is no shock hazard with low-voltage detectors. Coating systems greater than 20 mils (508 μ) are normally checked using high-voltage detectors, which present a shock hazard to the user. The wire brushes for these detectors are available in a number of different sizes and shapes. Coating manufacturers should be consulted to determine if a high-voltage detector can be used and, if so, what voltage level should be used. The use of excessive voltage may damage an otherwise sound coating.

WARNING

High-voltage detectors present a shock hazard and must be used as prescribed in the equipment manufacturer's instructions. Avoid any contact with the metal components of these instruments during operation. In addition, these detectors should not be used in areas containing flammable or explosive vapors.

In use, the earth ground lead of the detector is normally attached to the bare metal of the coating substrate and the wet sponge or wire brush is drawn over the coating. Discontinuities are signaled by an alarm on the detector. In addition, sparking may be observed at the location of the detected discontinuity. For existing above-ground pipelines or uncovered sections of existing buried pipelines where the substrate metal is not accessible without damaging the coating, the earth ground lead is connected to a driven ground rod or connected to bare cable laid in water adjacent to the pipeline. In all cases, the coating area to be tested should be cleaned of any contaminants such as oil, grease, dust, dirt, or debris.

NOTE

Holiday/pinhole detectors can usually be used to check coatings applied over concrete substrates because of the normally conductive water content of concrete.

LOW-VOLTAGE DETECTOR

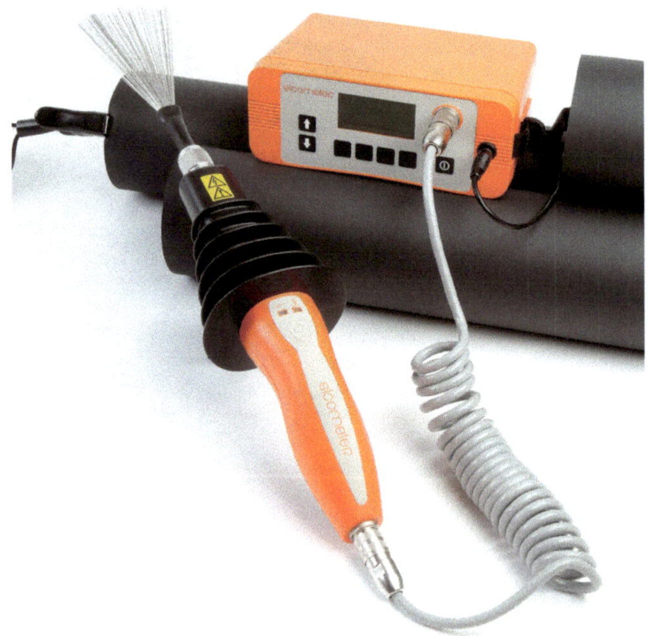

HIGH-VOLTAGE DETECTOR

207F42.EPS

Figure 42 Low- and high-voltage holiday/pinhole detectors.

10.1.4 Hardness Testers

Hardness resistance of a coating is one of several mechanical resistance tests that can be conducted on a coating. All hardness testers are destructive to coatings and most hardness tests are conducted on test coupons covered on one side with the coating to be tested. One of the simplest hardness tests that can be used in the field is the pencil hardness test called the Wolff-Wilborn test. It is typically conducted with a test kit that can be used in accordance with the requirements in *ASTM D3363*. The kit includes a wheeled tester that holds a special pencil. The wheeled tester is designed to maintain a constant force of 7.5N

(Newtons) on the point of the pencil lead and to maintain the correct angle for the pencil, which increases the reproducibility of the test. The kit also contains a set of special pencils in 14 hardnesses from 6B to 6H along with two special pencil sharpeners and a sandpaper block that are used to prepare the lead points of the pencils. The ASTM test is conducted on a horizontal surface, and the tester, equipped with the hardest pencil (6H) first, is pushed forward from the rear a distance of about 0.25 inch (3 mm). Then, the movement path is visually inspected for scratches or gouges made by the pencil. The test is progressively repeated, spaced about 0.5 inch (12 mm) away from each previous test, using the next softest pencil each time. The end point for the ASTM tests is as follows:

- Gouge hardness = The hardest pencil that leaves the coating uncut over a 0.25 inch (3 mm) length
- Scratch hardness = The hardest pencil that will not scratch the coating

The Barcol impressor is an indentation hardness tester that meets *ASTM D2583* for use with rubber, plastic, or fiberglass-reinforced linings and soft metals. It measures the depth of an indentation in Barcol impresser scale numbers. In some applications, conversion tables can be used to translate the Barcol scale numbers to other common hardness scales. Manufacturer-supplied calibration disks are used to calibrate the tester at the extreme ends of its measurement range. In use, the indenter point must always be perpendicular to the surface being tested. Normally, test panels of the linings are used when taking a measurement. If necessary, both sides of the support leg at the rear of the tester must be shimmed to make the point perpendicular. To make a measurement, press firmly on the tester until the indicator stops changing and note the Barcol number at the peak reading indication. On softer materials, the indication may normally fall off somewhat from the peak reading. Make sure not to damage the precision indenter point by hitting or dragging it on any surface.

10.1.5 Solvent Resistance Test

Solvent resistance tests (also called solvent rub tests) are conducted by soaking a cotton ball or rag in an appropriate solvent (sometimes a MEK solvent) and rubbing an area of the coating surface back and forth (called a double rub) a given number of times. The effect on the surface is then evaluated in terms of gloss loss, softening, or degree of rub-off as compared to a control standard for the coating that is known to be fully cured and that has been subjected to the same test with the same solvent by the same operator. With this method, a good qualitative determination of the degree of cure can be made. This is important for coatings such as epoxies, vinyl esters, polyesters, alkyds, and urethanes because they should reach certain levels of solvent resistance prior to being topcoated and prior to being placed into service. However, the level of solvent resistance by itself does not necessarily indicate full cure, and some coatings become solvent-resistant before they become sufficiently cured for service. The time to reach a specific level of solvent resistance can be altered by temperature, film thickness, air movement, and, for water-borne or water-reactive coatings, humidity. The test solvent effects on the coating varies considerably with the coating type and the solvent used. Coating manufacturers may specify the solvent, the number of double rubs, and the specific test results needed.

As defined in the *ASTM D5402* standard, solvent rub tests are commonly used to determine the extent of curing for organic coatings that undergo chemical change during the cure process. No specific number of rubs is defined by the standard. The *ASTM D4752* standard is used to determine the extent of cure for ethyl silicate (inorganic) zinc-rich primers using a 50 double rub test and a MEK solvent.

10.2.0 Overcoating and Coating Repair

Most coatings require a drying and curing time to achieve a solvent-free and solvent-resistant film. The time varies by the type and composition of

On Site

Gauges for Non-Ferrous Substrates

In addition to the ferrous DFT measurements, many models of electronic gauges are available that can perform DFT measurements on non-ferrous metal substrates using eddy currents at the surface of the coating. Other similar electronic gauges using ultrasound principles are available to check DFT on non-metallic substrates such as masonry, concrete, wood, and plastic.

the coating. Drying and curing time is normally dependent on humidity, temperature, and film thickness. Manufacturers normally list drying and curing times in a schedule on the container label and in the PDS for a coating material. The times are given for a certain humidity and film thickness at a number of various ambient temperatures. The times relate to the period required before the coating can be touched, handled, recoated (maximum and minimum times) as well as the length of time required for total cure. If a coating is recoated before the specified recoating time, the previous coat may soften and lift, or solvent entrapment, delamination (delayering), blistering, and other failures may occur. If too much time has occurred between coats, the surface of the previous coat may be too hard to permit adhesion, and delamination or other failures may occur. In this case, the previous coating should be allowed to fully cure and overcoating, if necessary, must be done in accordance with the manufacturer's instructions.

A problem that can occur with coatings is that a blush or white coating appears on coated surfaces shortly after application. It is caused by water in coating films that have been applied during periods of cool temperatures or high humidity (in excess of 85 percent RH). If allowed by specification, a slower-drying solvent or retarder may help to reduce the amount of blushing; however, for best results, coating operations should be postponed or containment with environmental control should be used. Coatings using polyamine, and to lesser extent, polyamide curing agents, tend to develop amine blush if moisture accumulates on the surface or high humidity exists while the curing takes place. Amine blush is a hazy or oily residue on the surface caused by unreacted amine reacting with carbon dioxide and moisture to form a carbamate. The reaction is aggravated by increased film thickness and excess carbon dioxide from portable gas heaters. This type of blushing may not be readily apparent and can interfere with adhesion of succeeding coats if not detected and/or removed.

Amine blush can only be positively detected by laboratory analysis. To perform the laboratory analysis, surface scrapings of the coating surface at spot locations across the surface must be submitted to the laboratory. However, amine blush can also be detected on surfaces using an amine blush check kit. These are very sensitive kits that are pH-based and reliably detect amine blush. Unfortunately, they may also give a positive indication for other types of surface contamination.

Because amine blush is readily soluble in water, pressure washing or wiping using water with a mild detergent along with clear water rinsing can remove it. As a result, it may be simpler to just clean amine-cured coating surfaces before recoating them.

Generally, when overcoating a fully cured new coating or an older existing coating with surface discontinuities (pinholes, holidays, delamination, blisters, corrosion, etc.), the coating must be repaired and prepared for the new coating layer. This involves masking off the entire area to be overcoated, removing any discontinuities down to the substrate by chemical or mechanic means, and feathering the edges of these areas back into the existing sound coating. The spot repairs are cleaned and primed as required. Then, the existing coating is usually abraded to allow adhesion of the overcoating material and solvent, and/or pressure washed to remove any contaminants. The entire existing coating is primed, if required, and then overcoated.

10.3.0 Coating Application Failures and Remedies

Common coating application failures that occur during or shortly after application, along with their remedies, are summarized in *Table 3*. Most of these are detectable by inspection at the time of application or during the curing time after application completion.

11.0.0 Waste Management

A waste stream is any waste material generated at a particular location. The waste may be hazardous or non-hazardous. Waste needs to be disposed of, and depending on the type of waste, disposal can be costly. The first step in waste management is to control it at its source. Minimizing waste decreases costs in two ways. First, it saves money on materials. Second, it saves money on disposal fees. Minimizing waste is also good for the environment. Waste deliberately or accidentally released into the environment can pollute water and air, kill animals, destroy vegetation, and threaten human health. In addition, hefty fines may be associated with improper waste disposal. The following guidelines should be followed for proper waste control and disposal:

- Use only the amount of coating that is needed.
- Place recyclables in their proper container.
- Never mix hazardous and non-hazardous waste.
- Use care not to release waste into the environment.

Table 3 Common Coating Application Failures

Description	Cause	Remedy
SAGS. Also called runs or curtains. Excess flow of paint.	Spray gun too close to work; too much thinner; too much paint; or surface too hard or glossy to hold paint.	Before cure, brush out excess paint and modify spray conditions. After cure, sand and apply another coat.
ORANGE PEEL. Hills, valleys in paint resembling skin of orange.	Paint too viscous; gun too close to surface; solvent evaporated too fast; or air pressure too low for proper atomization.	Before cure, brush out excess paint and modify spray conditions. After cure, sand and apply another coat.
OVERSPRAY. Also called dry spray. Dry, flat, pebbly surface.	Particles reaching surface not wet enough to level because of too rapid solvent evaporation; gun too far from surface; or paint particles falling outside spray pattern.	Before cure, remove by dry brushing followed by solvent wiping. After cure, sand and apply another coat.
COBWEBBING. Thin, stringy paint, spider web-like particles.	Solvent evaporating too rapidly. Most common with fast evaporating lacquers, such as vinyls and chlorinated rubbers.	Use a slower evaporating solvent or apply when cooler.
CRATERING. Also called pitting. Small, uniform indentations in film.	Air pockets trapped in wet film during spraying.	Sand or blast to smooth finish and apply additional coats.
FISH EYES. Separation or pulling apart of wet film to expose underlying finish or substrate.	Application over oil, dirt, silicone, or incompatible coating.	Sand or blast remove; brush-apply a fresh coat plus topcoat.
BLUSHING. Flat finish with milky appearance.	Moisture condensation in high humidity with fast evaporating or unbalanced thinner in spray application.	Sand or blast remove; respray with retarder added to thinner.
UNEVEN GLOSS. Nonuniform sheen, shiny spots.	Nonuniform film thickness; moisture in film; temperature change during curing; or paint applied over soft or wet undercoat.	Allow to dry and apply another finish coat under acceptable conditions for moisture and humidity.
WRINKLING. Rough, crinkled surface.	Surface skinning over uncured paint because of too much thickness and/or too warm weather, especially with oil-base paints.	Scrape off wrinkles and apply thinner coat; avoid intense sunlight.
BLISTERING. Small to large broken or unbroken bubbles.	Solvent entrapment; oil, moisture or salt-contaminated surfaces; or cathodic disbonding.	Put blowers in enclosed areas to accelerate solvent release; clean surface contamination; provide proper levels of cathodic protection.
PINHOLING. Tiny, deep holes exposing substrate.	Insufficient paint spray atomization; coarse atomization; or settled pigment. Mist coat not applied to zone surface before final coat.	If uncured, brush out and apply additional coat. If cured, apply additional coat.

207T03.EPS

11.1.0 Non-Hazardous Waste Disposal

Non-hazardous waste is defined in accordance with company policy and local regulations. Always dispose of the non-hazardous waste as described by company policy and in compliance with applicable regulations.

11.2.0 Hazardous Waste Disposal

Hazardous waste is dangerous or potentially dangerous to human health or the environment. Hazardous wastes can be liquids, solids, contained gases, or sludge. They can be the byproducts of manufacturing processes or discarded products, such as solvents. These products need to be disposed of in accordance with the local, regional, or national law. Local or regional law can be stricter than national laws, but cannot be more lenient. The product MSDS indicates disposal methods in accordance with national law.

For example, the United States EPA has determined that a number of wastes are hazardous. These are defined in EPA document *Title 40, Protection of Environment Part 261—Identification and Listing of Hazardous Waste.* Other wastes can also be classified as hazardous, such as those that are capable of spontaneous combustion, those with a low flash point, like some solvents, and those that are toxic, like some paints.

Any person or site creating hazardous waste is called a generator. Generators may be large or small, based on how much waste they produce per month. Some generators produce a very small amount of hazardous waste each month. These companies may be exempt from following the rules that large and small generators need to follow, but they still need to properly dispose of hazardous waste. All large and small quantity generators are required to do the following:

- Obtain an EPA identification number from the state or authority having jurisdiction.
- Comply with the manifest system.
- Handle wastes properly before shipment; for example, packaging, labeling, marking, and accumulation time. Waste must not be mixed, including separate types of hazardous waste.
- Comply with record keeping and reporting requirements.

It is very important to know what waste is hazardous. Hazardous waste must be placed in the appropriate container in accordance with applicable regulations. This waste must be disposed of by a contractor authorized to dispose of it. Sometimes waste needs to be tested to determine its exact contents.

Should a hazardous material ever be spilled in an area, check the MSDS for cleanup instructions. Local government agencies may require notification of a spill, depending on the material and how large an area is involved. Some states may have more stringent rules about handling hazardous materials and cleaning up a spill. Check with a supervisor if there are any questions.

SUMMARY

This module described the environmental conditions required for applying coatings, quality control and inspection required during coating application, and coating application failures and remedies. Specialty spray systems and the application methods for these systems were discussed. Equipment descriptions and application techniques were provided for conventional air spray, HVLP, airless, and air-assisted airless spray equipment. The equipment and techniques to mix various types of coatings were also covered. The information found in coating specifications and on the PDS is vital to proper coating application.

Review Questions

1. TLV for a coating defines the coating's _____.
   a. vapor concentration needed for ignition and explosion
   b. amount and concentration of airborne vapor that can be safely inhaled during a specified period
   c. highest vapor concentration to which workers can be exposed over a stated period of time
   d. threshold vapor limit to which workers can be exposed for a specific work period

2. If foaming occurs after a coating is shaken, the coating should be _____.
   a. immediately remixed
   b. allowed to sit
   c. applied immediately
   d. discarded

3. The physical properties of a coating material are determined by _____.
   a. the EPA
   b. SSPC
   c. the manufacturer
   d. ASTM International

4. The coating film must _____ to its solid state.
   a. flow
   b. corrode
   c. oxidize
   d. dry or cure

5. Brush or roller application should only be used when the _____.
   a. areas are very large
   b. surfaces are flat
   c. PDS permits
   d. job is big enough

6. The WFT for a product can be found in the _____.
   a. MSDS
   b. PDS
   c. OSHA
   d. EPA

7. If any conflicts between specifications and the PDS arise, _____.
   a. they must be resolved
   b. the PDS takes priority
   c. follow the manufacturer's suggestions
   d. they are not important

8. A manufacturer's document that recommends application methods is a(n) _____.
   a. PDS
   b. MSDS
   c. coating specification
   d. quality control document

9. Coatings for steel normally require a surface temperature at least _____.
   a. equal to the ambient air temperature
   b. 5°F above the dew point
   c. 5°F below the dew point
   d. 50°F above the dew point

10. The RH can be calculated by using a(n) _____.
    a. anemometer
    b. single thermometer
    c. sling psychrometer
    d. manometer

11. Areas around bolts and corners can be accurately coated with a _____.
    a. squeegee
    b. roller
    c. brush
    d. trowel

12. The majority of coating failures are due to faulty surface preparation or application.
    a. True
    b. False

13. To test for the presence of oil or grease on concrete, apply water to the surface, and see if it _____.
    a. evaporates
    b. beads
    c. runs off
    d. steams

14. To evaluate existing coatings, check the film thickness, the condition of the substrate, and the _____.
    a. surface temperature
    b. psychrometer reading
    c. quality of adhesion
    d. MSDS

15. The time required from the beginning of the component mixing process until applying the coating is called _____.
    a. pot life
    b. introduction
    c. curing
    d. sweat-in time

16. Epoxies have some of the most complex coating _____.
    a. ingredients
    b. preparation requirements
    c. restrictions
    d. materials

17. Viscosity is affected by the amount of solvent the coating material contains and the _____.
    a. pigment color
    b. humidity when mixed
    c. temperature of the material
    d. viscometer aperture

18. The most common types of viscosity cups are _____.
    a. SAE and metric
    b. Ford, Zahn, and DIN
    c. ISO and imperial
    d. Number 4 and Number 2

19. The compressed air supply for a sprayer must be _____.
    a. at least 1,500 psi (103.42 bar)
    b. at least 100°F (37.78°C)
    c. clean, dry, and free of oil
    d. at full pressure before hookup

20. The general sequence for coating surfaces is to coat the _____.
    a. whole surface at once
    b. edges, inside corners, and outside corners first
    c. flat parts first
    d. edges last

21. Curing time is affected by the _____.
    a. color of the coating
    b. temperature of the substrate surface
    c. method of application
    d. specifications

22. When ambient temperatures are too high, coatings may _____.
    a. not evaporate the solvent
    b. run or sag
    c. not properly wet the surface
    d. cure too slowly

23. The daily cycle of coating work on ferrous metals is called the _____.
    a. recycling coat
    b. curing cycle
    c. ambient cycle
    d. blast and coat cycle

24. The appearance of flash rust can be described as _____.
    a. rust sheen
    b. "the blast is lost"
    c. "the blasted surface"
    d. the blast is found

25. The blast surface can be retained for a relatively long period by controlling the _____.
    a. humidity
    b. exposure to sunlight
    c. traffic in the work area
    d. environmental contamination

26. The two basic types of dehumidifiers are mechanical refrigerant and _____.
    a. quantum electronic
    b. RH control
    c. absorption-type desiccant
    d. natural dehydration

27. Refrigerant dehumidifiers are most commonly used in _____.
    a. tank work
    b. cold weather
    c. fall, winter, and spring
    d. summer

28. If allowed by the specification, a method to reduce blushing is to _____.
    a. use a fast-drying solvent
    b. use a retarder
    c. raise the RH
    d. lower the ambient temperature

29. A repair procedure for coating sags is to _____.
    a. brush out excess paint, and after cure, sand and apply another coat
    b. sand or blast to a smooth finish and apply additional coats
    c. scrape off and apply thinner coats while avoiding intense sunlight
    d. sand or blast to remove, and respray coating with a fast-drying thinner

30. A waste stream is _____.
    a. unauthorized dumping of waste into a stream
    b. any waste generated at a particular location
    c. hazardous material from industrial work
    d. a polluted stream near an industrial area

Trade Terms Introduced in This Module

Atomizing: Reducing to fine particles or spray.

Blast and coat cycle: A daily cycling between surface preparation operations and coating applications operations.

Horn: One of two protrusions that blows air and material in a spread pattern from the nozzle of a conventional spray gun.

Plasma: An ionized gas created by passing a gas such as argon, nitrogen, helium, or hydrogen, through an electric arc. The atomic components of the gas break apart and then recombine, producing a tremendous amount of energy.

Short-term exposure limit (STEL): The highest concentration of an airborne material to which workers can be exposed over a prescribed period of time.

Threshold limit value (TLV): The upper limit for the concentration of solvents in the air.

Time-weighted average (TWA): The amount of time a worker can be exposed to a material over a certain period—usually given in hours allowed to work per day or week.

Upper explosive limit (UEL): The highest amount of vapor concentration in the air that could explode. Vapor concentrations above the UEL have displaced so much oxygen that there is not enough oxygen in the air to explode.

Appendix A

Material Safety Data Sheets

MATERIAL SAFETY DATA SHEET

B65W511
04 00

Section 1 -- PRODUCT AND COMPANY IDENTIFICATION

PRODUCT NUMBER

B65W511

HMIS CODES	
Health	2
Flammability	2
Reactivity	0

PRODUCT NAME
POLY-LON* 1900 Polyester Polyurethane (Part A), Extra White

MANUFACTURER'S NAME
THE SHERWIN-WILLIAMS COMPANY
101 Prospect Avenue N.W.
Cleveland, OH 44115

EMERGENCY TELEPHONE NO.
(216) 566-2917

DATE OF PREPARATION
01-APR-07

INFORMATION TELEPHONE NO.
(216) 566-2902

Section 2 -- COMPOSITION/INFORMATION ON INGREDIENTS

% by WT	CAS No.	INGREDIENT	UNITS	VAPOR PRESSURE
11	110-43-0	Methyl n-Amyl Ketone		
		ACGIH TLV	50 ppm	3.855 mm
		OSHA PEL	100 ppm	
18	108-65-6	1-Methoxy-2-Propanol Acetate		
		ACGIH TLV	Not Available	1.8 mm
		OSHA PEL	Not Available	
3	1332-58-7	Kaolin		
		ACGIH TLV	2 mg/m3 as Resp. Dust	
		OSHA PEL	10 mg/m3 Total Dust	
		OSHA PEL	5 mg/m3 Respirable Fraction	
35	13463-67-7	Titanium Dioxide		
		ACGIH TLV	10 mg/m3 as Dust	
		OSHA PEL	10 mg/m3 Total Dust	
		OSHA PEL	5 mg/m3 Respirable Fraction	

Section 3 -- HAZARDS IDENTIFICATION

ROUTES OF EXPOSURE
INHALATION of vapor or spray mist.
EYE or SKIN contact with the product, vapor or spray mist.

EFFECTS OF OVEREXPOSURE
EYES: Irritation.
SKIN: Prolonged or repeated exposure may cause irritation.
INHALATION: Irritation of the upper respiratory system.
May cause nervous system depression. Extreme overexposure may result in unconsciousness and possibly death.

SIGNS AND SYMPTOMS OF OVEREXPOSURE
Headache, dizziness, nausea, and loss of coordination are indications of excessive exposure to vapors or spray mists.
Redness and itching or burning sensation may indicate eye or excessive skin exposure.

Continued on page 2

207A01A.EPS

MEDICAL CONDITIONS AGGRAVATED BY EXPOSURE

May cause allergic respiratory and/or skin reaction in susceptible persons or sensitization. This effect may be delayed several hours after exposure.

CANCER INFORMATION

For complete discussion of toxicology data refer to Section 11.

Section 4 -- FIRST AID MEASURES

EYES: Flush eyes with large amounts of water for 15 minutes. Get medical attention.

SKIN: Wash affected area thoroughly with soap and water. Remove contaminated clothing and launder before re-use.

INHALATION: If any breathing problems occur during use, LEAVE THE AREA and get fresh air. If problems remain or occur later, IMMEDIATELY get medical attention.

INGESTION: Do not induce vomiting. Get medical attention immediately.

Section 5 -- FIRE FIGHTING MEASURES

FLASH POINT	LEL	UEL
102 F PMCC	1.1	13.1

FLAMMABILITY CLASSIFICATION

Combustible, Flash above 99 and below 200 F

EXTINGUISHING MEDIA

Carbon Dioxide, Dry Chemical, Foam

UNUSUAL FIRE AND EXPLOSION HAZARDS

Closed containers may explode when exposed to extreme heat.

Application to hot surfaces requires special precautions.

During emergency conditions overexposure to decomposition products may cause a health hazard. Symptoms may not be immediately apparent. Obtain medical attention.

SPECIAL FIRE FIGHTING PROCEDURES

Full protective equipment including self-contained breathing apparatus should be used.

Water spray may be ineffective. If water is used, fog nozzles are preferable. Water may be used to cool closed containers to prevent pressure build-up and possible autoignition or explosion when exposed to extreme heat.

Section 6 -- ACCIDENTAL RELEASE MEASURES

STEPS TO BE TAKEN IN CASE MATERIAL IS RELEASED OR SPILLED

Remove all sources of ignition. Ventilate the area.

Remove with inert absorbent.

Section 7 -- HANDLING AND STORAGE

STORAGE CATEGORY

DOL Storage Class II

Continued on page 3

207A01B.EPS

B65W511 page 3

PRECAUTIONS TO BE TAKEN IN HANDLING AND STORAGE

Contents are COMBUSTIBLE. Keep away from heat and open flame. Consult NFPA Code. Use approved Bonding and Grounding procedures.

Keep container closed when not in use. Transfer only to approved containers with complete and appropriate labeling. Do not take internally. Keep out of the reach of children.

Section 8 -- EXPOSURE CONTROLS/PERSONAL PROTECTION

PRECAUTIONS TO BE TAKEN IN USE

NO PERSON SHOULD USE THIS PRODUCT, OR BE IN THE AREA WHERE IT IS BEING USED, IF THEY HAVE CHRONIC (LONG-TERM) LUNG OR BREATHING PROBLEMS OR IF THEY EVER HAD A REACTION TO ISOCYANATES.

Use only with adequate ventilation.

Avoid contact with skin and eyes. Avoid breathing vapor and spray mist.

Wash hands after using.

This coating may contain materials classified as nuisance particulates (listed "as Dust" in Section 2) which may be present at hazardous levels only during sanding or abrading of the dried film. If no specific dusts are listed in Section 2, the applicable limits for nuisance dusts are ACGIH TLV 10 mg/m3 (total dust), 3 mg/m3 (respirable fraction), OSHA PEL 15 mg/m3 (total dust), 5 mg/m3 (respirable fraction).

VENTILATION

Local exhaust preferable. General exhaust acceptable if the exposure to materials in Section 2 is maintained below applicable exposure limits. Refer to OSHA Standards 1910.94, 1910.107, 1910.108.

RESPIRATORY PROTECTION

Where overspray is present, a positive pressure air supplied respirator (TC19C NIOSH/MSHA approved) should be worn. If unavailable, a properly fitted organic vapor/particulate respirator approved by NIOSH/MSHA for protection against materials in Section 2 may be effective. Follow respirator manufacturer's directions for use. Wear the respirator for the whole time of spraying and until all vapors and mists are gone. NO PERSONS SHOULD BE ALLOWED IN THE AREA WHERE THIS PRODUCT IS BEING USED UNLESS EQUIPPED WITH THE SAME RESPIRATOR PROTECTION RECOMMENDED FOR THE PAINTERS.

When sanding or abrading the dried film, wear a dust/mist respirator approved by NIOSH/MSHA for dust which may be generated from this product, underlying paint, or the abrasive.

PROTECTIVE GLOVES

To prevent skin contact, wear gloves which are recommended by glove supplier for protection against materials in Section 2.

EYE PROTECTION

Wear safety spectacles with unperforated sideshields.

OTHER PROTECTIVE EQUIPMENT

Use barrier cream on exposed skin.

OTHER PRECAUTIONS

This product must be mixed with other components before use. Before opening the packages, READ AND FOLLOW WARNING LABELS ON ALL COMPONENTS.

Intentional misuse by deliberately concentrating and inhaling the contents can be harmful or fatal.

Continued on page 4

207A01C.EPS

B65W511 page 4

Section 9 -- PHYSICAL AND CHEMICAL PROPERTIES

PRODUCT WEIGHT	11.78 lb/gal	1411 g/l
SPECIFIC GRAVITY	1.42	
BOILING POINT	284 - 308 F	140 - 153 C
MELTING POINT	Not Available	
VOLATILE VOLUME	47 %	
EVAPORATION RATE	Slower than ether	
VAPOR DENSITY	Heavier than air	
SOLUBILITY IN WATER	N.A.	

VOLATILE ORGANIC COMPOUNDS (VOC Theoretical - As Packaged)

3.56 lb/gal	427 g/l	Less Water and Federally Exempt Solvents
3.56 lb/gal	427 g/l	Emitted VOC

Section 10 -- STABILITY AND REACTIVITY

STABILITY -- Stable

CONDITIONS TO AVOID
None known.

INCOMPATIBILITY
None known.

HAZARDOUS DECOMPOSITION PRODUCTS
By fire: Carbon Dioxide, Carbon Monoxide

HAZARDOUS POLYMERIZATION
Will not occur

Section 11 -- TOXICOLOGICAL INFORMATION

CHRONIC HEALTH HAZARDS

No ingredient in this product is an IARC, NTP or OSHA listed carcinogen.

Prolonged overexposure to solvent ingredients in Section 2 may cause adverse effects to the liver, urinary and blood forming systems.

Persons sensitive to isocyanates will experience increased allergic reaction on repeated exposure.

Rats exposed to titanium dioxide dust at 250 mg./m3 developed lung cancer, however, such exposure levels are not attainable in the workplace.

Reports have associated repeated and prolonged overexposure to solvents with permanent brain and nervous system damage.

TOXICOLOGY DATA

CAS No.	Ingredient Name				
110-43-0	Methyl n-Amyl Ketone				
		LC50	RAT	4HR	Not Available
		LD50	RAT		1670 mg/kg
108-65-6	1-Methoxy-2-Propanol Acetate				
		LC50	RAT	4HR	Not Available
		LD50	RAT		8500 mg/kg
1332-58-7	Kaolin				
		LC50	RAT	4HR	Not Available
		LD50	RAT		Not Available
13463-67-7	Titanium Dioxide				
		LC50	RAT	4HR	Not Available
		LD50	RAT		Not Available

Continued on page 5

207A01D.EPS

B65W511 page 5

Section 12 -- ECOLOGICAL INFORMATION

ECOTOXICOLOGICAL INFORMATION
No data available.

Section 13 -- DISPOSAL CONSIDERATIONS

WASTE DISPOSAL METHOD
Waste from this product may be hazardous as defined under the Resource Conservation and Recovery Act (RCRA) 40 CFR 261.
Waste must be tested for ignitability to determine the applicable EPA hazardous waste numbers.
Incinerate in approved facility. Do not incinerate closed container. Dispose of in accordance with Federal, State/Provincial, and Local regulations regarding pollution.

Section 14 -- TRANSPORT INFORMATION

US Ground (DOT)
May be Classed as a Combustible Liquid for U.S. Ground.
UN1263, PAINT, 3, PG III, (ERG#128)

Bulk Containers may be Shipped as:
UN1263, PAINT, COMBUSTIBLE LIQUID, PG III, (ERG#128)

Canada (TDG)
May be Classed as a Combustible Liquid for Canadian Ground.
UN1263, PAINT, CLASS 3, PG III, (ERG#128)

IMO
UN1263, PAINT, CLASS 3, PG III, (39 C c.c.), EmS F-E, S-E

Section 15 -- REGULATORY INFORMATION

SARA 313 (40 CFR 372.65C) SUPPLIER NOTIFICATION

CAS No.	CHEMICAL/COMPOUND	% by WT	% Element

No ingredients in this product are subject to SARA 313 (40 CFR 372.65C) Supplier Notification.

CALIFORNIA PROPOSITION 65
WARNING: This product contains chemicals known to the State of California to cause cancer and birth defects or other reproductive harm.
TSCA CERTIFICATION
All chemicals in this product are listed, or are exempt from listing, on the TSCA Inventory.

Continued on page 6

207A01E.EPS

B65W511 page 6

Section 16 -- OTHER INFORMATION

This product has been classified in accordance with the hazard criteria of the Canadian Controlled Products Regulations (CPR) and the MSDS contains all of the information required by the CPR.

The above information pertains to this product as currently formulated, and is based on the information available at this time. Addition of reducers or other additives to this product may substantially alter the composition and hazards of the product. Since conditions of use are outside our control, we make no warranties, express or implied, and assume no liability in connection with any use of this information.

207A01F.EPS

MATERIAL SAFETY DATA SHEET

B65V500
07 00

Section 1 -- PRODUCT AND COMPANY IDENTIFICATION

PRODUCT NUMBER

B65V500

HMIS CODES	
Health	2*
Flammability	1
Reactivity	1

PRODUCT NAME
POLY-LON* 1900 Polyester Polyurethane Hardener (Part B)

MANUFACTURER'S NAME
THE SHERWIN-WILLIAMS COMPANY
101 Prospect Avenue N.W.
Cleveland, OH 44115

EMERGENCY TELEPHONE NO.
(216) 566-2917

DATE OF PREPARATION
01-APR-07

INFORMATION TELEPHONE NO.
(216) 566-2902

Section 2 -- COMPOSITION/INFORMATION ON INGREDIENTS

% by WT	CAS No.	INGREDIENT	UNITS	VAPOR PRESSURE
0.7	822-06-0	Hexamethylene Diisocyanate (max.)		
		ACGIH TLV	0.005 ppm	0.05 mm
		OSHA PEL	Not Available	
100	28182-81-2	Hexamethylene Diisocyanate Polymer		
		ACGIH TLV	Not Available	
		OSHA PEL	Not Available	

Section 3 -- HAZARDS IDENTIFICATION

ROUTES OF EXPOSURE
EYE or SKIN contact with product.

EFFECTS OF OVEREXPOSURE
EYES: Irritation.
SKIN: Prolonged or repeated exposure may cause irritation.
INHALATION: Irritation of the upper respiratory system.

SIGNS AND SYMPTOMS OF OVEREXPOSURE
Redness and itching or burning sensation may indicate eye or excessive skin exposure.

MEDICAL CONDITIONS AGGRAVATED BY EXPOSURE
May cause allergic respiratory and/or skin reaction in susceptible persons or sensitization. This effect may be delayed several hours after exposure.

CANCER INFORMATION
For complete discussion of toxicology data refer to Section 11.

Continued on page 2

207A01G.EPS

Section 4 -- FIRST AID MEASURES

EYES: Flush eyes with large amounts of water for 15 minutes. Get medical attention.

SKIN: Wash affected area thoroughly with soap and water.

INHALATION: If any breathing problems occur during use, LEAVE THE AREA and get fresh air. If problems remain or occur later, IMMEDIATELY get medical attention.

INGESTION: Do not induce vomiting. Get medical attention immediately.

Section 5 -- FIRE FIGHTING MEASURES

FLASH POINT	LEL	UEL
>200 F PMCC	N.A.	N.A.

FLAMMABILITY CLASSIFICATION
Not Applicable

EXTINGUISHING MEDIA
Carbon Dioxide, Dry Chemical, Foam

UNUSUAL FIRE AND EXPLOSION HAZARDS
Closed containers may explode (due to the build-up of pressure) when exposed to extreme heat.

During emergency conditions overexposure to decomposition products may cause a health hazard. Symptoms may not be immediately apparent. Obtain medical attention.

SPECIAL FIRE FIGHTING PROCEDURES
Full protective equipment including self-contained breathing apparatus should be used.

Water spray may be ineffective. If water is used, fog nozzles are preferable. Water may be used to cool closed containers to prevent pressure build-up and possible autoignition or explosion when exposed to extreme heat.

Section 6 -- ACCIDENTAL RELEASE MEASURES

STEPS TO BE TAKEN IN CASE MATERIAL IS RELEASED OR SPILLED
All personnel in the area should be protected as in Section 8.

Cover spill with absorbent material. Deactivate spilled material with a 10% ammonium hydroxide solution (household ammonia). After 10 minutes, collect in open containers and add more ammonia. Cover loosely. Wash spill area with soap and water.

Section 7 -- HANDLING AND STORAGE

STORAGE CATEGORY
DOL Storage Class IIIB

PRECAUTIONS TO BE TAKEN IN HANDLING AND STORAGE
Keep container closed when not in use. Transfer only to approved containers with complete and appropriate labeling. Do not take internally. Keep out of the reach of children.

Continued on page 3

207A01H.EPS

===

Section 8 -- EXPOSURE CONTROLS/PERSONAL PROTECTION

PRECAUTIONS TO BE TAKEN IN USE

NO PERSON SHOULD USE THIS PRODUCT, OR BE IN THE AREA WHERE IT IS BEING USED, IF THEY HAVE CHRONIC (LONG-TERM) LUNG OR BREATHING PROBLEMS OR IF THEY EVER HAD A REACTION TO ISOCYANATES.

Use only with adequate ventilation.

Avoid contact with skin and eyes. Avoid breathing vapor and spray mist.

Wash hands after using.

This coating may contain materials classified as nuisance particulates (listed "as Dust" in Section 2) which may be present at hazardous levels only during sanding or abrading of the dried film. If no specific dusts are listed in Section 2, the applicable limits for nuisance dusts are ACGIH TLV 10 mg/m3 (total dust), 3 mg/m3 (respirable fraction), OSHA PEL 15 mg/m3 (total dust), 5 mg/m3 (respirable fraction).

VENTILATION

Local exhaust preferable. General exhaust acceptable if the exposure to materials in Section 2 is maintained below applicable exposure limits. Refer to OSHA Standards 1910.94, 1910.107, 1910.108.

RESPIRATORY PROTECTION

Where overspray is present, a positive pressure air supplied respirator (TC19C NIOSH/MSHA approved) should be worn. If unavailable, a properly fitted organic vapor/particulate respirator approved by NIOSH/MSHA for protection against materials in Section 2 may be effective. Follow respirator manufacturer's directions for use. Wear the respirator for the whole time of spraying and until all vapors and mists are gone. NO PERSONS SHOULD BE ALLOWED IN THE AREA WHERE THIS PRODUCT IS BEING USED UNLESS EQUIPPED WITH THE SAME RESPIRATOR PROTECTION RECOMMENDED FOR THE PAINTERS.

When sanding or abrading the dried film, wear a dust/mist respirator approved by NIOSH/MSHA for dust which may be generated from this product, underlying paint, or the abrasive.

PROTECTIVE GLOVES

To prevent skin contact, wear gloves which are recommended by glove supplier for protection against materials in Section 2.

EYE PROTECTION

Wear safety spectacles with unperforated sideshields.

OTHER PROTECTIVE EQUIPMENT

Use barrier cream on exposed skin.

OTHER PRECAUTIONS

This product must be mixed with other components before use. Before opening the packages, READ AND FOLLOW WARNING LABELS ON ALL COMPONENTS.

===

Section 9 -- PHYSICAL AND CHEMICAL PROPERTIES

PRODUCT WEIGHT	9.41 lb/gal	1127 g/l
SPECIFIC GRAVITY	1.13	
BOILING POINT	Not Applicable	
MELTING POINT	Not Available	
VOLATILE VOLUME	0 %	
EVAPORATION RATE	N.A.	
VAPOR DENSITY	N.A.	
SOLUBILITY IN WATER	N.A.	

Continued on page 4

207A01I.EPS

VOLATILE ORGANIC COMPOUNDS (VOC Theoretical - As Packaged)

0.00 lb/gal	0 g/l	Less Water and Federally Exempt Solvents
0.00 lb/gal	0 g/l	Emitted VOC

Section 10 -- STABILITY AND REACTIVITY

STABILITY -- Stable

CONDITIONS TO AVOID

None known.

INCOMPATIBILITY

Contamination with Water, Alcohols, Amines and other compounds which react with isocyanates, may result in dangerous pressure in, and possible bursting of, closed containers.

HAZARDOUS DECOMPOSITION PRODUCTS

By fire: Carbon Dioxide, Carbon Monoxide

HAZARDOUS POLYMERIZATION

Will not occur

Section 11 -- TOXICOLOGICAL INFORMATION

CHRONIC HEALTH HAZARDS

No ingredient in this product is an IARC, NTP or OSHA listed carcinogen.

Persons sensitive to isocyanates will experience increased allergic reaction on repeated exposure.

TOXICOLOGY DATA

CAS No.	Ingredient Name				
822-06-0	Hexamethylene Diisocyanate (max.)				
		LC50	RAT	4HR	Not Available
		LD50	RAT		738 mg/kg
28182-81-2	Hexamethylene Diisocyanate Polymer				
		LC50	RAT	4HR	Not Available
		LD50	RAT		Not Available

Section 12 -- ECOLOGICAL INFORMATION

ECOTOXICOLOGICAL INFORMATION

No data available.

Section 13 -- DISPOSAL CONSIDERATIONS

WASTE DISPOSAL METHOD

Waste from this product is not hazardous as defined under the Resource Conservation and Recovery Act (RCRA) 40 CFR 261.

Incinerate in approved facility. Do not incinerate closed container. Dispose of in accordance with Federal, State/Provincial, and Local regulations regarding pollution.

Continued on page 5

207A01J.EPS

B65V500 page 5

Section 14 -- TRANSPORT INFORMATION

US Ground (DOT)
Not Regulated for Transportation.

Canada (TDG)
Not Regulated for Transportation.

IMO
Not Regulated for Transportation.

Section 15 -- REGULATORY INFORMATION

SARA 313 (40 CFR 372.65C) SUPPLIER NOTIFICATION

CAS No.	CHEMICAL/COMPOUND	% by WT	% Element

No ingredients in this product are subject to SARA 313 (40 CFR 372.65C) Supplier Notification.

TSCA CERTIFICATION
All chemicals in this product are listed, or are exempt from listing, on the TSCA Inventory.

Section 16 -- OTHER INFORMATION

This product has been classified in accordance with the hazard criteria of the Canadian Controlled Products Regulations (CPR) and the MSDS contains all of the information required by the CPR.

The above information pertains to this product as currently formulated, and is based on the information available at this time. Addition of reducers or other additives to this product may substantially alter the composition and hazards of the product. Since conditions of use are outside our control, we make no warranties, express or implied, and assume no liability in connection with any use of this information.

207A01K.EPS

Appendix B

Product Data Sheet

Sherwin Williams

Industrial & Marine Coatings

5.23

POLY-LON® 1900

POLYESTER POLYURETHANE

Part A	**B65-500**	**Series**
Part B	**B65V500**	**Hardener**

PRODUCT INFORMATION

Revised 6/05

Product Description

POLY-LON 1900 POLYESTER POLYURETHANE is a heavy duty, high performance, two component, exterior/interior, VOC compliant, high solids, polyester-aliphatic urethane. When properly cured, this dries to a super tough, "wet look", high gloss, flexible finish with maximum gloss retention, color retention, and chalk resistance. Designed to withstand aggressive industrial environments and provide excellent durability against severe weather conditions, prolonged exterior exposure, abrasion, impact, and general chemical attack.

- Retains its exterior appearance over a wide range of chemical, weather, and mechanical conditions.
- Superior exterior color and gloss retention.

Recommended Uses

For use over prepared metal and masonry surfaces in industrial environments such as:

- Tank exteriors
- Pipelines
- Structural steel
- Bridges
- Marine vessels
- Rolling stock
- Conveyors
- Refineries
- Walls
- Floors

Conforms to AWWA D102-03 OCS #5 & #6.

- Suitable for use in USDA inspected facilities.

Product Characteristics

Finish: High Gloss

Color: Wide range of colors available

Volume Solids: 65% ± 2%, mixed, may vary by color

Weight Solids: 76% ± 2%, mixed, may vary by color

VOC (EPA Method 24): mixed

Unreduced:	<340 g/L; 2.80 lb/gal
Reduced 10%:	<388 g/L; 3.23 lb/gal

May vary by Color

Mix Ratio: 3:1 by volume, 4 gallon mix

Recommended Spreading Rate per coat:

Wet mils:	3.0 - 4.5
Dry mils:	2.0 - 3.0
Coverage:	360 - 545 sq ft/gal approximate

NOTE: Brush or roll application may require multiple coats to achieve maximum film thickness and uniformity of appearance.

Drying Schedule @ 3.0 mils wet @ 50% RH:

	@50°F	@ 77°F	@100°F
To touch:	16 hours	2 hours	30 minutes
To handle:	24 hours	10 hours	2 hours
To recoat:			
minimum:	24 hours	12 hours	2 hours
maximum:	3 days	48 hours	24 hours
To cure:	7 days	7 days	5 days

If maximum recoat time is exceeded, abrade surface before recoating. Drying time is temperature, humidity, and film thickness dependent.

Pot Life:	5 hours	4 hours	45 minutes

Sweat-in-Time: None required

Shelf Life: 12 months, unopened
Store indoors at 40°F to 100°F.

Flash Point: 102°F TCC, mixed

Reducer/Clean Up: Reducer #132, R7K132

Performance Characteristics

System Tested: (unless otherwise indicated)
Substrate: Steel
Surface Preparation: SSPC-SP10
2 cts. Epolon II Primer @ 2.5 mils dft/ct
1 ct. Poly-Lon 1900 @ 2.0 mils dft

Abrasion Resistance:
Method: ASTM D4060, CS17 wheel, 1000 cycles, 1 kg load
Result: 76 mg loss (average of 5 trials)

Adhesion:
Method: ASTM D3359 Method B
Result: 5B, 100% Retention
Method: ASTM D4541
Result 1200 psi

Direct Impact Resistance:
Method: ASTM D2794
Result: 100 in. lb.

Dry Heat Resistance:
Method: ASTM D2485
Result: 200°F, 250°F intermittent

Exterior Durability:
Method: 2 years at 45° South
Result: Excellent, 87% gloss retention

Flexibility: (urethane only)
Method: ASTM D522, 180° bend, 1/4" mandrel
Result: Passes

Humidity Resistance:
Method: ASTM D4585, 100°F, 2000 hours
Result: No blistering, cracking, softening or delamination

Pencil Hardness:
Method: ASTM D3363
Result: 6H

Salt Fog Resistance:
Method: ASTM B117, 1000 hours
Result: Rating 10 per ASTM D610 for rusting, less than 1/16" creepage at scribe. No blistering, cracking, softening, or delamination of the film.

Meets the requirements of SSPC Paint No. 36, Levels 2 & 3

continued on back

207A02A.EPS

Industrial & Marine Coatings

SHERWIN WILLIAMS

5.23

POLY-LON® 1900

POLYESTER POLYURETHANE

PART A	B65-500	SERIES
PART B	B65V500	HARDENER

PRODUCT INFORMATION

RECOMMENDED SYSTEMS

Steel:
1-2 cts. Epolon II Epoxy Primer @ 2.0 - 4.0 mils dft/ct
1-2 cts. Poly-Lon 1900 Polyester Polyurethane @ 2.0 - 3.0 mils dft/ct

Steel:
1 ct. Zinc Clad II Plus @ 3.0 - 5.0 mils dft
1 ct. Epolon II Epoxy Primer @ 2.0 - 4.0 mils dft
1-2 cts. Poly-Lon 1900 Polyester Polyurethane @ 2.0 - 3.0 mils dft/ct

Steel:
1 ct. Epoxy Mastic Aluminum II @ 6.0 mils dft
1-2 cts. Poly-Lon 1900 Polyester Polyurethane @ 2.0 - 3.0 mils dft/ct

Galvanizing:
1-2 cts. Epolon II Epoxy Primer @ 2.0 - 4.0 mils dft/ct
1-2 cts. Poly-Lon 1900 Polyester Polyurethane @ 2.0 - 3.0 mils dft/ct

Concrete/Masonry:
1 ct. Kem Cati-Coat HS Epoxy Filler/Sealer @ 10.0 - 20.0 mils dft
1-2 cts. Poly-Lon 1900 Polyester Polyurethane @ 2.0 - 3.0 mils dft/ct

The systems listed above are representative of the product's use. Other systems may be appropriate.

SURFACE PREPARATION

Surface must be clean, dry, and in sound condition. Remove all oil, dust, grease, dirt, loose rust, and other foreign material to ensure adequate adhesion.

Refer to product Application Bulletin for detailed surface preparation information.

Minimum recommended surface preparation:

* Iron & Steel:	SSPC-SP6/NACE 3, 1-2 mils profile
* Galvanizing:	SSPC-SP1
* Concrete & Masonry:	SSPC-SP13/NACE 6, or ICRI 03732, CSP 1-3

* Primer required

TINTING

Tint Part A with 844 Colorant at 200% tint strength. Five minutes minimum mixing on a mechanical shaker is required for complete mixing of color.

APPLICATION CONDITIONS

Temperature:	40°F minimum, 100°F maximum (air, surface, and material) At least 5°F above dew point
Relative humidity:	75% maximum

Refer to product Application Bulletin for detailed application information.

ORDERING INFORMATION

Packaging:	1 gallon mix:	4 gallon mix:
Part A:	.75 gallons	3 gallons
Part B:	1 quart	1 gallon
(premeasured components)		

Weight per gallon: 11.4 ± 0.2 lb mixed, may vary with color

SAFETY PRECAUTIONS

Refer to the MSDS sheet before use.

Published technical data and instructions are subject to change without notice. Contact your Sherwin-Williams representative for additional technical data and instructions.

DISCLAIMER

The information and recommendations set forth in this Product Data Sheet are based upon tests conducted by or on behalf of The Sherwin-Williams Company. Such information and recommendations set forth herein are subject to change and pertain to the product offered at the time of publication. Consult your Sherwin-Williams representative to obtain the most recent Product Data Information and Application Bulletin.

WARRANTY

The Sherwin-Williams Company warrants our products to be free of manufacturing defects in accord with applicable Sherwin-Williams quality control procedures. Liability for products proven defective, if any, is limited to replacement of the defective product or the refund of the purchase price paid for the defective product as determined by Sherwin-Williams. NO OTHER WARRANTY OR GUARANTEE OF ANY KIND IS MADE BY SHERWIN-WILLIAMS, EXPRESSED OR IMPLIED, STATUTORY, BY OPERATION OF LAW OR OTHERWISE, INCLUDING MERCHANTABILITY AND FITNESS FOR A PARTICULAR PURPOSE.

207A02B.EPS

Industrial & Marine Coatings

5.23A

POLY-LON® 1900

POLYESTER POLYURETHANE

PART A	B65-500	SERIES
PART B	B65V500	HARDENER

APPLICATION BULLETIN

Revised 6/05

SURFACE PREPARATION

Surface must be clean, dry, and in sound condition. Remove all oil, dust, grease, dirt, loose rust, and other foreign material to ensure adequate adhesion.

Iron & Steel

Remove all oil and grease from surface by Solvent Cleaning per SSPC-SP1. Minimum surface preparation is Commercial Blast Cleaning per SSPC-SP6/NACE 3. For better performance, use Near White Metal Blast Cleaning per SSPC-SP10/NACE 2. Blast clean all surfaces using a sharp, angular abrasive for optimum surface profile (1-2 mils). Prime any bare steel the same day as it is cleaned or before flash rusting occurs.

Galvanized Steel

Allow to weather a minimum of six months prior to coating. Solvent Clean per SSPC-SP1. When weathering is not possible, or the surface has been treated with chromates or silicates, first Solvent Clean per SSPC-SP1 and apply a test patch. Allow paint to dry at least one week before testing adhesion. If adhesion is poor, brush blasting per SSPC-SP7 is necessary to remove these treatments. Rusty galvanizing requires a minimum of Hand Tool Cleaning per SSPC-SP2, prime the area the same day as cleaned or before flash rusting occurs.

Poured Concrete

New

For surface preparation, refer to SSPC-SP13/NACE 6, or ICRI 03732, CSP 1-3. Surfaces must be clean, dry, sound and offer sufficient profile to achieve adequate adhesion. Minimum substrate cure is 28 days at 75°F. Remove all form release agents, curing compounds, salts, efflorescence, laitance, and other foreign matter by sandblasting, shotblasting, mechanical scarification, or suitable chemical means. Refer to ASTM D4260. Rinse thoroughly to achieve a final pH between 8.0 and 10.0. Allow to dry thoroughly prior to coating.

Old

Surface preparation is done in much the same manner as new concrete, however, if the concrete is contaminated with oils, grease, chemicals, etc., they must be removed by cleaning with a strong detergent. Refer to ASTM D4258. Form release agents, hardeners, etc. must be removed by sandblasting, shotblasting, mechanical scarification, or suitable chemical means. If surface deterioration presents an unacceptably rough surface, Kem Cati-Coat HS Epoxy Filler/Sealer is recommended to patch and resurface damaged concrete. Fill all cracks, voids and bugholes with ArmorSeal Crack Filler.

Always follow the standard methods listed below:

ASTM D4258 Standard Practice for Cleaning Concrete.
ASTM D4259 Standard Practice for Abrading Concrete.
ASTM D4260 Standard Practice for Etching Concrete.
ASTM F1869 Standard Test Method for Measuring Moisture Vapor Emission Rate of Concrete.
SSPC-SP 13/Nace 6 Surface Preparation of Concrete.
ICRI 03732, Concrete Surface Preparation

APPLICATION CONDITIONS

Temperature:	40°F minimum, 100°F maximum (air, surface, and material) At least 5°F above dew point
Relative humidity:	75% maximum

APPLICATION EQUIPMENT

The following is a guide. Changes in pressures and tip sizes may be needed for proper spray characteristics. Always purge spray equipment before use with listed reducer. Any reduction must be compatible with the existing environmental and application conditions.

Reducer/Clean Up .... Reducer #132, R7K132

Airless Spray

Pressure .... 2400 - 3000 psi
Hose .... 3/8" ID
Tip .... .013" - .017"
Filter .... 60 mesh
Reduction .... As needed up to 10% by volume

Conventional Spray

Gun .... Binks 95
Cap 63P
Tip 66
Atomization Pressure ... 50 - 60 psi
Fluid Pressure .... 20 - 30 psi
Reduction .... As needed up to 10% by volume

Brush

Brush .... Natural Bristle
Reduction .... Not recommended

Roller

Cover .... 1/4" woven with phenolic core
Reduction .... Not recommended

If specific application equipment is not listed above, equivalent equipment may be substituted.

Polyurethane 5.23A

continued on back

207A02C.EPS

Industrial & Marine Coatings

SHERWIN WILLIAMS

5.23A

POLY-LON® 1900

POLYESTER POLYURETHANE

PART A	B65-500	SERIES
PART B	B65V500	HARDENER

APPLICATION BULLETIN

APPLICATION PROCEDURES

Surface preparation must be completed as indicated.

Mix contents of each component thoroughly with power agitation. Make certain no pigment remains on the bottom of the can. Then combine three parts by volume of Part A with one part by volume of Part B. Thoroughly agitate the mixture with power agitation. Allow the material to sweat-in as indicated. Re-stir before using.

If reducer is used, add only after both components have been thoroughly mixed, after sweat-in.

Apply paint at the recommended film thickness and spreading rate as indicated below:

Recommended Spreading Rate per coat:

Wet mils:	3.0 - 4.5
Dry mils:	2.0 - 3.0
Coverage:	360 - 545 sq ft/gal approximate

NOTE: Brush or roll application may require multiple coats to achieve maximum film thickness and uniformity of appearance.

Drying Schedule @ 3.0 mils wet @ 50% RH:

	@50°F	@ 77°F	@100°F
To touch:	16 hours	2 hours	30 minutes
To handle:	24 hours	10 hours	2 hours
To recoat:			
minimum:	24 hours	12 hours	2 hours
maximum:	3 days	48 hours	24 hours
To cure:	7 days	7 days	5 days

If maximum recoat time is exceeded, abrade surface before recoating. Drying time is temperature, humidity, and film thickness dependent.

Pot Life:	5 hours	4 hours	45 minutes

Sweat-in-Time: none required

Application of coating above maximum or below minimum recommended spreading rate may adversely affect coating performance.

PERFORMANCE TIPS

Stripe coat all crevices, welds, and sharp angles to prevent early failure in these areas.

When using spray application, use a 50% overlap with each pass of the gun to avoid holidays, bare areas, and pinholes. If necessary, cross spray at a right angle.

Spreading rates are calculated on volume solids and do not include an application loss factor due to surface profile, roughness or porosity of the surface, skill and technique of the applicator, method of application, various surface irregularities, material lost during mixing, spillage, overthinning, climatic conditions, and excessive film build.

Excessive reduction of material can affect film build, appearance, and adhesion.

Do not apply the material beyond recommended pot life.

Do not mix previously catalyzed material with new.

In order to avoid blockage of spray equipment, clean equipment before use or before periods of extended downtime with Reducer #132, R7K132.

Mixed coating is sensitive to water. Use water traps in all air lines. Moisture contact can reduce pot life and affect gloss and color.

E-Z Roll Urethane Defoamer is acceptable for use. See data page 5.99 for details.

Quik-Thane Urethane Accelerator is acceptable for use. See data page 5.97 for details.

Refer to Product Information sheet for additional performance characteristics and properties.

CLEAN UP INSTRUCTIONS

Clean spills and spatters immediately with Reducer #132, R7K132. Clean tools immediately after use with Reducer #132, R7K132. Follow manufacturer's safety recommendations when using any solvent.

SAFETY PRECAUTIONS

Refer to the MSDS sheet before use.

Published technical data and instructions are subject to change without notice. Contact your Sherwin-Williams representative for additional technical data and instructions.

DISCLAIMER

The information and recommendations set forth in this Product Data Sheet are based upon tests conducted by or on behalf of The Sherwin-Williams Company. Such information and recommendations set forth herein are subject to change and pertain to the product offered at the time of publication. Consult your Sherwin-Williams representative to obtain the most recent Product Data Information and Application Bulletin.

WARRANTY

The Sherwin-Williams Company warrants our products to be free of manufacturing defects in accord with applicable Sherwin-Williams quality control procedures. Liability for products proven defective, if any, is limited to replacement of the defective product or the refund of the purchase price paid for the defective product as determined by Sherwin-Williams. NO OTHER WARRANTY OR GUARANTEE OF ANY KIND IS MADE BY SHERWIN-WILLIAMS, EXPRESSED OR IMPLIED, STATUTORY, BY OPERATION OF LAW OR OTHERWISE, INCLUDING MERCHANTABILITY AND FITNESS FOR A PARTICULAR PURPOSE.

207A02D.EPS

BREAKDOWN OF A CONVENTIONAL AIR SPRAY GUN

PARTS LIST

1	AIR CAP	1	22	NEEDLE VALVE	1
2	FLUID NOZZLE	1	23	RING	1
3	HEAD AND BODY ASSEMBLY	1	24	FLUID CONTROL ASSEMBLY	1
4	Head	1	25	Screw	1
5	Screw	1	26	Spring	1
6	Body	1	27	Ring	1
7	SCREW Trigger	1	28	Body	1
8	STUD Trigger	1	29	CONNECTION	1
9	SIDE PORT CONTROL ASSEMBLT	1	30	TRIGGER	1
10	Stem	1	31	AIR VALVE ASSEMBLY	1
11	Spring	1	32	Nut	1
12	Washer	1	33	Packing (kit of 5)	1
13	Packing (kit of 5)	1	34	Body	1
14	Spindle	1	35	Stem	1
15	Gasket	1	36	SPRING	1
16	Housing	1	37	GASKET	1
17	Knob	1	38	SLEEVE	1
18	Washer	1	39	RETAINER	1
19	Screw	1	40	WIPER	1
20	GASKET	1	41	PACKING	1
21	HOUSING	1	42	NUT	1

207A03B.EPS

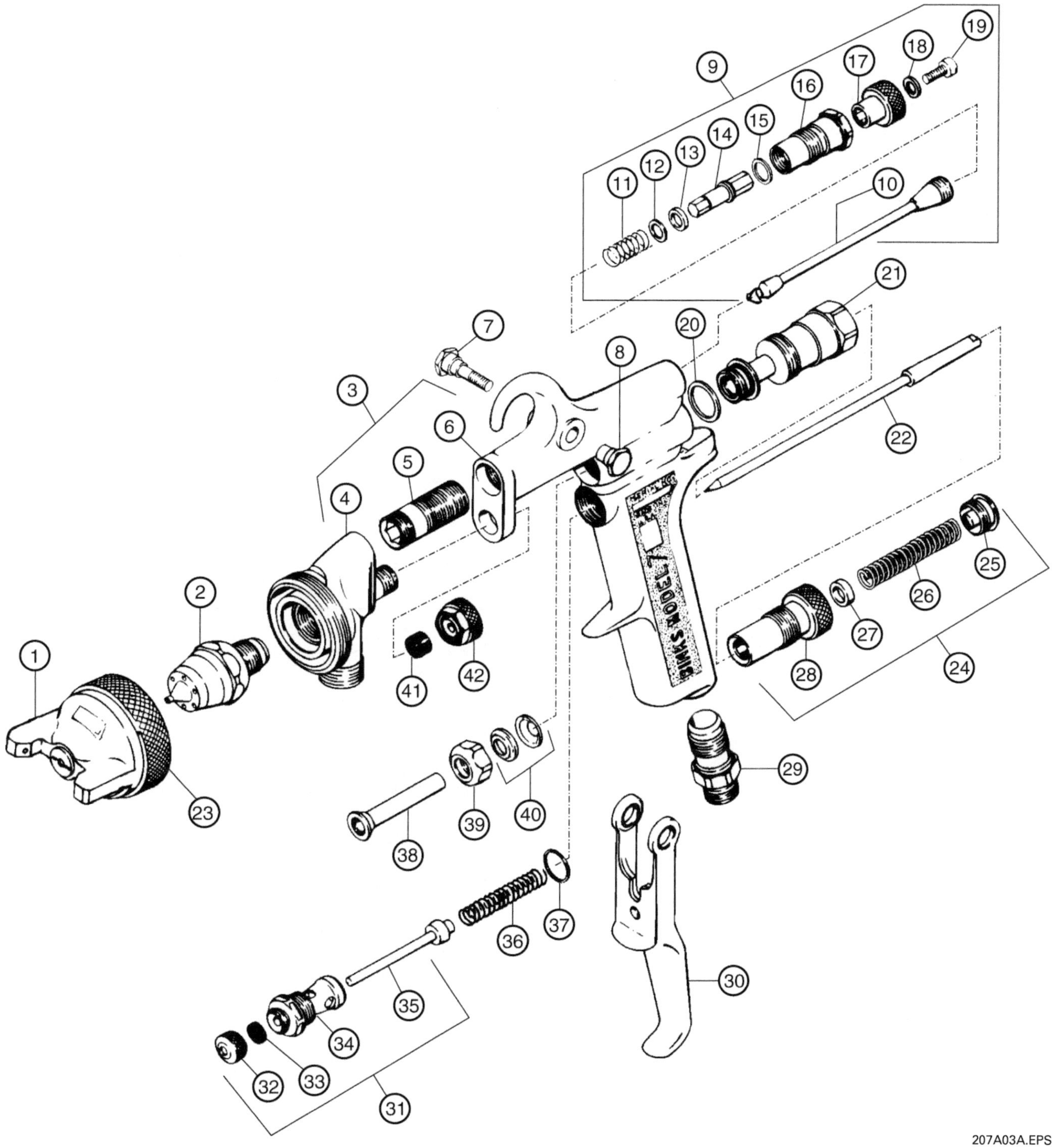
1
2
3
4
5
6
7
8
9
10
11
12
13
14
15
16
17
18
19
20
21
22
23
24
25
26
27
28
29
30
31
32
33
34
35
36
37
38
39
40
41
42

207A03A.EPS

Breakdown of a Typical Airless Spray Gun

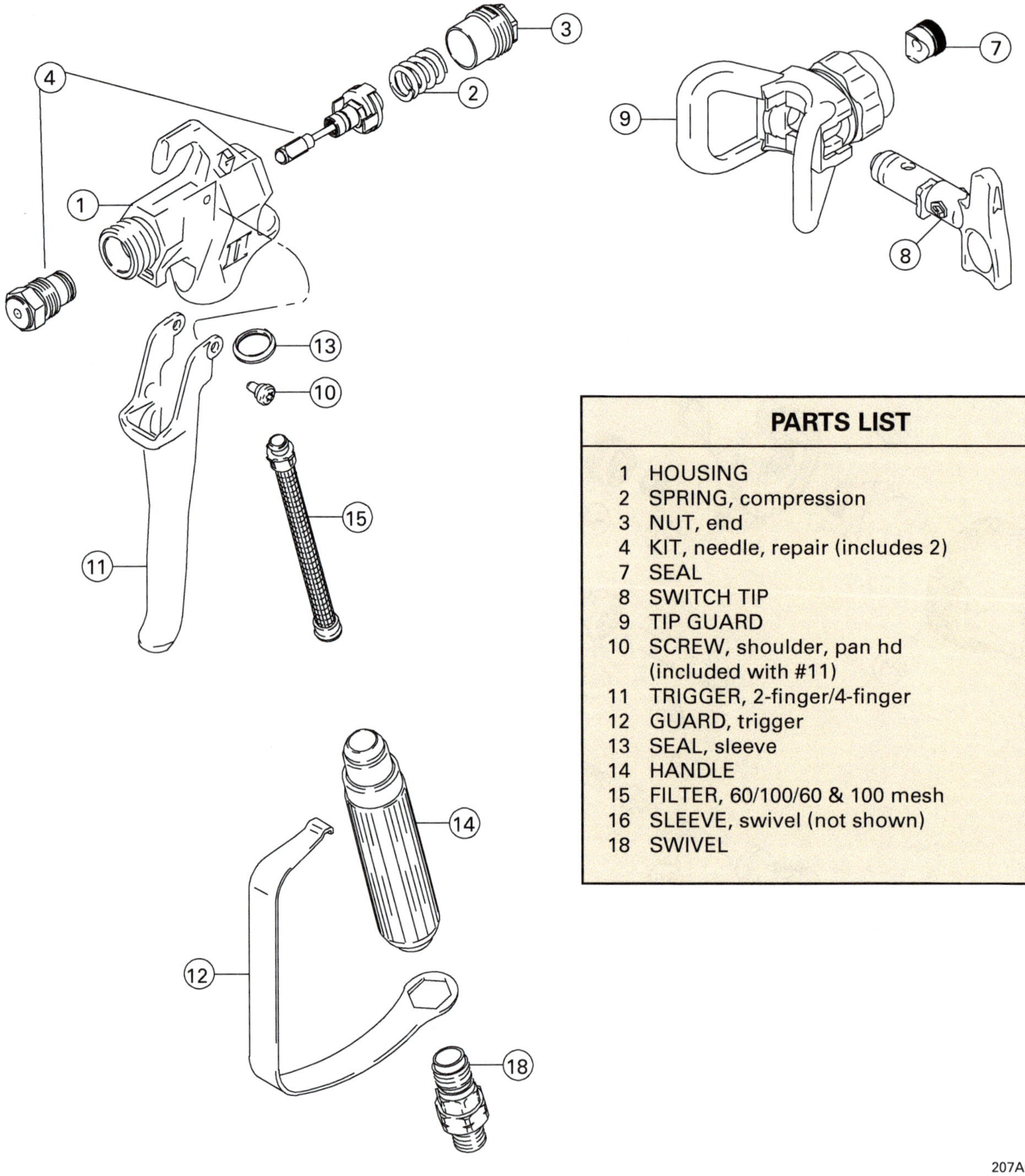

PARTS LIST		
1	HOUSING	1
2	SPRING, compression	1
3	NUT, end	1
4	KIT, needle, repair (includes 2)	1
7	SEAL	1
8	SWITCH TIP	1
9	TIP GUARD	1
10	SCREW, shoulder, pan hd (included with #11)	2
11	TRIGGER, 2-finger/4-finger	1
12	GUARD, trigger	1
13	SEAL, sleeve	1
14	HANDLE	1
15	FILTER, 60/100/60 & 100 mesh	1
16	SLEEVE, swivel (not shown)	1
18	SWIVEL	1

207A03C.EPS

Breakdown of a Typical Air-Assisted Airless Spray Gun

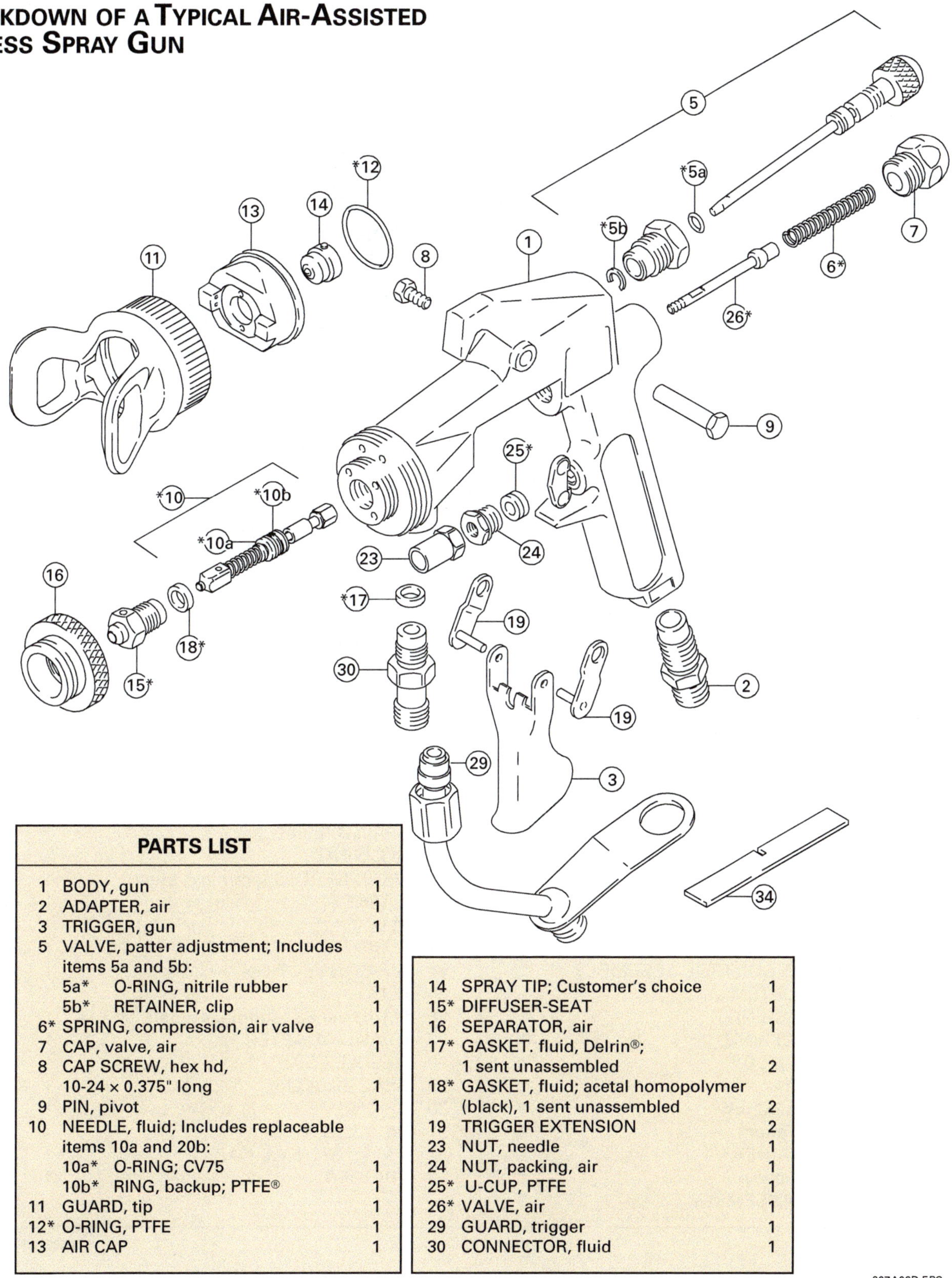

PARTS LIST

1	BODY, gun	1
2	ADAPTER, air	1
3	TRIGGER, gun	1
5	VALVE, patter adjustment; Includes items 5a and 5b:	
	5a* O-RING, nitrile rubber	1
	5b* RETAINER, clip	1
6*	SPRING, compression, air valve	1
7	CAP, valve, air	1
8	CAP SCREW, hex hd, 10-24 × 0.375" long	1
9	PIN, pivot	1
10	NEEDLE, fluid; Includes replaceable items 10a and 20b:	
	10a* O-RING; CV75	1
	10b* RING, backup; PTFE®	1
11	GUARD, tip	1
12*	O-RING, PTFE	1
13	AIR CAP	1

14	SPRAY TIP; Customer's choice	1
15*	DIFFUSER-SEAT	1
16	SEPARATOR, air	1
17*	GASKET. fluid, Delrin®; 1 sent unassembled	2
18*	GASKET, fluid; acetal homopolymer (black), 1 sent unassembled	2
19	TRIGGER EXTENSION	2
23	NUT, needle	1
24	NUT, packing, air	1
25*	U-CUP, PTFE	1
26*	VALVE, air	1
29	GUARD, trigger	1
30	CONNECTOR, fluid	1

207A03D.EPS

BREAKDOWN OF A TYPICAL HVLP SPRAY GUN

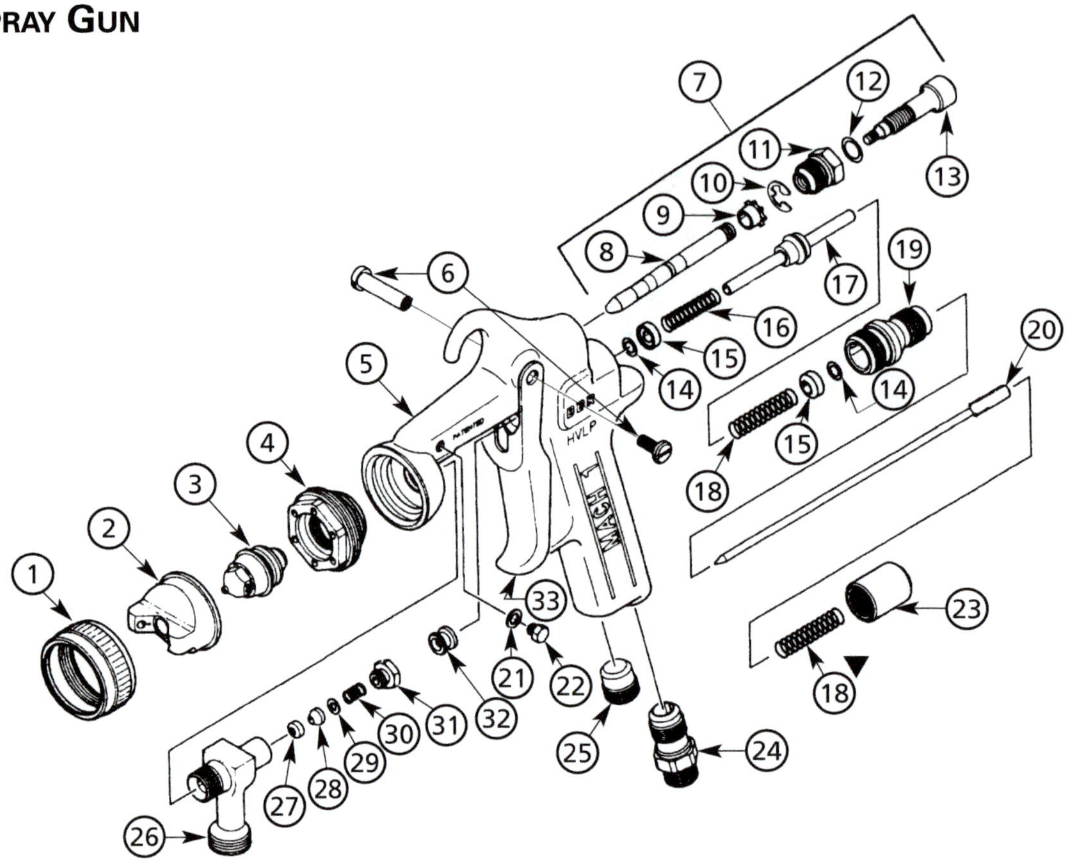

PARTS LIST					
1	RETAINING RING (metal)	1	18	SPRING (blue)	2
2	AIR NOZZLE	1	19	HOUSING	1
3	FLUID NOZZLE	1	20	FLUID NEEDLE (stainless steel)	1
4	HEAD INSERT	1	21	GASKET	1
5	GUN BODY ASSEMBLY	1	22	PLUG	1
6	TRIGGER STUD & SCREW KIT	1	23	FLUID CONTROL KNOB	1
7	SIDE PORT CONTROL ASSEMBLY	1	24	AIR CONNECTION	1
8	CONTROL STEM	1	25	PLUG	1
9	GUIDE BUSHING	1	26	FLUID INLET (stainless steel)	1
10	RETAINING RING	1	27	GLAND ADAPTER	1
11	CONTROL BODY	1	28	NEEDLE SEAL	1
12	O-RING	1	29	SEAL BACKUP	1
13	CONTROL SCREW	1	30	SPRING	1
14	O-RING (Teflon)	2	31	PACKING NUT	1
15	SEAL RETAINER	2	32	VALVE SPINDLE CAP	1
16	SPRING (yellow)	1	33	TRIGGER	1
17	SPINDLE ASSEMBLY	1			

207A03E.EPS

Additional Resources

This module presents thorough resources for task training. The following resource material is suggested for further study.

ASTM D4227, Standard Practice for Qualification of Coating Applicators for Application of Coatings to Concrete Surfaces. Latest Edition. West Conshohocken, PA: ASTM International.

ASTM D4228, Standard Practice for Qualification of Coating Applicators for Application of Coatings to Steel Surfaces, Latest Edition. West Conshohocken, PA: ASTM International.

NACE International, for corrosion control resources and information. www.nace.org

SSPC-PA 2, Measurement of Dry Coating Thickness With Magnetic Gages. Pittsburgh, PA: Society for Protective Coatings, 2004.

The Protective Coating User's Handbook. Dr. Louis D. Vincent. Houston, Texas: NACE Press, 2004.

Figure Credits

Graco Inc., Module opener, 207F08, 207F12, 207F16, 207F17 (bottom), 207F18–207F22, 207F24–207F27, 207T01, 207T02A, 207T02B, 207F30–207F34, Appendix D, Appendix E

©iStockphoto.com/stayorgo (Jeffrey Zavitski), 207F01

©iStockphoto.com/kajetan, 207F02 (chemical-resistant gloves)

North Safety Products USA, 207F02 (heat-resistant gloves), 207F03 (full facepiece respirator), 207F05

Scott Health and Safety, 207F03 (half-mask respirator), 207F04

Red Devil Inc., 207F06

Miracle Paint Rejuvenator Company, Inc., 207F07 (five-gallon unit)

Greg Londrigan, Hougen Mfg., 207F07 (one-gallon unit)

Rogers & Associates Consulting Services, 207F09

Norcross Corporation, www.viscosity.com, 617-969-7020, 207F10

Topaz Publications, Inc., 207F11, 207F14 (electronic thermometer)

Sherwin-Williams Co., 207F13, Appendix A, Appendix B

TSI Incorporated, 207F14 (anemometer)

Extech Instruments, 207F14 (digital psychrometer, electronic hygrometer)

Photo courtesy of Paul N. Gardner Company, Inc., 207F15, 207F42 (low-voltage detector)

ITW Finishing Innovations, 207F17 (top three), 207F23, 207F28, 207F29, Appendix C

Bry-Air, Inc., www.bry-air.com, 207F35

TEMP-AIR, Inc., 207F36–207F39

Altex Coatings Limited, 207F40

DeFelsko Corporation, 207F41

Elcometer, Inc., 800-521-0635, www.elcometer.com, 207F42 (high-voltage detector)

NACE International, 207T03

NCCER CURRICULA — USER UPDATE

NCCER makes every effort to keep its textbooks up-to-date and free of technical errors. We appreciate your help in this process. If you find an error, a typographical mistake, or an inaccuracy in NCCER's curricula, please fill out this form (or a photocopy), or complete the online form at **www.nccer.org/olf**. Be sure to include the exact module ID number, page number, a detailed description, and your recommended correction. Your input will be brought to the attention of the Authoring Team. Thank you for your assistance.

Instructors – If you have an idea for improving this textbook, or have found that additional materials were necessary to teach this module effectively, please let us know so that we may present your suggestions to the Authoring Team.

NCCER Product Development and Revision

13614 Progress Blvd., Alachua, FL 32615

Email: curriculum@nccer.org

Online: www.nccer.org/olf

❑ Trainee Guide ❑ Lesson Plans ❑ Exam ❑ PowerPoints Other ____________________

Craft / Level: Copyright Date:

Module ID Number / Title:

Section Number(s):

Description:

Recommended Correction:

Your Name:

Address:

Email: Phone:

Glossary

Accident: An unplanned event preceded by an unsafe act, unsafe condition, or both, which disrupts activities and often results in personal injury or property damage.

Acute effect: An immediate reaction to chemical exposure, such as nausea or death.

Acute exposure: A short-term, usually intense, exposure to a chemical.

Aggregate: Crushed rock or gravel that is screened to size for use in concrete.

Anchorage: A secure point of attachment for lifelines, lanyards, or deceleration devices.

Apparatus: An assembly of machines used together to perform a particular job.

Arc: The flow of electrical current from one point to another through a gas such as air.

Area: The space enclosed by the outline of a two-dimensional shape.

Atmospheric hazard: A potential danger in the air or a condition of poor air quality.

Atomizing: Reducing to fine particles or spray.

Barrier coating: A coating that keeps moisture and oxygen away from a substrate.

Best practices plan: A safety program that includes the best available practices and equipment to ensure worker safety.

Blast and coat cycle: A daily cycling between surface preparation operations and coating applications operations.

Body harness: Straps that may be secured about the worker in a manner that will distribute the fall-arrest forces over at least the thighs, pelvis, waist, chest, and shoulders, with means for attaching it to other components of a personal fall-arrest system.

Bonding: Connecting conductive objects to establish electrical continuity.

Chronic effect: A delayed or long-term reaction to chemical exposure. Chronic effects can result in serious diseases, such as cancer.

Chronic exposure: A long-term, usually low-dose, exposure to a chemical.

Circumference: The line that encloses a two-dimensional shape.

Combustible: Air or materials that can explode and cause a fire.

Competent person: A company-appointed person who can identify working conditions or surroundings that are unsanitary, hazardous, or dangerous to employees and who has authorization to correct or eliminate these conditions promptly.

Compliance: A safety plan that meets the minimum OSHA safety standards.

Condensation-cure coating: A coating in which two or more organic molecules react, forming a new bond and giving off a molecule of water.

Confined space: An area that is large enough for a person to work in, but with limited ways of entering and exiting, and that is not meant for continuous human occupancy.

Connector: A device that is used to couple (connect) parts of the personal fall-arrest system and positioning device systems together. It may be an independent component of the system, such as a carabiner. It may also be an integral component of part of the system, such as a buckle or D-ring sewn into a body belt or body harness, or a snaphook spliced or sewn to a lanyard.

Coreaction-cure coating: A coating that requires the addition of an activator or catalyst to the base material. Also called a plural or multi-component coating.

Cylinder: A shape created by a circle moving in a straight line through space perpendicular to the surface of the circle.

Decibels (dBA): A measure of sound intensity or loudness. The higher the decibel level, the louder and more potentially damaging the sound is.

Desiccant: Material that attracts and retains water vapor.

Duty rating: American National Standards Institute (ANSI) rating assigned to ladders. It indicates the type of use the ladder is designed for (industrial, commercial, or household) and the maximum working load limit (weight capacity) of the ladder. The working load limit is the maximum combined weight of the user, tools, and any materials bearing down on the rungs of a ladder.

Electrically powered tools: Tools that use electrical current to operate.

Electrolyte: A nonmetal that is capable of supporting electrical current flow, such as a fluid in which salt has been dissolved.

Emissions: Materials that escape the containment area, including airborne plumes of material, abrasives, paint chips, debris, spills, or leaks of water.

Energy source: Any source of electrical, mechanical, hydraulic, pneumatic, chemical, thermal, or other energy.

Energy-isolating device: Any mechanical device that physically prevents the transmission or release of energy. These include, but are not limited to, manually operated electrical circuit breakers, disconnect switches, line valves, and blocks.

Entryway: The area of a containment structure by which workers may enter and leave a containment and ventilation system.

Exponent: A number or symbol placed to the right and above another number, symbol, or expression, denoting the power to which the latter is to be raised.

Feathering: Working the edges of the repaired area back, so there is a fairly smooth transition from the repair area to the sound coating.

Flame-resistant: Materials that resist burning for a while but that continue to burn once ignited.

Flame-retardant: Materials that stop burning once the source of fire has been removed.

Flash rusting: The rusting that occurs from the time the waterjetting process starts to the time the water used in the process dries.

Forced airflow: Air that is supplied to a containment area through the use of fans and blowers.

Free fall: The act of falling before a personal fall-arrest system begins to apply force to arrest the fall.

Ground fault circuit interrupter: A protective device that functions to de-energize a circuit or portion thereof within an established period of time when a current to ground exceeds some predetermined value. This value is less than that required to operate the overcurrent protective device of the supply circuit.

Grounded: Connected to ground or to a conductive body that extends the ground connection.

Guards: Devices that protect tool operators from dangerous moving parts such as blades, gears, and pulleys.

Hazard Communication (HazCom) Standard: A federal OSHA regulation requiring employers to educate and inform workers about chemical hazards on the job site (*29 CFR 1910.1200*).

Horn: One of two protrusions that blows air and material in a spread pattern from the nozzle of a conventional spray gun.

Hydration-cure coating: A coating that cures by a process called hydration, in which water combines chemically with the coating and its constituents.

Hydrocarbons: Petroleum products such as oil or grease.

Hygrometer: Humidity measuring device.

Hypothermia: A life-threatening condition caused by exposure to very cold temperatures.

Impact tools: Tools that must strike or be struck to accomplish their task. Impact tools include hammers, chisels, and taps.

Impenetrable: Materials that dust and wind will not flow through or penetrate.

Impermeable: Materials that water will not flow through or penetrate.

Inhibitive coating: A coating that contains pigments that react with absorbed moisture within the coating and then react with the steel surface, passivating the metal and interfering with the corrosion process.

Ions: Electrically charged particles.

Lanyard: A flexible line of rope, wire rope, or strap that generally has a connector at each end for connecting the body belt or body harness to a deceleration device, lifeline, or anchorage.

Leading edge: The edge of a floor, roof, or formwork for a floor or other walking/working surface (such as the deck) which changes location as additional floor, roof, decking, or formwork sections are placed, formed, or constructed. A leading edge is considered to be an unprotected side and edge during periods when it is not actively and continuously under construction.

Lifeline: A component consisting of a flexible line connected vertically to an anchorage at one end (vertical lifeline), or connected horizontally to an anchorage at both ends (horizontal lifeline), and which serves as a means for connecting other components of a personal fall-arrest system to the anchorage.

Lockout: The placement of a lockout device on an energy-isolating device, in accordance with an established procedure, ensuring that the energy-isolating device and the equipment being controlled cannot be operated until the lockout device is removed.

Lockout device: Any device that uses positive means such as a lock to hold an energy-isolating device in a safe position, thereby preventing the energizing of machinery or equipment.

Makeup air: Air that replaces the air that has been exhausted from the containment enclosure.

Material safety data sheet (MSDS): A document that must accompany any hazardous material. The MSDS identifies the substance and gives the exposure limits, the physical and chemical characteristics, the kind of hazard it presents, precautions for safe handling and use, and specific control measures.

Microsiemens: Units of electrical conductivity.

Negative pressure: A condition in an enclosure in which exhaust air exceeds input air.

Noble: A term used to describe a metal that is resistant to corrosion.

Non-permit required confined space: A work space free of any atmospheric, physical, electrical, and mechanical hazards that can cause death or injury.

Opacity: A coating's ability to cover or hide a substrate.

Oxidation-cure coating: A coating containing vegetable oil that cures by a process called oxidation, in which coating ingredients react with the oxygen in the air.

Oxygen-deficient: An atmosphere in which there is not enough oxygen. This is usually considered less than 19.5 percent oxygen by volume.

Oxygen-enriched: An atmosphere in which there is too much oxygen. This is usually considered more than 23.5 percent oxygen by volume.

Patch test: A test used in overcoating operations in which the selected coating is applied over a small area of the existing coating to see if they are compatible.

Permit required confined space: A confined space that has real or possible hazards. These hazards can be atmospheric, physical, electrical, or mechanical.

Perpendicular: At a right angle to the plane of a line or surface.

Personal fall-arrest system: A system used to stop an employee in a fall from a working level. It consists of an anchorage, connectors, and a body harness, and may include a lanyard, deceleration device, lifeline, or suitable combinations of these. As of January 1, 1998, using a body belt for fall arrest is prohibited.

Pi: The constant that defines the relationship between the diameter and circumference or the radius and area of a circle.

Plasma: An ionized gas created by passing a gas such as argon, nitrogen, helium, or hydrogen, through an electric arc. The atomic components of the gas break apart and then recombine, producing a tremendous amount of energy.

Platform: A work surface elevated above lower levels. Platforms can be constructed using individual wood planks, fabricated planks, fabricated decks, and fabricated platforms.

Pneumatic tools: Tools that use pressurized air to operate.

Power tools: Tools that require a power source, such as electricity, hydraulics, or pneumatics, to operate.

Radius: A straight line from the center of a circle to a point on the edge of the circle.

Sacrificial coating: A coating that corrodes to preserve the metal it coats.

Scaffolding: A temporary supported framework or suspended platform or work area designed to support workers, materials, and equipment at elevated or otherwise inaccessible job sites.

Self-retracting lifeline: A deceleration device containing a drum-wound line that can be slowly extracted from, or retracted onto, the drum under slight tension during normal employee movement, and which, after the onset of a fall, automatically locks the drum and arrests the fall.

Short-term exposure limit (STEL): The highest concentration of an airborne material to which workers can be exposed over a prescribed period of time.

Sieve test: A test performed on abrasives as they are received to see if they are sized to specification.

Silica: A white or colorless crystalline compound used to manufacture a wide variety of materials, especially glass and concrete.

Support structure: The framework upon which the containment material is mounted.

Surfactants: Additives used in oil-and-water emulsions that break down the surface tension of the liquid.

Tagout: The placement of a tagout device on an energy-isolating device, in accordance with an established procedure, to indicate that the energy-isolating device and the equipment being controlled may not be operated until the tagout device is removed.

Tagout device: Any prominent warning device, such as a tag and a means of attachment, that can be securely fastened to an energy-isolating device in accordance with an established procedure. The tag indicates that the machine or equipment to which it is attached is not to be operated until the tagout device is removed in accordance with the energy-control procedure.

Threshold limit value (TLV): The upper limit for the concentration of solvents in the air.

Tie coat: A coating that is compatible with both the existing coating and the new coating.

Time-weighted average (TWA): The amount of time a worker can be exposed to a material over a certain period—usually given in hours allowed to work per day or week.

Two-point swing scaffolding: A manual or power-operated platform supported by hangers at two points suspended from overhead supports in a way that allows it to be raised or lowered to the working position.

Upper explosive limit (UEL): The highest amount of vapor concentration in the air that could explode. Vapor concentrations above the UEL have displaced so much oxygen that there is not enough oxygen in the air to explode.

Volume: The space enclosed by a three-dimensional shape.

Index

A

I

N

O

Q

R

T

U

V

W

X

Z